Guide to Brain-Computer Music Interfacing

Eduardo Reck Miranda
Julien Castet
Editors

Guide to Brain-Computer Music Interfacing

Springer

Editors
Eduardo Reck Miranda
Interdisciplinary Centre for Computer Music Research (ICCMR)
Plymouth University
Plymouth
UK

Julien Castet
Immersion
Bordeaux
France

ISBN 978-1-4471-7210-9 ISBN 978-1-4471-6584-2 (eBook)
DOI 10.1007/978-1-4471-6584-2

Springer London Heidelberg New York Dordrecht

Softcover reprint of the hardcover 1st edition 2014

Printed on acid-free paper

Springer is part of Springer Science+Business Media (www.springer.com)

Foreword

Brain-Computer Music Interfacing (BCMI): One Place Where Science and Music May Meet in Deep Theoretical Territory

There is little doubt that as we gain facility in the intense, disciplined practices required to probe the origins of musical impulses, pathways of emergence can be observed and experienced by tracing backward down trails left in the normally outward flowering of forms from the brain, from their initial neural geneses, upward through their manifestations in neural network *holarchies*,[1] and finally to their manifestations in localized, time-space expressions. Along these pathways, the *languaged* forms of the presumably testable theoretical models of science and the investigative, speculative models of *propositional music* converge.

Propositional music involves a point of view about composing in which composers might build proposed models of worlds, universes, evolution, brains, consciousness or whole domains of thought and life, and then proceed to make dynamical musical embodiments of these models, inviting us to experience them in spontaneously emerging sonic forms (Rosenboom 2000a). For musicians who are interested in deep conceptual and theoretical investigations, BCMI is a natural attractor and a predictable outgrowth from mid-twentieth century explosions in interdisciplinary thinking, Cybernetics, Artificial Intelligence, Linguistics, Systems Theory, self-organization, morphogenesis, algorithmic music, and so on. Following early explorations in BCMI from that time, we are now experiencing a new flowering of what might be called *antidisciplinary thinking* in the arts and sciences, which, among other things, reexamines fundamental distinctions among scientific and cultural languages (Beech 2013).

Some extant model paradigms that employ BCMI in artistic creation can legitimately claim to be new *musical propositions*. Others focus more on direct mappings of neurological data onto acoustic parameters of sound or components of traditionally familiar musical structures. Both may reveal fruitful investigative

[1] The term *holarchy* is used to refer to structures that have both top-down and bottom-up dynamical aspects.

pathways; and there is a big difference between them. Direct mapping, often called *sonification*, draws on the profoundly integrative powers of auditory perception—possibly enhanced by musical experience and training—to *hear* relationships and find clues to hidden orders and unsuspected patterns. With careful, *active imaginative* listening, we can learn to distinguish the features of complexity and parse subtle relationships among differentiable, complex entities. Our auditory perception and cognition toolkits can be fine-tuned in these respects to extraordinary degrees. Later, we can probe these newly perceived patterns and quantify them with other investigative tools. Aesthetic propositions, on the other hand, may probe the very nature of thought, itself—in this case musical thought—and its symbiotic inhabiting of the brain in co-creative, adaptive, emergent, feedback-driven phenomena. Circling back to scientific exploration, these musical BCMI propositions may fuel important investigations into the nature of prediction, presumed causal relationships, ways to understand the global functions of the brain and other insights leading to paradigm shifts, from which even new practical applications may follow. Here again, the integrative powers of musical perception and musical cognition may be brought to bear on deep aesthetic investigations, which may, in return, offer important insights and clues for scientific explorations and theoretical modeling. Musical BCMI propositions, if we are alert, may take us back to first principles again and again, questioning our understanding of evolution and categorization. Recalling Charles Sanders Peirce's *doctrine* of dicisigns, we might soon discover a *biomusical semiotics* emerging from *neuro-musical propositions*. See (Stejernfelt 2014) for an analysis of the doctrine of *dicisigns*.

We should be careful, though, to avoid the potential misconception of BCMI as *MCMI (Mind-Computer Music Interfacing)*. At the moment, we don't really know any more about what the mind *is* than we know about what energy *is*. Richard Feynman reminds us that all we really know about energy is that it is related to some quantity we can calculate that doesn't change when *things happen* (conservation law) (Feynman 1995). Similarly, we don't really have a good characterization of what mind *is*, except we claim to be able to sense its presence. We do have many interesting speculative theories about the mind's emergence (Swan 2013), and some may pan out. It may be that minds have evolved to be able to *know*—perhaps an illusion—what *other minds* are thinking and feeling.[2] Perhaps the concept of *mind* equates to what we believe we can *know* as being relatively constant when we are observing relationships among entities that we differentiate as *individually unique* and label as being *conscious*. Is mind a constraint-based, emergent phenomena that might have begun already with proto-life forms (Deacon 2013)?

And what about intelligence? Nobody seems to really know clearly what it is, but everyone believes they can tell when it is *absent*. Are intelligences differentiable and knowable in advance of their encounter? Perhaps intelligence might be also

[2] I've been thinking about this a lot after discussing it with cognitive scientist (also a musician), Scott Makeig, Director of the Swartz Center for Computational Neuroscience at the University of California, San Diego (UCSD).

considered as a *field*, or *intellisphere*, perhaps even on inter-stellar scales (Rosenboom 2003b). Stjernfelt invokes the term *cognitive field* in describing Peirce's undoing of the misleading dualisms that dangerously pocket this intellectual terrain, particularly when considering causal modeling (Stjernfelt 2014). Yet, how do we describe what's going on in the brain in holistic terms, the interacting of neural atoms in a macro-form? Here again we have propositional language issues. There is imprecision in propositions, though imprecise meanings also have value. The term, *field*, has a precise meaning, as in *vector field*, but also imprecise ones, as in the nature of *chreods* or *zones of influence* (Thom 1975). Quantum paradigms have taught us that the universe is not a precision instrument. Imprecision and approximation cannot be overcome. Rather, though, they have value in permitting guided explorations into the fringes of thought.

So BCMI is striving to balance its need for precision in developing practical applications with its also critical need to explore imprecise paradigms, which often enable breakthroughs in thought and vision. BCMI is destined to open new doors, as long as we remain open to the unpredictable. We need to gain better understanding of global complexity in both brains and music. BCMI may offer useful tools for this. Music is fundamentally about time, and therefore about *qualities of change*. The spatiotemporal evolution of holistic brain phenomena is also about qualities of change. This is a good match for unveiling the future.

My own work with BCMI began in the 1960s (Rosenboom 1972, 1976a, 1976b, 1977, 1990, 1997, 2000b, 2000c, 2003a, 2006). Recently, this work has resurfaced, re-energized by new, accessible technology, and significant advances in methods for analyzing brain signals. The most recent example is a new composition called *Ringing Minds*, created in 2014 in collaboration with Tim Mullen and Alex Khalil at the Schwartz Center for Computational Neuroscience in the University of California at San Diego (UCSD). *Ringing Minds* builds on techniques for extracting principal oscillation patterns (POPs or *eigenmodes*) from maximally independent sources of intracranial electroencephalogram (EEG) data (Mullen et al. 2012). These tools were originally developed for epilepsy research. Mullen adapted them to analyze EEG data from an ensemble of five brain music performers. We treated the data as if it were arising from a collective brain, a five-person brain, and extracted eigenmode data in which we could identify distributions of these resonant modes across the collective brain. We also extracted event-related potentials (ERPs) by averaging simultaneous signals spatially, across the five brains, instead of across time with a single brain. In this way, we hoped to extract pointers to attention shifts in the five-performer group resulting from musical events emanating from two improvising musicians on stage. I played electric violin, and Khalil built and played the *lithophone*, an instrument that looks like a xylophone, but made with stone slabs struck with a hammer. I built a software-based electronic music instrument for this work, the central core of which is a very large array of complex resonators that can respond to the collective EEG eigenmode data in a way that generates a vast sound field of ringing components. The control parameters of the instrument can also be varied in real-time performance. The instrument is, in effect, a *compositional model* inspired by the analytical model working on the EEG signals from the five-brain performing group. The *model* thus becomes an *instrument*.

Ringing Minds investigates many things. Among them are complex relationships manifesting among the components of a sound environment—the resonator field—and a group of individuals, who may interact with this environment via natural listening patterns and/or use biofeedback techniques to try to influence that environment. With careful, active imaginative listening to the results of this fine-grained resonant field, one can imagine witnessing both local and global processes interacting and perceive small-scale, quantum processes zooming out into larger scale arenas of human perceptibility. The *local* is mirrored in the *global* and is impressed on the environment, which bears witness to both the emergence of coherence among components and the loss of coherence. This propositional neuromusical model is analogous to an intriguing propositional physical model that relates objective and subjective quantum states to their *environment as witness* (Ollivier et al. 2004).

This new volume, *Guide to Brain-Computer Music Interfacing*, offers a wonderful collection of BCMI tools with which adventuresome explorers may pursue both practical and propositional models in both neuromusic and music neuroscience. In an introductory chapter, Eduardo Reck Miranda contextualizes what follows under an umbrella with three major challenges: extracting meaningful control information from brain signals, designing effective generative music techniques that respond to this information, and effectively improving peoples lives. That's a tall order, but achievable with today's tools. Important techniques for prosthetics, device control, and handsfree interaction with computers follow. Very important work on event-related potentials (ERPs), especially with P300 waves, is included. P300 analysis was critical to my 1970s work on detecting possible neural correlates of attention shifts associated with musical features and forms (Rosenboom 1977, 1990, 1997, 2000b, 2000c). In that early work a combination of template matching and signal averaging techniques—then implemented with real-time, hardware computing devices—was used to shorten the normally long latencies separating targeted features in musical (or raw acoustic) forms from the points in time when their ERP concomitants could be reliably observed. A means for calculating predicted expectancy quantifications for events in sequences was also employed and used to predict when and where ERPs with strong P300 components were likely to occur. These predictions identified time points where analysis algorithms would be triggered. Feedback about successful and unsuccessful predictions was then used to influence sound synthesis and compositional form-generating algorithms.

Later in this volume, questions are explored about semiotic BCI—recalling Peirce again—and the use of machine learning to dig into relationships among music and emotions. This is complex and delicate territory rife with presumptions about musical emotions that need clarification; and some of the tools offered here may be helpful in that quest. Excellent tutorials on signal extraction, brain electric fields, passive BCI, and applications for genetic algorithms are offered along with historical surveys. In a penultimate chapter, Miranda and colleagues return to describe how BCMI research has received motivation from health and medical sectors as well as the entertainment industry and to advocate for the importance of "the potential impact on musical creativity of better scientific understanding of the brain, and the development of increasingly sophisticated technology to scan its activity." This book opens many doors.

Valencia, US David Rosenboom

References

Beech A (2013) Science and its fictions. Notes from a talk on scientific Method, California Institute of the Arts, private communication, Valencia, CA

Deacon TW (2013) Incomplete nature, how mind emerged from matter. W. W. Norton & Co, New York

Feynman RP (1995) Six easy pieces. Addison-Wesley Pub. Co, Reading, MA

Mullen T, Worrell G, Makeig S (2012) Multivariate principal oscillation pattern analysis of ICA sources during seizure. Proceedings of the 34th Annual International Conference of the IEEE, EMBS San Diego, CA

Ollivier H, Poulin D, Zurek WH (2004) Objective properties from subjective quantum states: environment as witness. Phys Rev Lett PRL 93:22040-1–220401-4

Rosenboom D (1972) Methods of producing sounds or light flashes with alpha brain waves for artistic purposes. Leonardo 5, 1. In: Malina FJ (ed) (1973) Kinetic art. Dover Pub, New York, pp 152–156. Space Design 10 (Japanese translation) Tokyo: Kajima Institute Pub. Co, Tokyo, 1974

Rosenboom D (1976a) Brainwave music. LP record. Aesthetic Research Centre Records #ST1002, Toronto

Rosenboom D (1976b) Biofeedback and the arts, results of early experiments. Aesthetic Research Centre of Canada Publications, Vancouver

Rosenboom D (1977) On being invisible. LP record. Music Gallery Editions #MGE-4, Toronto

Rosenboom D (1990) The performing brain. Computer Music Journal 14(1):48–66

Rosenboom D (1997) Extended musical interface with the human nervous system: assessment and prospectus. In: Revised electronic monograph. Original, Leonardo Monograph Series, 1, San Francisco, 1990 http://www.davidrosenboom.com/media/extended-musical-interface-humannervous-system-assessment-and-prospectus

Rosenboom D (2000a) Propositional music: on emergent properties in morphogenesis and the evolution of music, essays, propositions, commentaries, imponderable forms and compositional method. In: Zorn J (ed) Arcana, musicians on music, Granary Books/Hips Road, New York, pp 203–232

Rosenboom D (2000b) Invisible gold, classics of live electronic music involving extended musical interface with the human nervous system. Audio CD. Pogus Productions, Chester, New York, P 21022-2

Rosenboom D (2000c) On being invisible II (Hypatia speaks to Jefferson in a dream). On: Transmigration music. Audio CD. Consortium to Distribute Computer Music, vol 30, #CRC 2940. Centaur Records, Inc., Baton Rouge, LA

Rosenboom D (2003a) Propositional music from extended musical interface with the human nervous system. In: Avanzini G et al (eds) The neurosciences and music. Annals of the New York Academy of Sciences, vol 999. New York Academy of Sciences, New York, pp 263–271

Rosenboom D (2003b) Collapsing distinctions: interacting within fields of intelligence on interstellar scales and parallel musical models. David Rosenboom Publishing, Valencia, CA http://www.davidrosenboom.com/media/collapsing-distinctions-interactingwithin-fields-intelligence-interstellar-scales-and

Rosenboom D (2006) Brainwave music 2006. Audio CD. EM Records #EN1054CD, Osaka, Japan

Stjernfelt F (2014) Natural propositions, the actuality of Peirce's doctrine of dicisigns. Docent Press, Boston, MA

Swan L (2013) Origins of mind. Biosemiotics 8. Springer, Dordrecht

Thom R (1975) Structural stability and morphogenesis. W.A. Benjamin Inc., Reading, MA

Preface

The idea of using brainwaves to make music dates back from the 1960s, when composers such as Alvin Lucier, Richard Teitelbaum, and David Rosemboom, to cite but three, looked into generating music with the electroencephalogram, abbreviated as EEG.

Lucier placed electrodes on his own scalp, amplified the signals, and relayed them through loudspeakers that were "directly coupled to percussion instruments, including large gongs, cymbals, tympani, metal ashcans, cardboard boxes, bass and snare drums" (Lucier 1980). The low frequency vibrations emitted by the loudspeakers set the surfaces and membranes of the percussion instruments into vibration. Teitelbaum used various biological signals including the EEG and ECG (electrocardiogram) to control electronic synthesisers (Teitelbaum 1976). Rosemboom subsequently looked into designing more sophisticated systems inspired by Cybernetics, exploring the concept of biofeedback in real-time music making (Rosenboom 1990).

Those pioneering composers left an important legacy of concepts and practices. However, apart from very few sparse initiatives here and there, the idea seems to have faded into oblivion until the end of the twentieth century. We reckon that one of the reasons for this stagnation is that EEG equipment was not as widely available as it is today. Moreover, techniques for analyzing EEG signals were not as well developed as they are today, and consequently we lacked sophisticated handling and understanding of the EEG.

A notable development for musicians was the appearance of a piece of equipment called BioMuse in the 1990s, manufactured by Benjamin Knapp and Hugh Lusted (1996). BioMuse provided a portable kit for digitally processing bio-signals such as the EEG, muscle movement, heartbeat, and so on. It was able to convert these signals into MIDI data, which facilitated the implementation of MIDI controllers using the EEG.

Within the last two decades or so, we have witnessed the emergence of the field of Brain-Computer Interfacing, or BCI (also referred to as Brain-Machine Interfacing, or BMI). Research into BCI is aimed at the development of technology to enable people control machines by means of commands expressed by signals, such as the EEG, detected directly from their brain. Most of this research is developed within Biomedical Engineering and is aimed at giving severely paralyzed patients the ability to control artificial limbs, wheel chairs, robotic equipment, machines, and so on. Obviously, in these cases, the user must be able to actually control these devices voluntarily and as precisely as possible. The user needs to produce specific patterns of EEG to command a machine and such a machine needs to interpret those patterns and do what the user wants it to do.

Continuing progress in BCI research combined with the emergence of more affordable EEG equipment are fostering a renaissance of approaches to making music with brain signals: the field of Brain-Computer Music Interfacing, abbreviated as BCMI, is now well established (Miranda 2010).

The field of BCMI has developed in tandem with the field of BCI. As with BCI, in BCMI the notion of active control of a system is an important aspect (Miranda et al. 2005; Miranda et al. 2003). However, the notion of control in an artistic application can, and should, be approached with flexibility. There might be cases where a composer might want to avoid explicit control altogether. Nevertheless, in order to make progress, the science and engineering behind BCMI research should be aimed at the development of control methods as well as approaches for mapping EEG information into musical information. In practice, composers may of course choose to ignore all of these, depending on what they want to achieve.

A number of low cost EEG equipment have been appearing in the market, most of which are commercialized in association with some sort of system for aiding meditation, relaxation, and so on. Whereas these have given musicians wider access to such technology, at the same time, however, pressures to manufacture them at low cost mean that the great majority of these systems fail to relay a reliable EEG signal for processing. This is an important fact we should all bear in mind, including those who are not so concerned with active control. Even in those cases where we might not wish to harness the EEG signal for explicit control of a music system, we do need a reliable EEG signal nevertheless. Otherwise we might end up making music with signals that are anything but the actual EEG. Therefore, the essential ingredients for making progress in the field of BCMI are: reliable hardware, powerful techniques for EEG signal processing, and creative methods for rendering the EEG signal into music. *Guide to Brain-Computer Music Interfacing* brings a number of chapters reporting on developments for the last two ingredients.

This book emerged from a workshop on EEG and music composition that took place in 2011 at the University of Bordeaux, France, supported by the French Association for Musical Informatics (Association Française d'Informatique Musicale, AFIM). The workshop included presentations that were entirely technical, focusing on hardcore EEG analysis, and ones that focused on practical musical applications. This is reflected in this book, but in addition to chapters developed from papers presented at the workshop, we also commissioned chapters from experts on topics that were not covered by the workshop.

We would like to thank all authors for their valuable contributions and Springer for the opportunity to publish this book.

Eduardo Reck Miranda
Julien Castet

References

Knapp B, Lusted H (1996) Controlling computers with neural signals. Sci Am 275(4):82–87
Lucier A (1980) Chambers. Wesleyan University Press, Middletown, CT
Miranda ER (2010) Plymouth brain-computer music interfacing project: from EEG audio mixers to composition informed by cognitive neuroscience. Int J Arts and Tech 3(2/3):154–176
Miranda ER, Roberts S, Stokes M (2005) On generating EEG for controlling musical systems. Biomed Tech 49(1):75–76
Miranda ER, Sharman K, Kilborn K, et al (2003) On Harnessing the Electroencephalogram for the Musical Braincap. Computer Music Journal 27(2):80–102
Rosenboom D (1990) The Performing Brain. Computer Music Journal 14(1):48–65
Teitelbaum R (1976) In Tune: Some Early Experiments in Biofeedback Music (1966–1974). In: Rosenboom D (ed) Biofeedback and the Arts: results of early experiments, Aesthetic Research Centre of Canada, Vancouver

Contents

Contributors

Marie Chavent IMB, UMR CNRS 5251, INRIA Bordeaux Sud-Ouest, University of Bordeaux, Bordeaux, France

Mitsuko Aramaki Laboratoire de Mécanique et d'Acoustique (LMA), CNRS UPR 7051, Aix-Marseille University, Marseille, France

Emmanuel Bigand Department of Cognitive Psychology, University of Burgundy, Dijon, France

Marco Congedo GIPSA-lab, CNRS and Grenoble University, Grenoble, France

Ian Daly School of Systems Engineering, University of Reading, Reading, UK

Myriam Desainte-Catherine University of Bordeaux, Bordeaux, Talence, France; CNRS, LaBRI, UMR 5800, Bordeaux, Talence, France

Joel Eaton Interdisciplinary Centre for Computer Music Research (ICCMR), Plymouth University, Plymouth, UK

Frédérique Faïta-Aïnseba University of Bordeaux, Bordeaux, France

Laurent George INRIA, Rennes, France

Mick Grierson Embodied Audiovisual Interaction Group (EAVI), Goldsmiths Digital Studios, Department of Computing, Goldsmiths College, London, UK

Zoran Josipovic Psychology Department, New York University, New York, NY, USA

Christian Jutten GIPSA-lab, CNRS and Grenoble University, Grenoble, France

Chris Kiefer Embodied Audiovisual Interaction Group (EAVI), Goldsmiths Digital Studios, Department of Computing, Goldsmiths College, London, UK

Richard Kronland-Martinet Laboratoire de Mécanique et d'Acoustique (LMA), CNRS UPR 7051, Aix-Marseille University, Marseille, France

Joseph Larralde University of Bordeaux, LaBRI, UMR 5800, Bordeaux, Talence, France

Anatole Lécuyer INRIA, Rennes, France

Pierrick Legrand IMB, UMR CNRS 5251, INRIA Bordeaux Sud-Ouest, University of Bordeaux, Bordeaux, France

Dan Lloyd Department of Philosophy, Program in Neuroscience, Trinity College, Hartford, CT, USA

Fabien Lotte Inria Bordeaux Sud-Ouest/LaBRI, Talence Cedex, France

Jean-Arthur Micoulaud-Franchi Laboratoire de Neurosciences Cognitives (LNC), CNRS UMR 7291, Aix-Marseille University, Marseille, France

Eduardo Reck Miranda Interdisciplinary Centre for Computer Music Research (ICCMR), Plymouth University, Plymouth, UK

Slawomir J. Nasuto School of Systems Engineering, University of Reading, Reading, UK

Ramaswamy Palaniappan Department of Engineering, School of Science and Engineering, University of Wolverhampton, Telford, UK

Etienne B. Roesch School of Systems Engineering, University of Reading, Reading, UK

Sandra Rousseau GIPSA-lab, CNRS and Grenoble University, Grenoble, France

Konstantinos Trochidis Department of Cognitive Psychology, University of Burgundy, Dijon, France

Leonardo Trujillo Instituto Tecnológico de Tijuana, Tijuana, BC, Mexico

Laurent Vézard IMB, UMR CNRS 5251, INRIA Bordeaux Sud-Ouest, University of Bordeaux, Bordeaux, France

Jean Vion-Dury Laboratoire de Neurosciences Cognitives (LNC), CNRS UMR 7291, Aix-Marseille University, Marseille, France

Pierre-Henri Vulliard University of Bordeaux, LaBRI, UMR 5800, Bordeaux, Talence, France

James Weaver School of Systems Engineering, University of Reading, Reading, UK

Duncan Williams Interdisciplinary Centre for Computer Music Research (ICCMR), Plymouth University, Plymouth, UK

Sølvi Ystad Laboratoire de Mécanique et d'Acoustique (LMA), CNRS UPR 7051, Aix-Marseille University, Marseille, France

Brain–Computer Music Interfacing: Interdisciplinary Research at the Crossroads of Music, Science and Biomedical Engineering

1

Eduardo Reck Miranda

Abstract

Research into brain–computer music interfacing (BCMI) involves three major challenges: the extraction of meaningful control information from signals emanating from the brain, the design of generative music techniques that respond to such information and the definition of ways in which such technology can effectively improve the lives of people with special needs and address therapeutic needs. This chapter discusses the first two challenges, in particular the music technology side of BCMI research, which has been largely overlooked by colleagues working in this field. After a brief historical account of the field, the author reviews the pioneering research into BCMI that has been developed at Plymouth University's Interdisciplinary Centre for Computer Music Research (ICCMR) within the last decade or so. The chapter introduces examples illustrating ICCMR's developments and glances at current work informed by cognitive experiments.

1.1 Introduction

Until recently, developments in electronic technologies have seldom addressed the well being of people with special needs within the health and education sectors. But now, brain–computer music interfacing (BCMI) research is opening up fascinating possibilities at these fronts. BCMI systems have the potential to be used as recreational devices for people with physical disability, to support music-based activity

E.R. Miranda (✉)
Interdisciplinary Centre for Computer Music Research (ICCMR), Plymouth University, Plymouth PL4 8AA, UK
e-mail: eduardo.miranda@plymouth.ac.uk

E.R. Miranda and J. Castet (eds.), *Guide to Brain-Computer Music Interfacing*,
DOI 10.1007/978-1-4471-6584-2_1

for palliative care, in occupational therapy, and indeed in music therapy, in addition to innovative applications in composition and music performance. It should be mentioned, however, that although I have an avid interest in developing assistive technology for medical and special needs, there are a number of potentially interesting artistic uses of BCMI technology beyond such applications.

Plymouth University's Interdisciplinary Centre for Computer Music Research (ICCMR) is a main protagonist of the field of BCMI. This chapter reviews the pioneering research we have been developing at ICCMR for over a decade. Our approach is hands-on orientated. We often start by dreaming scenarios followed by implementing proof-of-concept or prototype systems. Then, as we test these systems, we learn what needs to be further developed, improved, discarded, replaced and so on. These often lead to new dreamed scenarios and the cycle continues incrementally. In reality, as we shall see below, vision, practice and theory do not necessarily take place sequentially in our research.

This chapter begins with a brief discussion introduction to the field and approaches to BCMI. Then, it introduces two BCMI systems that my team and I have designed in response to two dreamed scenarios:

1. Would it be possible to play a musical instrument with signals from the brain? No hands used.
2. Would it be possible to build a BCMI system for a person with locked-in syndrome to make music?

Next, I briefly discuss what I have learned from building these systems and identify challenges for making further progress. I suggest that one of the pressing challenges of BCMI research is to gain a better understanding of how the brain processes music, with a view on establishing detectable meaningful musical neural mechanisms for BCMI control. Then, I present two experiments aimed at gaining some of such understanding: one addressing active listening and the other addressing tonal processing. Each experiment is followed by an introduction to work-in-progress prototypes that I developed in response to the dreamed scenarios that emerged from the experiments.

1.2 Background to BCMI

Human brainwaves were first measured in the mid of 1920s by Hans Berger (1969). Today, the EEG has become one of the most useful tools in the diagnosis of epilepsy and other neurological disorders. In the early 1970s, Jacques Vidal proposed to use the EEG to interface with a computer in a paper entitled *Toward Direct Brain-Computer Communication* (Vidal 1973). Many attempts at using the EEG as a means to interface with machines followed with various degrees of success; for instance, in early 1990s, Jonathan Wolpaw and colleagues developed a prototype of a system that enabled primitive control of a computer cursor by subjects with severe motor deficits (Walpaw et al. 1991).

As for using EEG in music, as early as 1934 a paper in the journal *Brain* had reported a method to listen to the EEG (Adrian and Matthews 1934). But it is now generally accepted that it was composer Alvin Lucier who composed the first musical piece using EEG in 1965: *Music for Solo Performer* (Lucier 1976). Composers such as Richard Teitelbaum (1976), Rosenboom (1976) and a few others followed with a number of interesting ideas and pieces of music.

The great majority of those early pioneers who have attempted to use the EEG to make music have done so by direct sonification of EEG signals. However, in 1990, David Rosenboom introduced a musical system whose parameters were driven by EEG components believed to be associated with shifts of the performer's selective attention (Rosenboom 1990). Rosenboom explored the hypothesis that it might be possible to detect certain aspects of our musical experience in the EEG signal. This was an important step for BCMI research as Rosenboom pushed the practice beyond the direct sonification of EEG signals, towards the notion of digging for potentially useful information in the EEG to make music with.

1.3 Approaches to Brain–Computer Music Interfacing

Research into brain–computer interfacing (BCI) is concerned with devices whereby users voluntarily control a system with signals from their brain. The most commonly used brain activity signal in BCI is the EEG, which stands for electroencephalogram. In such cases, users must steer their EEG in a way or another to control the system. This informs the hard approach to BCMI: a system whereby the user voluntarily controls music. However, it is arguable that voluntary control may not be always necessary for a music system. For instance, a music system may simply react to the mental states of the user, producing music that is not necessarily explicitly controlled. We shall refer to such systems as soft BCMI, as opposed to hard BCMI. In this chapter, however, we will give focus to hard BCMI: we are interested in active, voluntary control of music. An example of passive soft BCMI is introduced in Chap. 13.

A hard BCMI system requires users to produce patterns of brain signals voluntarily to control musical output and this often requires training. Therefore, playing music with a BCMI should normally require ability and learning. This can be attractive for many individuals; for example, as an occupational therapeutic tool for severe physical impairment.

In a previous paper, we identified two approaches to control the EEG for a BCI: *conscious effort* and *operant conditioning* (Miranda et al. 2011). Conscious effort induces changes in the EEG by engaging in specific cognitive tasks designed to produce specific EEG activity (Curran and Stokes 2003; Miranda et al. 2005). The cognitive task that is most often used in this case is motor imagery because it is relatively straightforward to detect changes in the EEG of a subject imagining the movement of a limb such as, for instance, the left hand (Pfurtscheller et al. 2007). Other forms of imagery, such as auditory, visual and navigation imagery, can be used as well.

Operant conditioning involves the presentation of a task in conjunction with some form of feedback, which allows the user to develop unconscious control of the EEG. Once the brain is conditioned, the user is able to accomplish the task without being conscious of the EEG activity that needs to be generated (Kaplan et al. 2005).

Somewhere in between the two aforementioned approaches is a paradigm referred to as evoked potentials, or event-related potentials, abbreviated as ERP. ERP occur from perception of a user to an external stimulus or set of stimuli. Typically, ERP can be evoked from auditory, visual or tactile stimuli producing auditory-, visual- and somatosensory-evoked potentials, respectively. ERP are the electrophysiological response to a single event and therefore is problematic to detect in EEG on a single trial basis, becoming lost in the noise of ongoing brain activity. But if a user is subjected to repeated stimulation at short intervals, the brain's response to each subsequent stimulus is evoked before the response to the prior stimulus has terminated. In this case, a steady-state response is elicited, rather than left to return to a baseline state (Regan 1989).

For users with healthy vision and eye movements, the steady-state visual-evoked potential (SSVEP) is a robust paradigm for a BCI. And it has the advantage that it does not require much training in order to be operated satisfactorily. Typically, the user is presented with images, or simple images, on a standard computer monitor representing actions available to perform with the BCI; these could be, for instance, letters or geometrical figures. In order to make a selection, users must simply direct their gaze at the image corresponding to the action they would like to perform. The images must have a pattern reversing at certain frequency. As the user's spotlight of attention falls over a particular image, the frequency of the pattern reversal rate can be detected in the user's EEG through basic spectral analysis. What is interesting here is that once the target signal is detected in the EEG, it is possible to classify not only a user's choice of image, but also the extent to which the user is attending it (Middendorf et al. 2000). Therefore, each target is not a simple binary switch, but can represent an array of options depending on the user's attention.

1.4 BCMI-Piano

The BCMI-Piano resulted from the first aforementioned dream: Would it be possible to play a musical instrument with signals from the brain? No hands needed.

Initially, we looked into translating aspects of the EEG onto musical notes played on a synthesiser. However, this strategy proved to be unsatisfactory. The system did not convey the impression that one was playing a musical instrument. The notes sounded as if they were generated randomly and the synthesised sounds lacked the auditory quality that one would expect to hear from an acoustic musical instrument.

In order to remediate this, we connected the system to a MIDI-enabled acoustic piano (Fig. 1.1). That is, an acoustic piano that can be played by means of MIDI signals. MIDI stands for musical instrument digital interface. It is a protocol developed in the 1980s, which allows electronic instruments and other digital

Fig. 1.1 With BCMI-Piano one can play music generated on the fly on a MIDI-enabled acoustic piano with brain signals. No hands needed

musical devices and software to communicate with each other. MIDI itself does not make sound. Rather, it encodes commands, such as 'note on' and 'note off'. In our case, MIDI commands controlled a mechanism built inside the piano that moved the hammers to strike the strings. This resulted in a system whereby a real piano is played with brain signals. The quality of the sound improved considerably. But still, we felt that the result was not convincingly musical: the output sounded almost as if the notes were generated randomly. If we were to demonstrate that it is possible to play music with brain signals, then system ought to do more than merely associate brainwave activity with notes. Ideally, the music should result from some form of musical thinking detectable in the EEG. But the task of decoding the EEG of a person thinking of a melody, or something along these lines, is just impossible with today's technology.

Thus, I came up with the idea of endowing the machine with some form of musical intelligence, which could be steered by the EEG. The big idea was to programme the system with the ability to compose music on the fly, obeying simple abstract generic commands, which might be conveyed by something detectable in the EEG. This would not necessarily correlate to any musical thought at all, but it would at least be a realistic point of departure. For instance, I was aware that it is relatively straightforward to detect a pattern in the EEG, called alpha rhythm, which is present in the EEG of a person with eyes closed and in a state of relaxation.

Thus, my team and I moved on to implement BCMI-Piano, a system that looks for information in the EEG signal and match the findings with assigned generative musical processes corresponding to distinct musical styles. We implemented an artificial intelligence system that is able to generate pieces of piano music in the style of classic composers, such as Schumann, Satie, Beethoven, Mozart and so on. For instance, if the system detects prominent alpha rhythms in the EEG, then it would activate assigned processes that generate music in the style of Robert Schumann's piano works. Conversely, if it detected an EEG pattern other than alpha rhythms, then it would generate music in the style of Ludwig van Beethoven's sonatas for piano.

After a few trials, we decided to use seven channels of EEG, which can be obtained with seven pairs of electrodes placed on the scalp, covering a broad area of the head. The signals were filtered in order to tear out signal interference (e.g. interference generated on electrodes near the eyes due to eye blinking) and added their signals together prior to executing the analyses. The system analyses the spectrum of the EEG and its complexity. The analyses yield two streams of control parameters for the generative music system: one, which carries information about the most prominent frequency band in the signal—popularly referred to as EEG rhythms—and another, which carries a measure of the complexity of the signal. The former was used to control algorithms that generated the music, and the latter was used to regulate the tempo and the loudness of the music.

The most prominent EEG frequency band is obtained with a standard fast Fourier transform (FFT) algorithm, and the measure of complexity is obtained with Hjorth's analysis (Hjorth 1970).

FFT analysis is well known in BCI research and will be discussed in more detail in other chapters of this volume. Basically, the system looks for two patterns of information in the spectrum of the EEG: alpha and beta rhythms. Alpha rhythms are strong frequency components in the signal between 8 and 13 Hz and beta rhythms are strong components between 14 and 33 Hz.

The less familiar Hjorth's analysis is a time-based amplitude analysis, which yields three measurements: activity, mobility and complexity. The signal is measured for successive epochs—or windows—of one to several seconds. Activity and mobility are obtained from the first and second time derivatives of amplitude fluctuations in the signal. The first derivative is the rate of change of the signal's amplitude. At peaks and troughs, the first derivative is zero. At other points, it will be positive or negative depending on whether the amplitude is increasing or decreasing with time. The steeper the slope of the wave, the greater will be the amplitude of the first derivative. The second derivative is determined by taking the first derivative of the first derivative of the signal. Peaks and troughs in the first derivative, which correspond to points of greatest slope in the original signal, result in zero amplitude in the second derivative and so forth.

Amplitude fluctuations in the epoch give a measure of activity. Mobility is calculated by taking the square root of the variance of the first derivative divided by the variance of the primary signal. Complexity is the ratio of the mobility of the first derivative of the signal to the mobility of the signal itself; for instance, a sine wave has a complexity equal to 1. Figure 1.2 shows an example of Hjorth analysis. A raw EEG signal is plotted at the top (C:1), and its respective Hjorth analysis is plotted below: activity (C:2), mobility (C:3) and complexity (C:4). The tempo of the music is modulated by the complexity measure.

BCMI-Piano's music algorithm was developed with the assistance of Bram Boskamp, then a postgraduate student at ICCMR. It generates the music using rules that are deduced automatically from a given corpus of examples. It deduces sequencing rules and creates a transition matrix representing the transition logic of what follows what. New musical pieces are generated in the style of the ones of the training corpus. Firstly, the system extracts blocks of music and deduces the rules

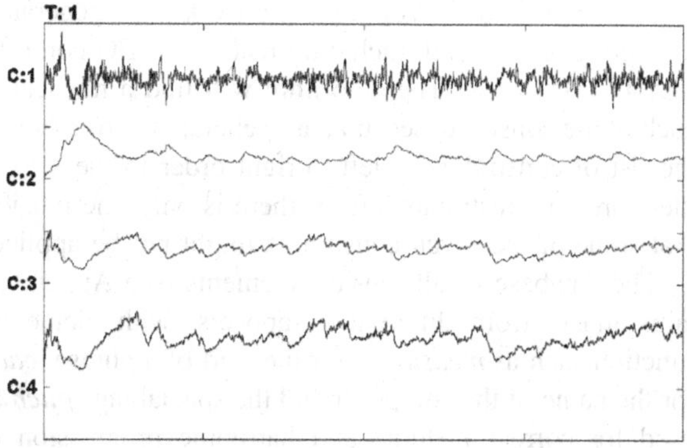

Fig. 1.2 A typical example of Hjorth analysis of an EEG signal

from the given examples. Then, it sequences those blocks in a domino-like manner based on the deduced rules (Miranda and Boskamp 2005).

Every time the system is about to produce a measure of music, it checks the power spectrum of the EEG at that moment and triggers the generative music instructions that are associated with the most prominent EEG rhythm in the signal. These associations are arbitrary and can be modified at will, which makes the system very flexible. The system is initialized with a reference tempo (e.g. 120 beats per minute), which is constantly modulated by Hjorth's measurement of complexity.

The EEG can influence the algorithm that generates the music in a well-defined way. We implemented a statistical predictor, which uses the deducted rules to generate short musical phrases with a beginning and an end that also allows for real-time steering with EEG information. The system generates musical sequences by defining top-level structures of sequences—referred below as sentences—and methods of generating similarity relationships or contrast relationships between elements. Consider the following example in LISP-like notation:

```
S -> (INC BAR BAR BAR BAR BAR HALF-CADENCE 8BAR-COPY)
```

From this top-level, the system retrieves rules for selecting a valid musical building block for each symbol (INC, BAR, etc.) and a rule for incorporating the EEG information in the generative process. For example:

```
INC -> ((EQUAL 'MEASURE 1)
  (EQUAL 'COMPOSER
  EEG-SET-COMPOSER))
BAR -> ((CLOSE 'PITCH 'PREV-PITCH-LEADING)
  (CLOSE 'PITCH-CLASS
      'PREV-PITCH-CLASS-LEADING)
  (EQUAL 'COMPOSER
  EEG-SET-COMPOSER))
```

Above is a definition of a network that generates a valid sentence with a beginning and an end, including real-time EEG control through the variable EEG-SET-COMPOSER. The algorithm will find a musical element in the database for each of the constraint sets that are generated above from INC and BAR, by applying the list of constraints in left-to-right order to the set of all musical elements until there are no constraints left, or there is only one musical element left. This means that some of the given constraints might not be applied.

The database of all musical elements (see Appendix for a short example) contains music from different composers, with elements tagged by their musical function such as *measure 1* for the start of a phrase, *cadence* for the end, *composer* for the name of the composer and the special tags *pitch* and *pitch-class* that are both used for correct melodic and harmonic progression or direction. The selection process is illustrated below.

The example database in the Appendix shows the main attributes that are used to recombine musical elements. P-CLASS (for *pitch-class*) is a list of two elements. The first is the list of start notes, transposed to the range of 0–11. The second is the list of all notes in this element (also transposed to 0–11). P is the *pitch* of the first (and highest) melodic note in this element, by matching this with the melodic note that the previous element was leading up to we can generate a melodic flow that adheres in some way to the logic of how the music should develop. The PCL (for *pitch-class leading*) elements contain the same information about the original next bar; this is used to find a possible next bar in the recombination process. Then, there are the INC, BAR and CAD elements. These are used for establishing whether those elements can be used for phrase starts (incipient), or cadence.

Simply by combining the musical elements with the constraint-based selection process that follows from the terminals of the phrase structure rewrite rules, we obtain a generative method that can take into account the EEG information. This generates musical phrases with building block connectivity like a domino game:

```
((EQUAL 'MEASURE 1)
(EQUAL 'COMPOSER EEG-SET-COMPOSER))
```

Assuming that there are also musical elements available from composers other than SCHU, the first constraint will limit the options to all *incipient* measures from all musical elements from all composers. The second constraint will then limit the options according to the current EEG analysis to the composer that is associated with the current EEG activity as follows:

```
((CLOSE 'PITCH 'PREV-PITCH-LEADING)
(CLOSE 'PITCH-CLASS
  'PREV-PITCH-CLASS-LEADING)
(EQUAL 'COMPOSER EEG-SET-COMPOSER))
```

In the given phrase structure, the rule that follows from BAR then defines the constraints put upon a valid continuation of the music. These constraints will limit the available options one by one and will order them according to the defined rule preferences. The CLOSE constraint will order the available options according to their closeness to the stored value. For example, after choosing:

```
(SCHU-1-1-MEA-1
  P-CLASS ((0 4) (0 3 4 6 7 9))
  P 76
  PCL ((2 7 11)(2 5 7 9 11))
  PL 83
  BAR INC
  CO SCHU)
```

as the beginning, PREV-PITCH-LEADING will have stored 83, and PREV-PITCH-CLASS-LEADING will have stored ((2 7 11) (2 5 7 9 11)). This will result in measure 2 and 4 being ranked highest according to both pitch and pitch-class, while measure 6 is also quite close according to pitch. This weighted choice will give a degree of freedom in the decision that is needed to generate pieces with an element of surprise. The music will not get stuck in repetitive loops, but it will find the closest possible continuation when no perfect match is available. We can still find a close match in this way if the third constraint eliminates all the obvious choices that are available, e.g. because a jump is requested to the musical elements of another composer, who might not use the same pitch-classes and pitches.

Figure 1.3 shows an example of resulting music with elements from the musical style of Schumann and Beethoven. In this example, the EEG jumped back and forth, from bar to bar, between the two styles. The harmonic and melodic distances are quite large from bar to bar, but they are the optimal choices in the set of chosen elements from the two composers.

After a few training sections, colleagues in the laboratory were able to increase and decrease the power of their alpha rhythms in relation to the beta rhythms practically at will, therefore being able to voluntarily switch between two styles of music. We noticed that the signal complexity measurement tended to be higher when beta rhythms were more prominent in the signal: the music in the style of Beethoven tended to be played slightly louder and faster than pieces in the style of Schumann.

At this point, I went on to address my second dreamed scenario: Would it be possible to build a BCMI system for a person with locked-in syndrome to make music?

As it turned out, after a critical evaluation of our system, informed by opinions and advice from health professionals and music therapists working with disable patients, I concluded that the BCMI-Piano was neither robust nor portable enough to be taken from the laboratory into the real world. The system comprised two laptops, two bulky hardware units for EEG amplification and too many dangling cables. Moreover, the skills required for placing the electrodes on the scalp and run

Fig. 1.3 An example output where the piece alternates between the styles of Schumann and Beethoven as the EEG jumps back and forth from bar to bar between alpha (between 8 and 13 Hz) and beta rhythms (between 14 and 33 Hz)

the various components of the system were time-consuming and far beyond the typical skills of a music therapist or carer. Also, it was generally agreed that, from the point of view of a user, the system would not give the feeling that they were really playing the piano or creating music. After all, it is the computer who composes the music; the user only switch between two modes of operation. I was advised that I should try to devolve the creative process to the user even if it is to create very simple music. I soon realised that this would require more options for control.

1.5 SSVEP-Music System

I teamed up with music therapist Wendy Magee and her colleagues at Royal Hospital for Neuro-disability, London, to develop a new system aimed at a trial with a locked-in syndrome patient, henceforth referred to Tester M. Tester M's only active movements following a stroke include eye movements, facial gestures and minimal head movements. She retained full cognitive capacity.

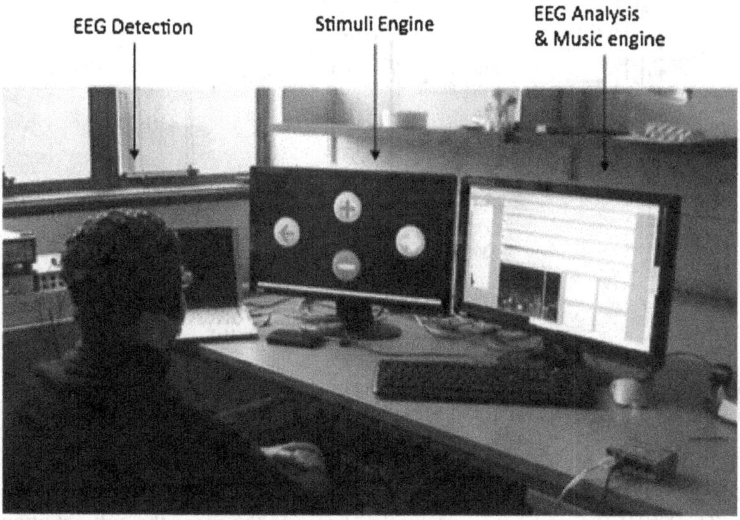

Fig. 1.4 Photograph of a subject operating the system

The two challenges that we had to address in this design were to devolve the creative process to the user and provide more options for control. Technically, a solution for the former would depend on the solution for the latter. Hence, we started by focusing on increasing the number of controls. To this end, we shifted from using EEG rhythms to adopting an evoked potential approach based on the SSVEP method that I mentioned earlier.

The new SSVEP-Music system was implemented in collaboration with Joel Eaton, a postgraduate research student at ICCMR, and John Wilson and Ramaswamy Palaniappan[1] of University of Essex (Miranda et al. 2011). Thanks to the SSVEP approach, we were able to implement four switches for control, as opposed to only one in BCMI-Piano. Moreover, each switch acted as a potentiometer for continuous control.

Figure 1.4 shows a photograph of a subject using the system. The monitor on the left hand side in Fig. 1.4 shows four images. These images flash at different frequencies, reversing their colours. Each image is associated with a musical task. Therefore, the system executes four different tasks, which the user can select by staring at the respective flashing image.

The SSVEP control signal can be used to generate the melody in a variety of ways, which can be customised. We provided a number of configurations that can be loaded into the system. For instance, suppose that the top image, shown on the monitor of the left hand side of the picture in Fig. 1.4, is associated with the task of generating a melody from an ordered set of five notes (Fig. 1.5). Let us say that this image flashes at a rate of 15 Hz. When one stares at it, the system detects that the

[1] Currently at University of Wolverhampton.

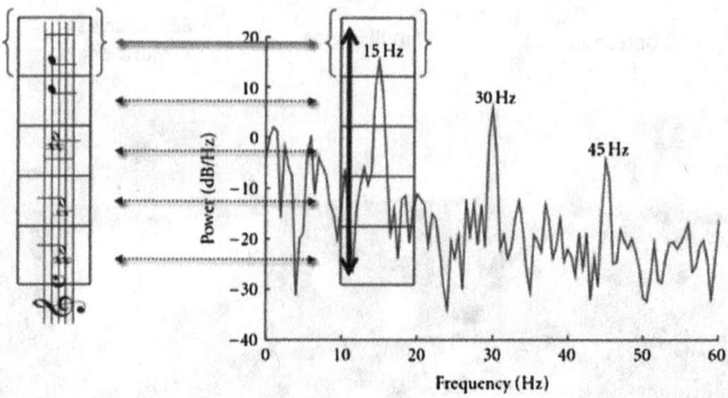

Fig. 1.5 Five bandwidths are established for the varying control signal, which are associated to the indices of an array of musical notes

subject is staring at this image and sends a command to generate the respective melody. The SSVEP paradigm is interesting because the more the subject attends to the flashing image, the more prominent is the amplitude of the EEG component corresponding to the brain's response to this stimulus. This produces a varying control signal, which is used to produce the melody and provide a visual feedback to the user: the size of the image increases or decreases in function of this control signal. In this way, we can steer the production of the notes by the intensity to which one attends to the respective image. One can bring the index down by looking away and bring the index up by staring at it again. Fine control of this variation can be achieved with practice, e.g. to repeat a single note many times or repeat a subgroup of notes. Each of the images could correspond to a distinct note sequence, and so on. This scheme opens up many possibilities for music control.

The SSVEP-Music system has proved to be rather successful because one can almost immediately produce musical notes with very little, or no training, simply by looking intently at the different images. As one learns to modulate the extent to which he or she is attending the images, more sophisticated musical control can be achieved, as if learning to play a musical instrument: the more one practices, the better one becomes at it.

Terster M trialled the system during a two-hour session. Being familiar with eye gaze technology for her alternative communication system, Tester M grasped the concept quickly and rapidly demonstrated her skills at playing the system with minimal practice. She was able to vary the intensity of her gaze, thus changing the amplitude of her EEG and vary the consequent melodic and dynamic output. Personal correspondence with Tester M following this trial communicated that she had enjoyed considerably using the system and that "…it was great to be in control again". This feedback is immensely gratifying and very encouraging. The possibilities for applying the system within group settings are immediately apparent and an exciting prospect for people with limited opportunities for participating as an equal partner in a group.

We are aware that some aspects of the system still require further refinement to make it more practical and viable for clinical applications. For instance, the frequency rate at which the images flash may limit using the system with people known to have epilepsy, a possible consequence following acquired brain injury.

The time required to place the electrodes on the scalp was reduced considerable: only three electrodes are required here. However, SSVEP-Music requires calibration to match the sensitivity of the system with the user's visual cortex response to the flashing images. This is outside typical clinical skills and can be time-consuming. The downside is that this calibration needs to be done before a session begins. Although this could be overcome with training, increasing the time burden of a clinical session is known to be a preventative factor influencing the uptake of technology by the health care sector.

Nevertheless, despite all the limitations, we succeeded in providing more control options and devolving the creative process to the user. And more importantly, we demonstrated that it is indeed possible to build a BCMI system for a person with locked-in syndrome to make music.

1.6 Discussion: Moving Forwards

In addition to the limitations discussed above, I am currently addressing two technical challenges, which I believe are important to move research into BCMI forwards:

a. Discovery of meaningful musical information in brain signals for control beyond the standard EEG rhythms.
b. Design of powerful techniques and tools for implementing flexible and sophisticated online generative music systems.

In order to address the former, I have been conducting brain scanning experiments aimed at gaining a better understanding of brain correlates of music cognition, with a view on discovering patterns of brain activity suitable for BCMI control. In the following section, I report on the results of two experiments: one on listening imagination and another on musical tonality.

As for the second challenge, it is important to ensure that the BCMI system offers an adequate musical repertoire or challenge to maintain the engagement of people who may have vastly sophisticated musical experiences and tastes. I have been looking into expanding the generative capabilities of the BCMI-Piano music algorithm by means of constraint satisfaction programming techniques.

1.7 Active Listening Experiment

This experiment was developed with Alex Duncan, a former postgraduate research student, and Kerry Kilborn and Ken Sharman, at University of Glasgow. The objective of the experiment was to test the hypothesis that it is possible to detect information in the EEG indicating when a subject is engaged in one of two mental tasks: *active listening* or *passive listening* (Miranda et al. 2003). In this context, the active listening task is to replay the experience of hearing some music, or part of that music, in the mind's ear. Conversely, passive listening is to listen to music without making any special mental effort. In day-to-day life experience, we are likely to be listening passively if we are relaxing to peaceful music or engaged in some other task while listening to music in the background.

Three non-musicians, young males, with age ranging between 18 and 25 years, participated in the experiment, which was divided into six blocks of trials, giving the participants the chance to relax. Each trial lasted for 8 sec and contained two parts: a rhythmic part, lasting for the entire trial, and a melodic riff part, lasting for the first half of the trial. A riff is a short musical passage that is usually repeated many times in the course of a piece of music. It was during the second half of each trial that the mental task was performed. The rhythmic part comprised four repetitions of a 1-bar rhythm loop. Two repetitions a 1-bar riff loop starting at the beginning of the trial and terminating halfway through were superimposed on the rhythmic part (Fig. 1.6).

In total, there were 15 unique riff loops: five played on a synthesised piano, five using an electronic type of timbre and five on an electric guitar. The music was in the style of a pop dance tune at 120 beats per minute, 4 beats per bar. The background rhythm looped seamlessly for the entire duration of each trial block. Blocks were named after the task the participant was instructed to perform on that block, and they were ordered as shown in Table 1.1. Each of the 15 riff parts was presented four times in each block in random order.

Participants were instructed to perform one of three mental tasks while listening to a continuous sequence of trials:

a. Active listening: listen to the looped riff that lasts for 2 bars, then immediately after it finishes, imagine that the riff continues for another 2 bars until the next trial begins.
b. Passive listening: listen to the entire 4-bar trial with no effort; just relax and focus on the continuing background part.
c. Counting task: listen to the looped riff that lasts for 2 bars, then immediately after it finishes, mentally count the following self-repeating sequence of numbers (i.e. mentally spoken): 1, 10, 3, 8, 5, 6, 7, 4, 2, 1, 10 and so forth.

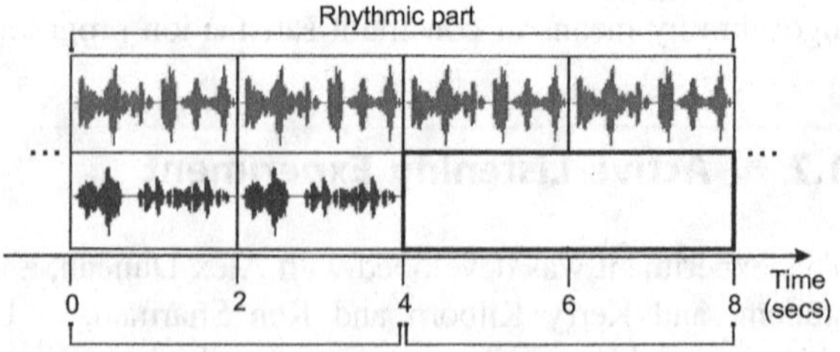

Fig. 1.6 Participants listened to 4-bar trials containing a looped riff lasting for 2 bars

Table 1.1 The experiment was divided into six blocks of trials. Blocks were named after the mental task the subjects were instructed to perform

Block	Subject 1	Subject 2	Subject 3
1	Active	Passive	Counting
2	Passive	Counting	Active
3	Counting	Active	Passive
4	Active	Passive	Counting
5	Passive	Counting	Active
6	Counting	Active	Passive

The classification task was to determine the class of 2-second multi-channel EEG segments, where class (1) = active listening, class (2) = passive listening and class (3) = counting task.

The counting task was included as a control task to determine whether the EEG features that might allow for the differentiation between the imagery and relaxed listening tasks are not merely a function of a concentrating versus a non-concentrating state of mind.

Only the last four seconds (i.e. the second half of each trial) were considered for analysis. These 4-second long segments were further divided into two 2-second long segments. Thus, each trial yielded two segments. There were 120 trials for each of the three conditions and each subject produced a total of 720 segments: 240 segments for each condition. The data are randomly partitioned into training set and testing set with split ratio of 9:1, resulting in 648 training segments and 72 testing segments.

We employed a linear auto-regression algorithm to represent the EEG data in a compressed form in terms of estimations of spectral density in time (Anderson and Sijercic 1996; Peters et al. 1997). Then, a classic single hidden-layer static neural network (multi-layer perceptron), with variable number of hidden units and up to three output units, was used for the classification task. The network was trained in batch mode for 50 epochs, using a scaled conjugate gradient algorithm, as described by Bishop (1995). The data were divided into two sets: a training set E and a test set T. The training set E was used to train the neural network to recognise the mental tasks of the elements that were left in T. In total, there were 768 inputs to the network. The network was reset, retrained and reassessed 10 times with different permutations of training and testing segments.

Classifications were made between 2-second long multi-channel segments belonging to pairs of conditions (for 2-way classification) and to all three conditions (for 3-way classification). The average classification scores, including confidence limits and standard deviation, for each subject are shown in Table 1.2.

Remarkably, the classification scores are above 90 % accuracy. We acknowledge that these results may not sound statistically robust because the experiment involved only three subjects. Nevertheless, they encouraged me to work towards the implementation of a BCMI on the assumption that the system would be capable to

Table 1.2 Average classification scores for the active listening experiment

Subject	Classification task	Mean	Min.	Max.	Deviation	Confidence
1	Active × passive	0.998	0.979	1.000	0.007	±0.007
	Active × counting	0.996	0.979	1.000	0.009	±0.009
	Passive × counting	0.994	0.979	1.000	0.010	±0.010
	Active × passive × counting	0.998	0.958	1.000	0.015	±0.016
2	Active × passive	0.994	0.979	1.000	0.010	±0.010
	Active × counting	0.973	0.896	1.000	0.031	±0.032
	Passive × counting	0.954	0.896	1.000	0.038	±0.039
	Active × passive × counting	0.951	0.903	0.986	0.023	±0.024
3	Active × passive	0.973	0.958	1.000	0.014	±0.014
	Active × counting	0.992	0.979	1.000	0.011	±0.011
	Passive × counting	0.994	0.958	1.000	0.014	±0.014
	Active × passive × counting	0.985	0.958	1.000	0.015	±0.016

establish if a subject is actively listening to music, or passively listening to it without any special mental effort. This notion is supported by a number of reports on experiments looking into musical imagination (Meister et al. 2004; Limb and Braun 2008; Miranda et al. 2005; Petsche et al. 1996).

1.7.1 Towards an Active Listening BCMI

The results from the above experiment encouraged me to look into the possibility of developing a BCMI whereby the user would be able to affect the music being generated in real time by focusing attention to specific constituents of the music. I designed a prototype, which produces two tracks of music of the same style of the music stimuli that was devised for the experiment: it comprises a rhythmic track and a solo track, which is generated by means of algorithms that transforms a given riff; they can transpose it, change rhythm, add a note, remove a note, play the riff backwards and so on.

Firstly, the neural network is trained to recognise when the incoming EEG corresponds to active or passive listening, as described in the experimental procedure. Needless to say, the person who controls the music here should be the same as the one who's EEG was used to train the system. The system works as follows: the rhythmic part is continuously played and a riff is played sporadically; an initial riff is given by default. Immediately after a riff is played, the system checks the subject's EEG. If it detects active listening behaviour, then the system applies some transformation on the riff that has just been played and plays it again. Otherwise, it does not do anything to the riff and waits for the subject's EEG response to the next sporadic riff. Sporadic riffs are always a repetition of the last played riff; in other words, it does not change until the system detects active listening behaviour.

In practice, I found it difficult to reliably detect active listening behaviour when a user is consciously trying to change the riffs online. Either more efficient EEG signal processing algorithms need to be employed or the paradigm is flawed, or both. More work is required to address this problem.

1.8 Neural Processing of Tonality Experiment

In Miranda et al. (2008) and Durrant et al. (2009), I introduced a functional magnetic resonance imaging (fMRI) study of tonality, which I developed with Simon Durrant, a former ICCMR research fellow, and Andre Brechmann, of the Leibniz Institute for Neurobiology, Germany. The objective of this experiment was to gain a better understanding of the neural substrates underlying the perception of tonality, with a view on developing a method to harness their behaviour to control a BCMI. We looked for differences in neural processing of tonal and atonal stimuli and also for neural correlates of distance around the circle of fifths, which describes how close one key is to another.

Tonality is concerned with the establishment of a sense of key, which in turn defines a series of expectations of musical notes. Within Western music, the octave is divided into twelve equal semitones, seven of which are said to belong to the scale of any given key. Within these seven tones, the first (or lowest) is normally referred to as the fundamental note of the chord and the one that the key is named after. A sense of key can be established by a single melodic line, with harmony implied, but can also have that harmony explicitly created in the form of chord progressions. Tonality defines clear expectations, with the chord built on the first tone (or degree) taking priority. The chords based on the fourth and fifth degrees also are important because their constituent members are the only ones whose constituent tones are entirely taken from the seven tones of the original scale and occurring with greater frequency than other chords. The chord based on the fifth degree is followed the majority of the time by the chord based on the first degree. In musical jargon, this is referred to as a dominant-tonic progression. This special relationship also extends to different keys, with the keys based on the fourth and fifth degrees of a scale being closest to an existing key by virtue of sharing all but one scale tone with that key. This gives rise to what is known as the circle of fifths, where a change—or modulation—from one key to another is typically to one of these other closer keys (Shepard 1982). Hence, we can define the closeness of keys based on their proximity in the circle of fifths, with keys whose first degree scale tones are a fifth apart sharing most of their scale tones, and being perceived as closest to each other (Durrant et al. 2009).

Sixteen volunteers, 9 females and 7 males, with age ranging between 19 and 31 years and non-musicians, participated in the experiment. Five experimental conditions were defined: *distant*, *close*, *same*, *initial* and *atonal* (that is, no key) conditions, respectively. As a result of the contiguity of groups, the first stimulus in each group followed the atonal stimulus in the previous group (except for the initial group), which was defined as the *initial* condition. The *distant* and *close* conditions therefore defined changes from one key to another (distant or close, respectively), whereas the *same* condition defined no change of key (i.e. the next stimulus was in the same key). The *atonal* condition defined a lack of key, which was included here as a control condition. The stimuli were ordered such that all tonal stimuli were used an equal number of times, and the conditions appeared in all permutations equally in order to control for order effects.

Each stimulus consisted of 16 isochronous events lasting 500 ms each, with each stimulus therefore lasting 8 s without gaps in between. Each event consisted of a chord recognised in Western tonal music theory, with each chord being in root position (i.e. the lowest note of the chord is also the fundamental note). The sense of key, or lack of it, was given by the sequence of 16 chords, rather than by individual chords. For a single run, stimuli were ordered into twenty-four groups of three stimuli with no gaps between stimuli or groups. The first stimulus in each group was always a tonal stimulus presented in the home key of C major, and the second was always a tonal stimulus that could either be in the distant key of F# major, the closely related key of G major or the same key of C major. In order to reset the listener's sense of relative key, the third stimulus in each group was always an atonal stimulus, that is, the chord sequences without any recognisable key (Fig. 1.7).

In order to draw the attention of the participants to the tonal structure of the stimulus stream, the behavioural task in the scanner was to click the left mouse button when they heard a change to a different key (*distant*, *close* and *initial*

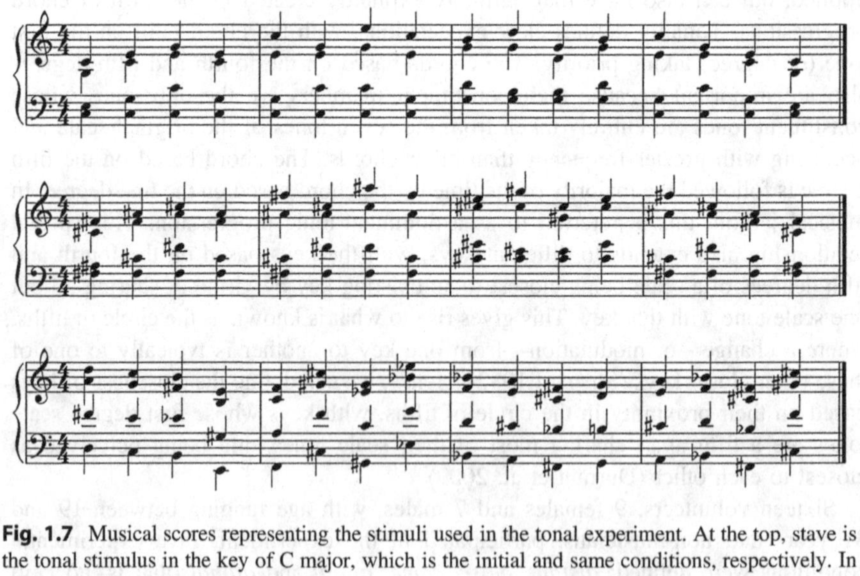

Fig. 1.7 Musical scores representing the stimuli used in the tonal experiment. At the top, stave is the tonal stimulus in the key of C major, which is the initial and same conditions, respectively. In the middle is the stimulus in the key of F# major, which is the distant condition. At the bottom is the stimulus in no obvious key

conditions), and right-click the mouse button when they heard a change to no key (*atonal* condition). As the participants were non-musicians, the task was explained as clicking in response to a given type of change so as to avoid misunderstandings of the meaning of the terms 'tonal', 'atonal' and 'key'. That is, they were instructed to indicate any change from one key to another by clicking on the left button of a mouse and a change towards a sequence with no key by clicking on the right button. Subjects were given an initial practice period in order to ensure that they understood the task.

The results of the behavioural tasks are shown in Table 1.3, which gives the percentage of trials that contained a left- or right-click for each condition. Second-level one-way analysis of variance (ANOVA) was performed for the left-click and right-click results, respectively, across all participants. *Distant*, *close* and *initial* conditions had a significantly higher number of left-click responses than for conditions *same* and *atonal*. Conversely, the *atonal* condition had a significantly higher amount of right mouse clicks than for *distant*, *same* and *atonal* conditions. These results confirm that the participants were able to perform the behavioural task satisfactorily and show that the participants had some awareness of the tonal structure of the stimuli.

As for the fMRI scanning, functional volumes were collected with 3 Tesla scanner using echo planar imaging. A more detailed description of the data acquisition procedures and analysis methods is beyond the scope of this chapter. In summary, each stimulus block lasted 8 s and was immediately followed by the next stimulus block. Analysis was performed with a general linear model (GLM) (Friston et al. 2006).

Group analysis revealed a cluster of fMRI activation around the auditory cortex (especially in the left hemisphere) showing a systematic increase in blood-oxygen-level-dependent (BOLD) amplitude with increasing distance in key. We have found a number of significant active neural clusters associated with the processing of tonality, which represent a diverse network of activation, as shown in Table 1.4 and Fig. 1.9.

We note the strong presence of medial structures, in particular *cingulate cortex* (label 5 in Fig. 1.8 and Table 1.4) and *caudate nucleus* (label 4 in Fig. 1.8 and Table 1.4) in response to key changes. Also significant is the bilateral activation for key changes of the *transverse temporal gyrus* also known as Heschl's gyrus (labels

Table 1.3 Behavioural results, showing the percentage of trials that contained a left- or right-click aggregated over all participants in the experiment

Condition	Left-click	Right-click
Distant	89.453	11.328
Close	83.594	0.7812
Same	26.563	3.5156
Initial	68.62	4.2969
Atonal	14.193	83.984

Table 1.4 Anatomical results of GLM analysis contrasting conditions with and without a key change

Anatomical name	*X*	*Y*	*Z*	Cluster
(1) Left transverse temporal gyrus	−51	−18	11	981
(2) Right insula	36	17	13	948
(3) Right lentiform nucleus	24	−1	1	750
(4) Right caudate	14	−4	22	1,443
(5) Left anterior cingulate	−1	41	11	2,574
(6) Left superior frontal gyrus	−12	50	36	2,241
(7) Right transverse temporal gyrus	51	−17	10	1,023

These active clusters preferentially favour key change stimuli. *X*, *Y* and *Z* are Talairach coordinates for plotting scans onto a standard template after normalisation of brain size and shape across the subjects

1 and 7 in Fig. 1.8 and Table 1.4), which contains the primary auditory cortex. The activation curves for the bilateral activation of the *transverse temporal gyrus* show strongest activity for the distant key changes, slightly less, but still significant

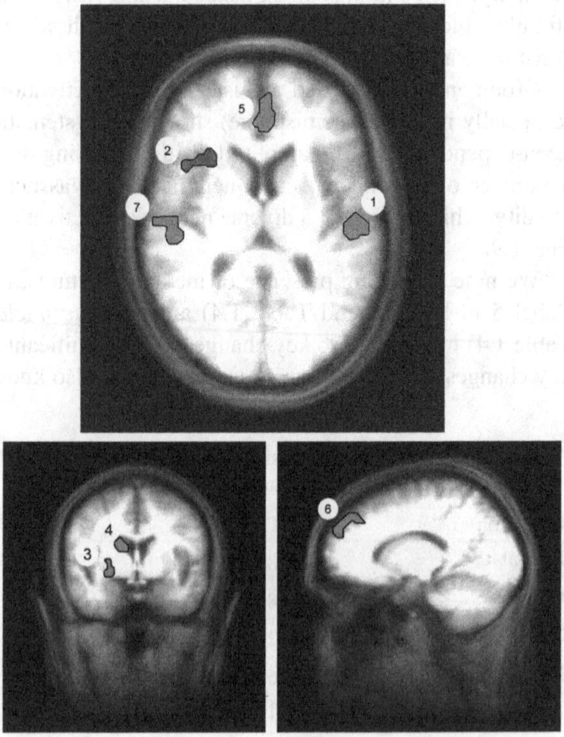

Fig. 1.8 Examples of clusters of activation for the contrast distant and close key versus same key, including bilateral activation of transverse temporal gyrus for which the activation curves are shown in Table 1.4

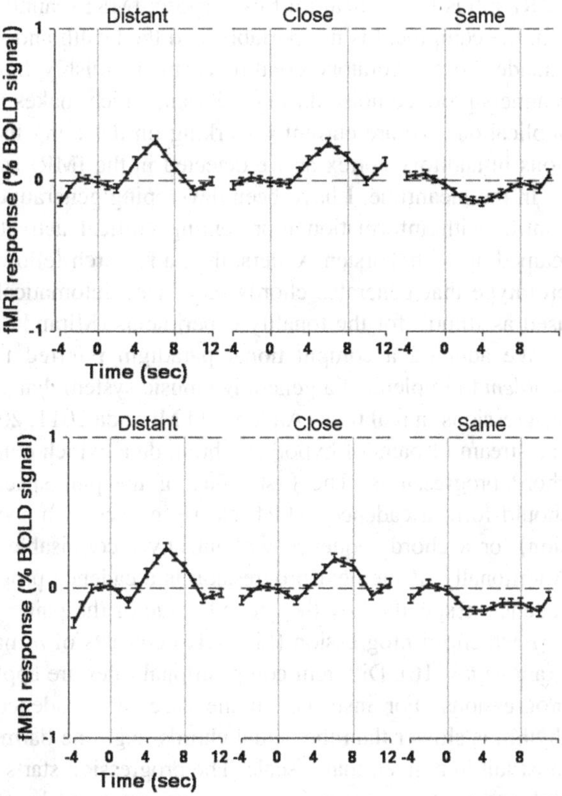

Fig. 1.9 Activation curves in *left* (*top graph*) and *right* (*bottom graph*) transverse temporal gyri for *distant* condition (*plot on the left side*), *close* condition (*plot in the middle*) and *same* condition (*plot on the right side*)

activity for the close key changes, and much less activity for no key changes (Fig. 1.9). It should be emphasised that this occurred across a variety of different stimuli, all of equal amplitude and with very similar basic auditory features, such as envelope and broad spectral content. Both left and right *transverse temporal gyri* showed very similar response curves highlighting the robust nature of these results. This might suggest that these areas may not be limited to low-level single note processing as commonly thought, but also are involved in some higher-order sequence processing. This is significant for my research as it could constitute a potential source of control information for a BCMI, associated with tonality and modulation. However, more testing needs to be developed in order to probe this.

1.8.1 Towards a BCMI for Controlling Tonality

The results of the tonality experiment suggest that it might indeed be possible to design a BCMI controlled with auditory cortex activity correlated to tonality. However, despite ongoing attempts at using fMRI for BCI (Weiskopf et al. 2004), fMRI still is impractical for this purpose: fMRI scanning is simply too expensive to run, the equipment is not portable, and the health and safety implications for usage outside strict laboratory conditions are fiendishly burdensome. Moreover, fMRI scanners produce noise during the scan, which makes it inconvenient for a musical application. We are currently working on detecting in the EEG equivalent activations in auditory cortex as we detected in the fMRI scans.

In the meantime, I have been developing generative music systems suitable for control with information representing cortical activations of tonal processing. I teamed up with Torsten Anders, then a research fellow at ICCMR, to implement a prototype that generates chords sequences automatically, in the style of the ones used as stimuli for the tonality experiments (Miranda et al. 2008).

We adopted a computational paradigm referred to as *constraint satisfaction problem* to implement a generative music system that generates sequences of chord progressions in real time (Anders and Miranda 2011, 2010). The input to the system is a stream of pairs of hypothetic brain data, which controls higher-level aspects of chord progressions. The first value of the pair specifies whether a progression should form a cadence, which clearly expresses a specific key (cadence progression), or a chord sequence without any recognisable key (key-free progression). Additionally, if the next progression is a cadence progression, then the key of the cadence is specified by the second value of the pair.

Each chord progression (Fig. 1.10) consists of n major or minor chords (in the example $n = 16$). Different compositional rules are applied to cadence and key-free progressions. For instance, in the case of a cadence, the underlying harmonic rhythm is slower than the actual chords (e.g. one harmony per bar), and all chords must fall in a given major scale. The progression starts and ends in the tonic chord, and intermediate root progressions are governed by Schoenberg's rules for tonal harmony (Schoenberg 1986). For a key-free, atonal progression, the rules established that all 12 chromatic pitch classes are used. For example, the roots of consecutive chords must differ and the set of all roots in the progression must express the chromatic total. In addition, melodic intervals must not exceed an octave. A custom dynamic variable ordering scheme speeds up the search process by visiting harmony variables (the root and whether it is major or minor), then the pitches' group (or classes) and finally the pitches themselves. The value ordering is randomized, so the system always produces different results.

As it is, the generative system design assumes that subjects would be able to produce the required control information in some way or another. In practice, however, it is unlikely that subjects would learn to produce bilateral activations of transverse temporal gyrus simply by imagining tonal progressions. The challenge here is to establish effective ways to embed in a realistic system design the theoretical understanding of neural correlates of tonal processing and generative musical algorithms. The research continues.

Fig. 1.10 Extract from a sequence of chord progressions generated by our constraints-based generative system. In this case, the system produced a sequence in C major, followed by a sequence in no particular key and then a sequence in A major

1.9 Concluding Remarks

There has been a tremendous progress in the field of BCI in the last decade or so, in particular on EEG signal processing aspects, which is the focus of a number of chapters in this volume. BCMI research is obviously benefiting from this progress. From the experienced I have gained with trialling the SSVEP-Music with a patient in a hospital setting, I learned the hard way that the hardware aspect of BCI lags behind the more theoretical advances in the field. The EEG electrode technology that is currently commercially available is adequate for medical diagnosis, but not for wearing on a more frequent and ad hoc basis. Dangling wires, electrodes cap, gel, required technical support for handling and so on need to disappear from the equation in order to pave the way for BCMI systems into the real world of health care. The equipment must be simple to switch on, set up and operate. Fortunately, the electronics industry is making continuing progress at this front: wireless electrodes that do not need gel are beginning to emerge in the market and more mobile, less conspicuous, good quality EEG amplifiers are becoming available—albeit good quality equipment still is not generally affordable.

From the musical side, my ICCMR team and I are continuously paving the way for the development of effective music algorithms for BCMI. I believe that an approach combining the technique developed for the BCMI-Piano and the constraints-based system built after the tonality experiment is a viable way to proceed, and I will continue working towards this goal.

The issue of harnessing the EEG for BCMI control with signals correlated to music cognition remains unresolved. It turns out that the most effective control methods are those that are not at all related to music, such as the SSVEP method. It is questionable whether the types of music cognition I touched upon in this chapter are the way forward or not. Much research is needed in order to make progress at this front.

1.10 Questions

1. Is voluntary control always necessary in BCI? Give some examples to illustrate your answer.
2. What is the difference between these three approaches of BCI control: conscious effort, operational conditioning and evoked potentials?
3. How does BCMI-Piano regulate the tempo and loudness of the music?
4. Why was the BCMI-Piano system not suitable for trial in a clinical context?
5. What are the differences between the SSVEP-Music and BCMI-Piano systems? Elaborate on advantages and disadvantages of both systems.
6. How many degrees of freedom are afforded by the SSVEP-Music system?
7. The chapter described a method to generate a melody using the SSVEP control signal. Could you envisage how this be done differently?
8. Why was the counting task included in the active listening experiment?
9. Given the state of the art of fMRI technology, would it be viable to build an fMRI-based BCMI?
10. How can fMRI technology help to advance research into the design of more sophisticated EEG-based BCMI systems?

1.11 Appendix: Database of Musical Elements

An excerpt from a database of musical elements where **CO** = composer (SCHU = Robert Schumann.), **P-CLASS** = pitch class, **P** = pitch, **PCL** = pitch-class leading, **PL** = pitch leading and **TPE** = type.

ID	SCHU-1-1-CAD
CO	SCHU
P-CLASS	((0 2 7)(0 2 4 5 7 11))
P	74
PCL	((0 4 9)(0 2 4 5 7 9 11))
PL	76
TPE	CAD

ID	SCHU-1-1-MEA-6
CO	SCHU
P-CLASS	((5 9)(0 5 7 9))
P	81
PCL	((0 2 7)(0 2 4 5 7 11))
PL	74
TPE	BAR

ID	SCHU-1-1-MEA-5
CO	SCHU
P-CLASS	((0 4)(0 4 7))
P	76
PCL	((5 9)(0 5 7 9))
PL	81
TPE	BAR

ID	SCHU-1-1-MEA-4
CO	SCHU
P-CLASS	((0 4)(0 3 4 6 7 9))
P	83
PCL	((0 4)(0 4 7))
PL	76
TPE	BAR

ID	SCHU-1-1-MEA-3
CO	SCHU
P-CLASS	((0 4)(0 3 4 6 7 9))
P	76
PCL	((2 7 11)(2 5 7 9 11))
PL	83
TPE	BAR

ID	SCHU-1-1-MEA-2
CO	SCHU
P-CLASS	((2 7 11)(2 5 7 9 11))
P	83
PCL	((0 4)(0 3 4 6 7 9))
PL	76
TPE	BAR

ID	SCHU-1-1-MEA-1
CO	SCHU
P-CLASS	((0 4)(0 3 4 6 7 9))
P	76
PCL	((2 7 11)(2 5 7 9 11))
PL	83
TPE	INC

References

Anders T, Miranda ER (2011) A survey of constraint programming systems for modelling music theories and composition. ACM Comput Surv 43(4):30

Anders T, Miranda ER (2010) Constraint application with higher-order programming for modeling music theories. Comp Music J 34(2):25–38

Anderson C, Sijercic Z (1996) Classification of EEG signals from four subjects during five mental tasks. In: Solving engineering problems with neural networks: proceedings of the conference on engineering applications in neural networks (EANN'96), pp 507–414

Adrian ED, Matthews BHC (1934) The Berger rhythm: potential changes from the occipital lobes in man. Brain 57(4):355–385

Berger H (1969) On the electroencephalogram of man. In: The fourteen original reports on the human electroencephalogram, electroencephalography and clinical neurophysiology. Supplement No. 28. Elsevier, Amsterdam

Bishop C (1995). Neural networks for pattern recognition. Oxford University Press

Curran EA, Stokes MJ (2003) Learning to control brain activity: a review of the production and control of EEG components for driving brain-computer interface (BCI) systems. Brain Cogn 51(3):326–336

Durrant S, Hardoon DR, Brechmann A, Shawe-Taylor J, Miranda ER, Scheich H (2009) GLM and SVM analyses if neural response to tonal and atonal stimuli: new techniques and a comparison. Connection Science (2)1:161–175

Friston KJ, Ashburner JT, Kiebel SJ, Nichols TE, Penny WD (2006) Statistical parametric mapping: the analysis of functional brain images. Academic Press, London

Hjorth B (1970) EEG analysis based on time series properties. Electroencephalogr Clin Neurophysiol 29:306–310

Kaplan A, Ya Kim JJ, Jin KS, Park BW, Byeon JG, Tarasova SU (2005) Unconscious operant conditioning in the paradigm of brain-computer interface based on color perception. Int J Neurosci 115:781–802

Limb CJ, Braun AR (2008) Neural substrates of spontaneous musical performance: an fMRI study of jazz improvisation. PLoS ONE 3(2):e1679. doi:10.1371/journal.pone.0001679

Lucier A (1976) Statement on: music for solo performer. In: Rosenboom D (ed) Biofeedback and the arts, results of early experiments. Aesthetic Research Center of Canada Publications, Vancouver

Meister IG, Krings T, Foltys H, Boroojerdi B, Muller M, Topper R, Thron A (2004) Playing piano in the mind—an FMRI study on music imagery and performance in pianists. Cogn Brain Res 19(3):219–228

Middendorf M, McMillan G, Calhoun G, Jones KS (2000) Brain-computer interfaces based on the steady-state visual-evoked response. IEEE Trans Rehabilitation Eng 8:211–2140

Miranda ER, Magee W, Wilson JJ, Eaton J, Palaniappan R (2011) Brain-computer music interfacing (BCMI): from basic research to the real world of special needs. Music Med. doi:10.1177/1943862111399290

Miranda ER, Durrant S, Anders T (2008) Towards brain-computer music interfaces: progress and challenges. In: Proceedings of International Symposium on Applied Sciences in Bio-Medical and Communication Technologies (ISABEL2008). Aalborg (Denmark)

Miranda ER, Boskamp B (2005) Steering generative rules with the eeg: an approach to brain-computer music interfacing, In: Proceedings of Sound and Music Computing 05. Salerno (Italy)

Miranda ER, Roberts S, Stokes M (2005) On generating EEG for controlling musical systems. Biomed Tech 49(1):75–76

Miranda ER, Sharman K, Kilborn K, Duncan A (2003) On harnessing the electroencephalogram for the musical braincap. Comp Music J 27(2):80–102

Peters BO, Pfurtscheller G, Flyvberg H (1997) Prompt recognition of brain states by their EEG signals. Theory Biosci 116:290–301

Petsche H, von Stein A, Filz O (1996) EEG aspects of mentally playing an instrument. Cogn Brain Res 3(2):115–123

Pfurtscheller G, Muller-Putz GR, Graimann B, Scherer R, Leeb R, Brunner C, Keinrath C, Townsend G, Nateem M, Lee FY, Zimmermann D, Höfler E (2007) Graz-brain-computer interface: state of researc. In: Dornheg G, Millán JM, Hinterberger T, McFarland D, Müller K-R (eds) Toward brain-computer interfacing. The MIT Press, Cambridge, pp 65–84

Regan D (1989) Human brain electrophysiology: evoked potentials and evoked magnetic fields in science and medicine. Elsevier, Amsterdam

Rosenboom D (ed) (1976) Biofeedback and the arts, results of early experiments. Aesthetic Research Center of Canada Publications, Vancouver

Rosenboom D (1990) Extended musical interface with the human nervous system. Leonardo Monograph Series No. 1. International Society for the Arts, Science and Technology, Berkeley

Shepard RN (1982) Structural representations of musical pitch. In Deutsch D (ed) The psychology of music. Oxford University Press, Oxford, pp 344–390

Schoenberg A (1986) Harmonielehre. Universal Edition. (7th Edition)

Teitelbaum R (1976) In Tune: Some Early Experiments in Biofeedback Music (1966–1974).In: Rosenboom D (ed) Biofeedback and the arts, results of early experiments. Aesthetic Research Center of Canada Publications, Vancouver

Vidal JJ (1973) Toward direct brain-computer communication. In: Mullins LJ (ed) Annual review of biophysics and bioengineering, pp 157–80

Walpaw J, McFerland D, Neat G, Forneris C (1991) An EEG-based brain-computer interface for cursor control. Electroencephalogr Clin Neurophisiol 78(3):252–259

Weiskopf N, Mathiak K, Bock SW, Scharnowski F, Veit R, Grodd W, Goebel R, Birbaumer N (2004) Principles of a brain-computer interface (BCI) based on real-time functional magnetic resonance imaging (fMRI). IEEE Trans Biomed Eng 51(6):966–970

Electroencephalogram-based Brain–Computer Interface: An Introduction

2

Ramaswamy Palaniappan

Abstract

Electroencephalogram (EEG) signals are useful for diagnosing various mental conditions such as epilepsy, memory impairments and sleep disorders. Brain–computer interface (BCI) is a revolutionary new area using EEG that is most useful for the severely disabled individuals for hands-off device control and communication as they create a direct interface from the brain to the external environment, therefore circumventing the use of peripheral muscles and limbs. However, being non-invasive, BCI designs are not necessarily limited to this user group and other applications for gaming, music, biometrics etc., have been developed more recently. This chapter will give an introduction to EEG-based BCI and existing methodologies; specifically those based on transient and steady state evoked potentials, mental tasks and motor imagery will be described. Two real-life scenarios of EEG-based BCI applications in biometrics and device control will also be briefly explored. Finally, current challenges and future trends of this technology will be summarised.

2.1 Introduction

Brain–computer interface (BCI) is a revolutionary field of science that is rapidly growing due to its usefulness in assisting disabled patients as it provides a direct mechanism of controlling external devices through simple manipulation of brain thoughts (Nicolas-Alonso and Gomez-Gil 2012). Disabled individuals here could

R. Palaniappan (✉)
Department of Engineering, School of Science and Engineering,
University of Wolverhampton, Telford, UK
e-mail: palani@wlv.ac.uk; palani@iee.org

E.R. Miranda and J. Castet (eds.), *Guide to Brain-Computer Music Interfacing*,
DOI 10.1007/978-1-4471-6584-2_2

be those that have lost most or all motor functions (known as 'locked in' syndrome) due to progressive neuromuscular diseases like amyotrophic lateral sclerosis (ALS) or muscular dystrophy or non-progressive such as stroke, traumatic brain injury and spinal cord injury. The BCI approaches for such individuals could be used in control of wheelchair, prosthesis, basic communication etc., as shown in Fig. 2.1. These users could use BCI to communicate with others to express their needs, feelings, etc. A simple example could be of a communication BCI system such as brain controlled word processing software.

However, in recent years, other industries have taken interest in this field where applications related to biometrics (Palaniappan 2008), games (Hasan and Gan 2012), cursor control (Wilson and Palaniappan 2011) etc., have emerged. Table 2.1 gives a non-exhaustive list of possible applications of BCI for both disabled and healthy individuals.

In general, there are two categories of BCI: invasive and non-invasive methods. Invasive BCI methods such as electrocorticogram (ECoG) have shown excellent performance in human (Langhenhove et al. 2008) and monkey (Borton et al. 2013). Nevertheless, non-invasive approaches based on electroencephalogram (EEG), magnetoencephalogram (MEG), positron emission topography (PET), functional magnetic resonance imaging (fMRI) and near-infrared spectroscopy (NIRs) are more popular as it is safer (minimal risk of infection etc.).

Among these non-invasive methods, EEG-based BCI is preferred due to it being practical (i.e. cheap and portable). We will focus on EEG-based BCI techniques here, specifically on transient visual evoked potential (better known as P300), motor imagery, steady-state visual evoked potential (SSVEP), mental tasks and briefly on slow cortical potential (SCP). Figure 2.2 shows a block diagram of the components involved in the processing of EEG data to implement a BCI.

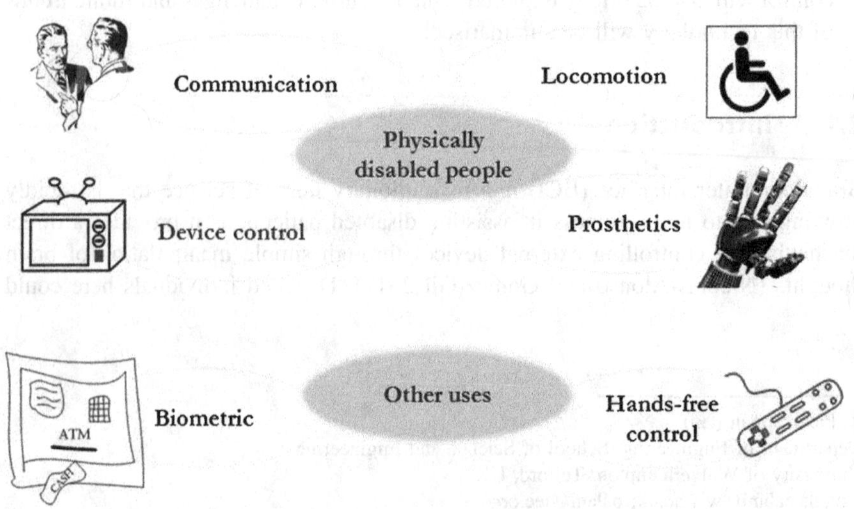

Fig. 2.1 Brain–computer interface applications

Table 2.1 Examples of possible BCI applications for disabled and healthy individuals

Disabled individuals	Healthy individuals
Restoring mobility—e.g. to control wheelchair movement	(*Mainly control of external devices*)
Environmental control—e.g. to control TV, power beds, thermostats, etc.	Mouse control in PC when fingers are on the keyboard
Prosthetics control (motor control replacement) —to control artificial limbs	Playing musical instruments by thoughts
Rehabilitative (assistive) control—to restore motor control (e.g.: strengthen/improve weak muscle)	Virtual reality
	Computer games (e.g. Mind Pacman)
	Flight/space control (pilots, astronauts)
	Biometrics

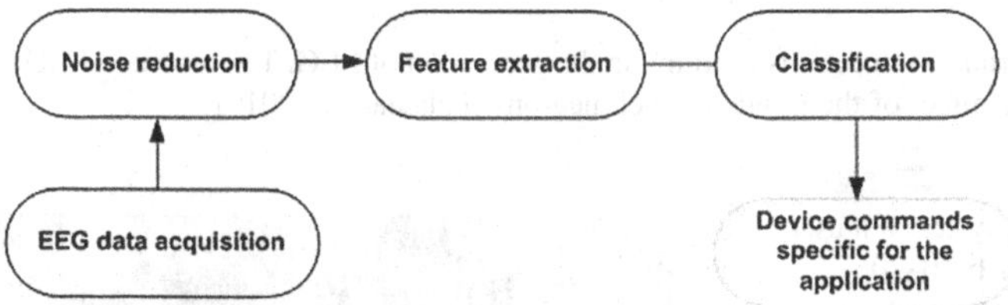

Fig. 2.2 EEG data processing for a BCI

2.2 Electroencephalogram

Acquiring electroencephalogram (EEG) is the first step in the BCI design. EEG is a type of oscillating electrical potential recorded from the scalp surface. It is generated by neuronal activations in the brain (as shown in Fig. 2.3) and is very small in amplitude (in the μV range) due to the attenuation caused by the skull and scalp. Evoked potentials are a specific type of EEG evoked during a stimulus like visual, auditory, etc.

EEG is usually recorded from a number of electrodes on the scalp. A standard electrode (channel) configuration is the 10–20 electrode system (Jasper 1958) of 19 active electrodes and two mastoids (reference) as shown in Fig. 2.4. However, it is common to extend this configuration and use higher number of channels such as 32, 64, 128 and even 256. The electrode locations are prefixed by a letter denoting the cortical area followed by a number (even for the right hemisphere and odd for the left). The prefix letter F stands for frontal, similarly C for central, P for parietal and O for occipital. The electrodes (normally made with Ag/AgCl) are used with gel to increase the conductance between scalp and electrodes but there are more recent advances in using dry electrodes made from gold. It is also common to re-reference the EEG using methods such as common averaging and Laplacian (Neuper and Klimesch 2006).

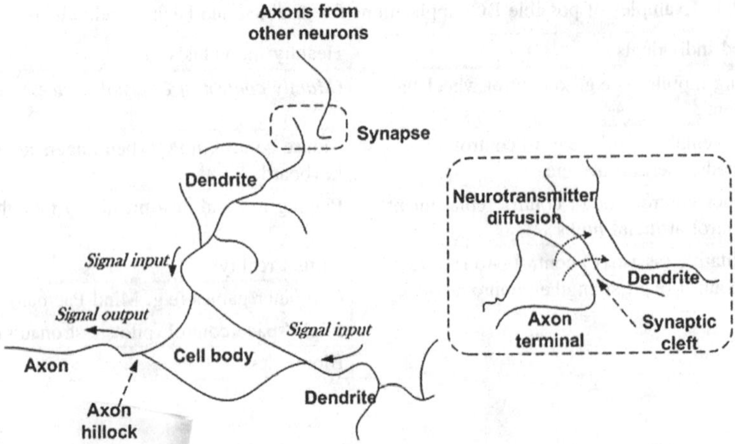

Fig. 2.3 Neuronal connections resulting in the generation of EEG. The recorded EEG is normally the cumulative effect of thousands of such neurons (Palaniappan 2010)

Fig. 2.4 BCI system **a** a user using BCI and **b** 10–20 electrode configuration

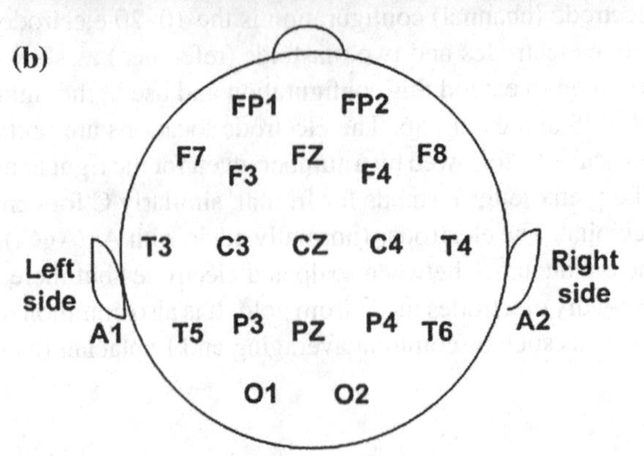

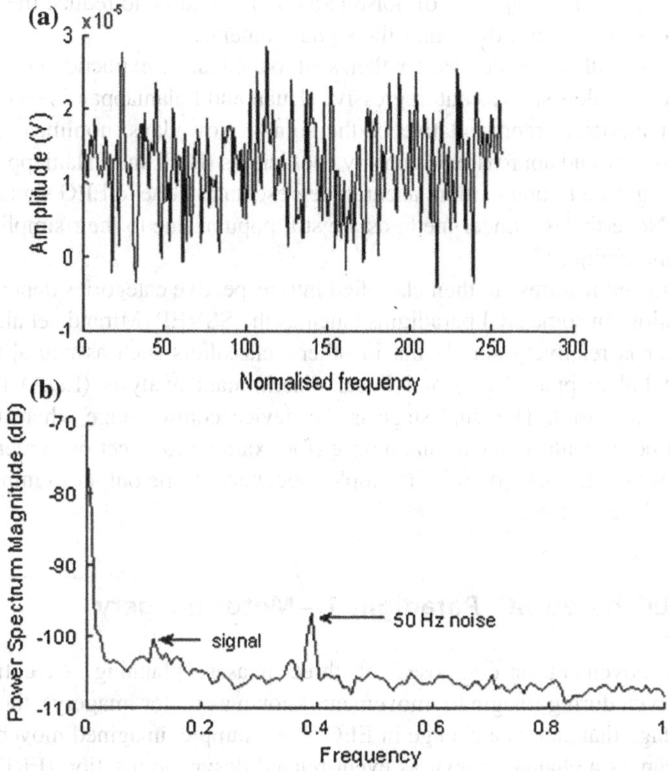

Fig. 2.5 **a** example of recorded EEG and **b** power spectral density of EEG showing signal and noise frequencies

As the EEG is of very small amplitude, it is normally amplified and converted to digital using an analog-to-digital converter. The digital conversion using sampling rates such as 256 Hz[1] is necessary to process the EEG signal using digital devices (like computers).

The EEG is normally obtained using certain BCI paradigms (to be discussed later), and the first-processing step is to reduce noise such as muscle artefacts, power line interference and other random noises from the EEG signals. Frequency-specific filtering using digital filters is commonly employed to filter the noise from the EEG; recently more sophisticated methods such as principal component analysis (PCA) and independent component analysis (ICA) have been employed. Figure 2.5a shows an example of the recorded EEG (using the SSVEP BCI paradigm) corrupted with power line interference. It can be seen the occurrence of this 50 Hz noise along with the signal frequency in the power spectral density plot of the

[1] With 256 Hz sampling rate, one second EEG will have 256 data points, other sampling rate up to 2,048 Hz is common.

EEG in Fig. 2.5b. The objective of noise reduction would be to reduce the noise as much as possible without distorting the signal contents.

The signals with noise reduced are then sent to the feature extraction stage, where mathematical models such as autoregressive (Huan and Palaniappan 2004) are used to extract parameters representative of the signal. Nowadays, nonlinear methods such as Lypunov and approximate entropy coefficients (Balli and Palaniappan 2013) are also being used to obtain more accurate representation due to EEG signals being nonlinear. Nevertheless, linear methods are still popular due to their simplicity and ease of computation.

The extracted features are then classified into respective categories depending on the application. In some BCI paradigms [such as the SSVEP (Miranda et al. 2011)], the classifier is relatively simple but in others, classifiers such as neural network (Huan and Palaniappan 2004) and linear discriminant analysis (LDA) (Asensio et al. 2011) are used. The final stage is the device control stage where the BCI output is used to control an external device (for example to select on-screen menus or move a wheelchair). Certain BCIs employ feedback of the output to improve the reliability of the system.

2.3 EEG-based BCI Paradigm 1—Motor Imagery

Voluntary movement is composed of three phases: planning, execution and recovery. Even during imaginary movement (known as motor imagery), there is the planning stage that causes a change in EEG. For example, imagined movements of left hand causes a change known as event-related desynchronisation (ERD) in the right motor cortex area, i.e. contralaterally to the imagined movement side and event-related synchronisation (ERS) in the left motor cortex area. Discrimination of these ERD/ERS can be used to design a BCI.

2.3.1 ERD/ERS

ERD and ERS generally occur in mu ($\sim$8–12 Hz) and beta ($\sim$13–20 Hz) frequency ranges. ERD is the EEG attenuation in primary and secondary motor cortices during preparatory stage which peaks at movement onset in the contralateral hemisphere while ERS is EEG amplification in ipsilateral hemisphere occurring during the same time. ERS appears to be an evolutionary built-in inhibitory mechanism, which explains why it is difficult to execute dissimilar tasks on both sides of the body simultaneously.[2]

In addition to mu and beta frequency ranges, sometimes there is also an increase in EEG energy in gamma (>30 Hz) frequency range. A simple electrode set-up for

[2] This can be demonstrated using an old trick. While sitting comfortably, lift right leg off the ground and rotate the right foot clockwise. Now, with right hand, draw number six in the air—what happens to the foot direction?

motor imagery will consist of two active channels in location C3 and C4 (i.e. motor cortex area), and EEG is obtained during an imagined movement (say either left or right hand). The EEG is filtered in mu and beta bands, and the energy of EEG from channels C3 and C4 are computed to decide on the movement class:

- if energy of $C3_{EEG}$ > energy of $C4_{EEG}$: left hand motor imagery
- if energy of $C4_{EEG}$ > energy of $C3_{EEG}$: right hand motor imagery
- if energy of $C3_{EEG} \approx$ energy of $C4_{EEG}$: no motor imagery

But this is a crude example and the actual EEG analysis involves several stages such as determining the appropriate electrode locations, spectral range and use of features such as band powers and classifiers to obtain accurate detection of the class of motor imagery.

2.4 EEG-based BCI Paradigm 2—SSVEP

SSVEP is a type of EEG that occurs when the visual stimulus flashes at a frequency higher than 6 Hz. It is maximal at the visual cortex, specifically in the occipital region. In this paradigm, a target block flickers with a certain frequency on screen (the flicker can also be achieved using LEDs) and the user looks at the flashes. The frequency following effect (sometimes known as photic response) of the brain causes EEG to oscillate in the frequency of the flickering object. The response is spontaneous and does not require any physical effort other than to gaze at the stimulus as required. In a similar manner, audio-based methods are explored but the results are not as accurate as the visual-based methods. The detection of the frequency of the EEG is sufficient to detect the focused object, though there is a recent study that showed the possibility of using SSVEP with eyes closed (Lim et al. 2013).

2.5 EEG-based BCI Paradigm 3—P300 VEP

P300 visual evoked potential (VEP) is another type of EEG that is evoked around 300–600 ms after visual stimulus onset (hence the term P300) and is maximal in midline locations (such as Fz, Cz and Pz). The potential is limited to 8 Hz, and hence, a low pass filter is normally used to filter VEP prior to analysis. It is evoked in a variety of decision-making tasks and in particular, when a target stimulus is identified, for example when a picture is recognised. A popular paradigm is the Donchin's speller matrix paradigm (Donchin et al. 2000) shown in Fig. 2.6. It consists of alphanumeric characters on screen and the rows and columns flash randomly. The row and column containing the target (focused) character will have a higher P300 amplitude compared to row or column that contains the unfocused character. However, this P300 amplitude is not normally detectable in a single trial due to contamination from higher background EEG and hence require averaging (or other forms of processing such as PCA and ICA) from a number of trials.

Fig. 2.6 Example of P300 VEP paradigm—Donchin's speller matrix

A	B	C	D	E	F
G	H	I	J	K	L
M	N	O	P	Q	R
S	T	U	V	W	X
Y	Z	1	2	3	4
5	6	7	8	9	_

The principle is based on the oddball paradigm where the frequency of the target stimulus is lower than the non-target stimulus. In this case, the target frequency is one sixth since only either one row or one column flashes at a time. A variation of this paradigm is where each alphanumeric character flashes thereby decreasing the frequency to one thirty-six—the lower this frequency, the higher is the P300 amplitude response, which allows easier detection, however resulting in slower response overall as it takes longer to complete the cycle.

2.6 EEG-based BCI Paradigm 4—Mental Task BCI

In this paradigm, users think of different mental tasks and since different tasks activate different areas of the brain, a set of multichannel EEG recordings will have distinct EEG patterns to differentiate the tasks, which could be used to design a BCI.

Examples of mental tasks used are (Keirn and Aunon 1990; Palaniappan 2006):

- Baseline task where users are asked to relax and think of nothing in particular;
- Computation task where users do nontrivial multiplication problems;
- Mental letter task composing where users mentally compose a letter to someone;
- Visual counting task where users visually imagine numbers written on a board with the previous number being erased before the next number is written;
- Geometric figure rotation task where users imagine a figure being rotated about an axis.

These mental tasks exhibit inter-hemispheric differences, and hence, the EEG pattern will be distinct (Keirn and Aunon 1990). For example, computation task involves the left hemisphere more while the visual task exhibits more activity in the right hemisphere. The detection of the inter-hemispheric activity can be done using asymmetry ratio where the powers of EEG channels in the left and right hemispheres are compared to decide the activated hemisphere, which can then be used to design a control interface.

2.7 EEG-based BCI 5—SCP BCI

SCP are low frequency potential shifts in EEG (around 1–2 Hz) and can last several seconds. It is possible to control SCP using feedback and reinforcement mechanism. Different tasks can be used to control either the positivity or negativity SCP. For example, cognitive tasks (or even inactive relaxed states) can generate positivity SCP while negativity SCP can be generated with tasks such as readiness/planning to move. Hence, it can be used to generate a binary signal, which can be used as a control mechanism. It is not as popular as the other EEG-based BCIs as it requires extensive training in order to give good performance.

2.8 EEG-based BCI—A Brief Comparison of the Paradigms

Comparing the different EEG-based BCIs, it can be seen that each method has its strengths and weaknesses. For example, motor imagery requires user training and also the response time is slower (the imaginary movement causes changes in EEG to show up typically after a few seconds) but this paradigm circumvents a visual interface and also be can run in the asynchronous mode, thereby allowing the user to turn the system ON/OFF and also use the control mechanism. Mental thoughts are similar in this regard but with the brain rapidly changing over time, such EEG-based BCIs will require frequent retraining.

SSVEP is very robust and requires only a single active channel but require users to gaze at flashing blocks, which is only practical for short periods of time (typically a few minutes). There is also the risk of triggering epilepsy if the flashing frequency is set to be too low. P300 VEP also suffers from this risk, though of a lesser degree. Of all the EEG-based BCIs, SCP requires the most extensive training and is less appealing for this reason but gives good performance.

2.9 Application 1—Biometrics (Password, PIN Generation)

The common biometric is fingerprint but in recent years, others such as DNA, hand geometry, palm print, face (optical and infrared), iris, retina, signature, ear shape, odour, keystroke entry pattern, gait, voice, etc., have been proposed. But all these biometrics can be compromised at some stage but biometrics based on BCI is more fraud resistant as thoughts cannot be forged!

The P300 BCI paradigm can be used to generate a sequence of passwords (or personal identification number, PIN) that can be used in ATMs and computer logins (Gupta et al. 2012). Instead of entering the password using a keypad, the

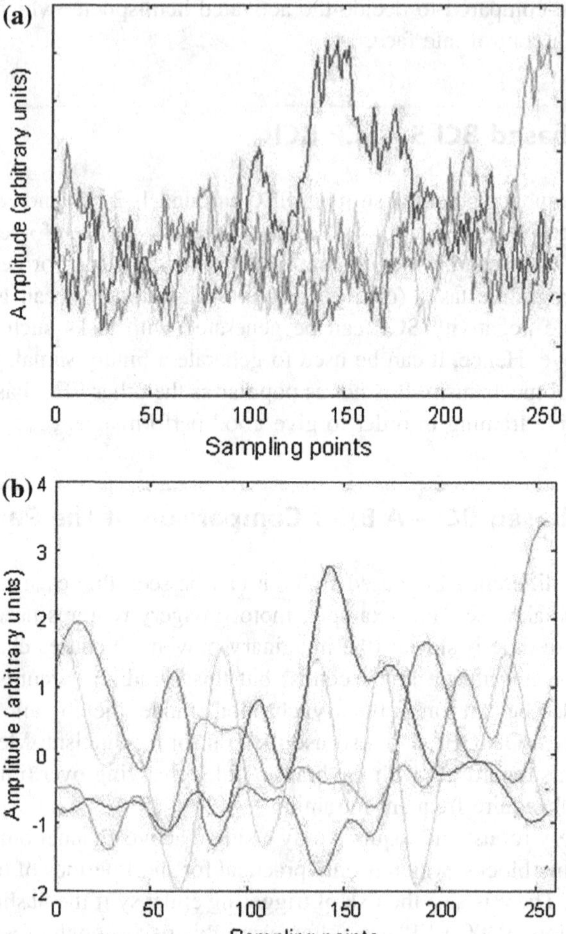

Fig. 2.7 Pass-colour biometric based on P300 VEP BCI **a** raw EEG **b** filtered EEG

alphanumeric characters will pop on the screen and when the character in the password appears on screen, this evokes the P300 potential which is not evoked when non-password characters appear. Similarly, colours can be used instead of alphanumeric characters (having the advantage of being language independent) to code a password (Gupta and Palaniappan 2013). For example, red-green-blue-red-yellow could be the 'pass-colour' for someone.

Figure 2.7a shows raw EEG signal, and Fig. 2.7b shows the filtered P300 signal from channel Cz where the bolder line shows the focused or target colour and the higher amplitude can be seen for the focused colour compared to the non-focused

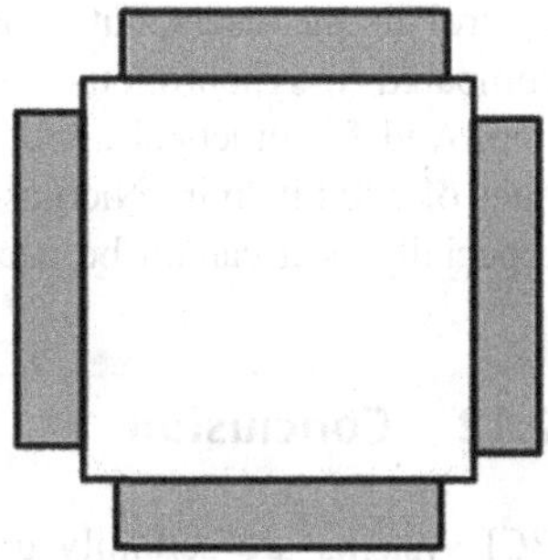

Fig. 2.8 SSVEP-based cursor—each shaded edge flickers with a certain distinct frequency

colours. The detection of the colours/characters through this mechanism overcomes problems like shoulder surfing.[3]

2.10 Application 2—Cursor Control

Most of the SSVEP-based BCIs focus on discrete control, for example selecting a menu on screen. In Wilson and Palaniappan (2011), an analog pointer was developed where instead of discrete control, the cursor on screen moved analogously based on the phase locking index (PLI) of the SSVEP frequency. The cursor was designed as shown in Fig. 2.8, where each edge flickers with either 15, 12, 10 or 8.57 Hz (the frequencies were chosen based on the refresh rate of the LCD screen of 60 Hz). The EEG was recorded from channel Oz in the visual cortex referenced to channel PO3. Depending on which block the user was looking at, the SSVEP will contain the respective frequency and its harmonics which can be detected using discrete Fourier transform (DFT) and other spectral analysis methods. Using this frequency measure, the cursor moved accordingly—the stronger the SSVEP response (measured by the PLI measure from DFT), the further the cursor moved on screen.

2.11 Challenges in BCI

The most difficult challenge at the moment for general BCI use is on the requirement of using gel to improve the conductance though the advances in electrode design (such as active electrodes) have reduced the set-up time considerably. Dry capacitive electrodes have been invented but the quality of the EEG signals is still poor. When it comes to patient usage, most of the advances are being tested on healthy, abled bodied subjects and the required adaptation for disabled people and in real noisy environments are not being studied extensively. Asynchronous (or self-paced) BCIs are more suitable for the disabled as these give additional ON/OFF

[3] Peeking over the shoulder to steal another person's password.

control to the users but proving to be difficult to obtain reliable accuracies as compared to synchronous systems. The response time of BCI systems need to be improved for practical applications—SSVEP BCI is relatively fast (with high bit rates of 100 bits/min (Nicolas-Alonso and Gomez-Gil 2012)) but not without issues especially as it cannot be used for long periods of time.

2.12 Conclusion

BCI systems are certainly useful for the disabled. However, in recent years, the focus has shifted from this original objective to other application areas like biometrics, games, virtual reality and indeed music (brain–computer music interface, or BCMI). EEG-based BCI still proves to be the most practical, portable and cost-effective. The current many advances in BCI technology—such as the advent of non-contact electrodes—will allow mind-controlled devices to become a reality in a decade or so, if not sooner. Imagine a thought-based speed dial: selecting a phone number to dial just by thinking/looking at photograph of the person using EEG from headphones—it could become a reality before we know it!

2.13 Questions

1. BCI approaches could be categorised as invasive or non-invasive. Discuss the advantages and disadvantages of each approach and list examples of approaches in each case.
2. Describe the different EEG-based BCI methods and comment on the practicality of each method.
3. Explore the appropriateness of other biological signals such as those based on electrocardiography, plethysmography, imagined speech etc., for use in non-muscular-based control systems.
4. Figure 2.6 shows a VEP paradigm based on P300 potential. It is based on oddball paradigm where the probability of target occurrence is lower than non-targets. Describe other ways where the characters can be flashed that will still evoke P300 potential in an appropriate manner for use as BCI speller.
5. Assuming the refresh rate of LCD screen of 60 Hz, list all the different possibilities of flicker frequencies for a SSVEP-based BCI system?
6. Two applications using EEG-based BCI have been discussed in this chapter. Describe other applications that might be appropriate using an EEG-based BCI.
7. What are the hurdles in the implementation of current BCI systems? Suggest possible solutions.
8. BCI system has been used in US judicial courts in place of polygraph (i.e. as a lie detector), and there is a common fear among the public that BCI technology can be exploited to read the mind. Based on the current level of technology, discuss if this is possible or just a sci-fi scenario.

9. An electroencephalophone (or sometimes known as encephalophone) uses BCI technology to generate or modulate sounds. Suggest how such as device could work using EEG signals.
10. What are the ethical, legal and societal issues surrounding BCI technology?

References

Asensio J, Gan JQ, Palaniappan R (2011) A study on temporal segmentation strategies for extracting common spatial patterns for brain computer interfacing. In: Proceedings of the 11th Annual Workshop on Computational Intelligence, Manchester, UK, 7–9 September 2011, pp 98–102

Balli T, Palaniappan R (2013) Approximate entropy as an indicator of nonlinearity in self paced voluntary finger movement EEG. Int J Med Eng Inf 5(2):103–116

Borton DA, Yin M, Aceros J, Nurmikko A (2013) An implantable wireless neural interface for recording cortical circuit dynamics in moving primates. J Neural Eng 10(2):026010

Donchin E, Spencer KM, Wijesinghe R (2000) The mental prosthesis: assessing the speed of a P300-based brain-computer interface. IEEE Trans Rehabil Eng 8(2):174–179

Gupta CN, Palaniappan R (2013) Using EEG and NIRS for brain-computer interface and cognitive performance measures. Int J Cogn Perform Support 1(1):69–81

Gupta CN, Palaniappan R, Paramesran R (2012) Exploiting the P300 paradigm for cognitive biometrics. Int J Cogn Biometrics 1(1):26–38

Hasan BAS, Gan JQ (2012) Hangman BCI: an unsupervised adaptive self-paced brain-computer interface for playing games. Comput Biol Med 42(5):598–606

Huan N-J, Palaniappan R (2004) Neural network classification of autoregressive features from electroencephalogram signals for brain-computer interface design. J Neural Eng 1(3):142–150

Jasper H (1958) The ten twenty electrode system of the international federation. Electroencephalographic Clin Neurophysiol 10:371–375

Keirn ZA, Aunon JI (1990) A new mode of communication between man and his surroundings. IEEE Trans Biomed Eng 37(12):1209–1214

Langhenhove AV, Bekaert M-H, N'Guyen J-P (2008) Using a brain-computer interface for rehabilitation: a case study on a patient with implanted electrodes. In: Proceedings of the 4th International Brain-Computer Interface Workshop and Training Course, Graz, Austria, 18–21 September 2008, pp 349–354

Lim J-H, Hwang H-J, Han C-H, Jung K-Y, Im C-H (2013) Classification of binary intentions for individuals with impaired oculomotor function: 'eyes close' SSVEP-based brain-computer interface (BCI). J Neural Eng 10(2):026021

Miranda E, Magee WL, Wilson JJ, Eaton J, Palaniappan R (2011) Brain-computer music interfacing: from basic research to the real world of special needs. Music Med 3(3):134–140

Neuper C, Klimesch W (2006) Event-related dynamics of brain oscillations. Prog Brain Res 159:1–448

Nicolas-Alonso LF, Gomez-Gil J (2012) Brain computer interfaces, a review. Sensors 12 (2):1211–1279

Palaniappan R (2006) Utilizing gamma band spectral power to improve mental task based brain computer interface design. IEEE Trans Neural Syst Rehabil Eng 14(3):299–303

Palaniappan R (2008) Two-stage biometric authentication method using thought activity brain waves. Int J Neural Syst 18(1):59–66

Palaniappan R (2010) Biological signal analysis. Bookboon (Ventus) Publishing, Denmark

Wilson JJ, Palaniappan R (2011) Analogue mouse pointer control via an online steady state visual evoked potential (SSVEP) brain–computer interface. J Neural Eng 8(2):025026

Contemporary Approaches to Music BCI Using P300 Event Related Potentials

3

Mick Grierson and Chris Kiefer

Abstract

This chapter is intended as a tutorial for those interested in exploring the use of P300 event related potentials (ERPs) in the creation of brain computer music interfaces (BCMIs). It also includes results of research in refining digital signal processing (DSP) approaches and models of interaction using low-cost, portable BCIs. We will look at a range of designs for BCMIs using ERP techniques. These include the P300 Composer, the P300 Scale Player, the P300 DJ and the P300 Algorithmic Improviser. These designs have all been used in both research and performance, and are described in such a way that they should be reproducible by other researchers given the methods and guidelines indicated. The chapter is not intended to be exhaustive in terms of its neuroscientific detail, although the systems and approaches documented here have been reproduced by many labs, which should be an indication of their quality. Instead, what follows is a basic introduction to what ERPs are, what the P300 is, and how it can be applied in the development of these BCMI designs. This description of ERPs is not intended to be exhaustive, and at best should be thought of as an illustration designed to allow the reader to begin to understand how such approaches can be used for new instrument development. In this way, this chapter is intended to be indicative of what can be achieved, and to encourage others to think of BCMI problems in ways that focus on the measurement and understanding of signals that reveal aspects of human cognition. With this in mind, towards the end of the chapter we look at the results of our most recent research in the area of P300 BCIs that may have an impact on the usability of future BCI systems for music.

M. Grierson (✉) · C. Kiefer
Embodied Audiovisual Interaction Group (EAVI), Goldsmiths Digital Studios, Department of Computing, Goldsmiths College, SE14 6NW London, UK
e-mail: m.grierson@gold.ac.uk

E.R. Miranda and J. Castet (eds.), *Guide to Brain-Computer Music Interfacing*,
DOI 10.1007/978-1-4471-6584-2_3

3.1 Music BCIs: The Challenges

EEG approaches have been applied in the domain of music for decades due to EEG's temporal resolution when compared to other brain information retrieval techniques, and also because it is considerably more cost effective and portable than the alternatives (e.g. fMRI[1], fNIRS[2]).

One would assume contemporary EEG-based BCMIs would draw from the state of the art in generalisable EEG. However, neurofeedback techniques that use spontaneous potentials (i.e. oscillations in raw EEG signals), or what are commonly called "Brainwave Frequencies" to understand and decode information from the brain are a popular method for consumer and research level BCMIs, largely because this approach is 'baked in' to various BCI devices that are commonly sold in the marketplace.

These approaches can be unreliable and challenging for users to control, and are therefore not as prevalent in other forms of EEG research as some other, more accurate methods. One such method is the ERP, or Event Related Potential technique, which is widely used in BCIs and psychology research (Grierson et al. 2011; Miranda et al. 2004; Chew and Caspary 2011; Nunez 2006).

3.2 P300 ERP Based Music BCIs

3.2.1 What Are ERPs?

ERPs are specific brain signals that can be detected in raw EEG data following the onset of a stimulus—for example, the beginning of a sound, or a change in the motion or visual appearance of an image or other sensory experience. There are many types of ERPs, each one thought to represent a different stage and/or process in the unconscious and conscious encoding of events by the brain. They can also be an indication of the way in which you are experiencing a particular stimulus—for example, such as whether you consciously noticed a particular event or not.

3.2.2 What Is the P300?

Under certain conditions, ERP components can be seen as amplitude curves and peaks derived from the raw EEG time series. They can be either positive or negative with respect to the baseline brain signal, and happen on or around particular points in time following a sensory input event. These two factors—their polarity and timing—provide the method by which they are labelled. For example, the N100 (or N1) is a negative peak approximately 100 ms after the onset of a sensory event that

[1] functional Magnetic Resonance Imaging.

[2] functional Near-Infrared Spectroscopy.

indicates the event is unpredictable in comparison to surrounding events, with the strength of the N100 component being roughly related to how random/unexpected the sensory event appears to the user. The N1 is often accompanied by the P2 or P200, which is a positive going response peaking at around 200 ms. Both these components are usually considered lower in amplitude and therefore slightly harder to detect than the P3, or P300, a positive-going potential, peaking at around 300 ms. It is thought that this component represents the moment at which an external event, being suitably unpredictable, causes a sensory signal considered important enough to cause a shift in attention in the brain. Basically, it is the point at which you consciously notice something happening—something out of the ordinary. As has already been stated, this component has a tendency to be higher in amplitude than many other parts of the raw EEG signal. Therefore it can sometimes be seen in the raw data signal as a peak. However, there are also other peaks in the signal, which makes detection complex (Fig. 3.1).

Importantly, the P300 is thought to consist of two potentials that interrelate: the P3a and P3b. The difference between them is key. The P3a is often called the "oddball" response, and is detectable under conditions where you experience a break in a pattern, or the onset of a new type of event in a stream of other events. However, the P3b is thought to occur when you intend to notice something in a stream of events, such as when you are searching for a specific target, such as a certain picture or letter.

In order to use the P300 ERP in the design of a BCMI, we will need to elicit the P300 through the use of unpredictable events (often called "oddballs"), or by some form of user-driven search task with rapid presentation of a range of choices. This requires significant expertise in digital signal processing (DSP) and audiovisual interaction paradigms, but is achievable by those with appropriate maths and programming skills using the basic methods described below and throughout this chapter.

To make life easier for those researchers who wish to utilise existing ERP detection software, there are some freely available P300 detection toolkits for Matlab (such as EEGLAB http://sccn.ucsd.edu/eeglab/), and also excellent alternative frameworks such as openVIBE and BCI2000. These toolkits support a range

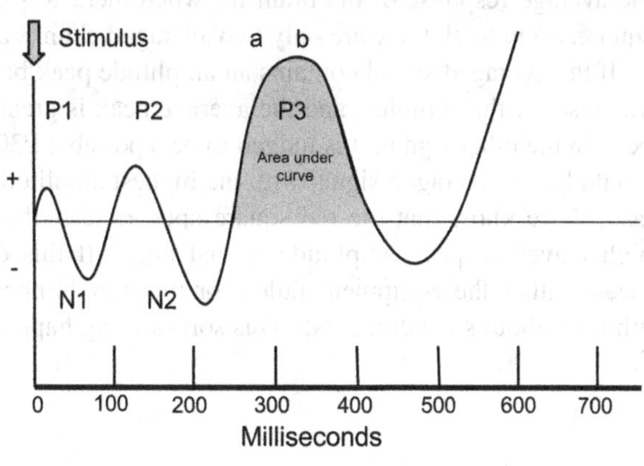

Fig. 3.1 A graph of common event related potential components following a stimulus. Note the inclusion of approximate timings for P3a and P3b, and the indication of the positive area underneath the P300 curve

of commercial and research grade (e.g. g.Tec MOBIlab) EEG hardware. The outputs of these can be easily connected to sound and music programming environments such as PD and Max through the use of UDP networking libraries like open sound control (OSC) to create BCMI systems. However, no matter what approach you take, it is crucial to understand the basic method and its limitations.

3.2.3 Detecting P300 ERPs Using the Averaging Method

Conventional methods for detecting ERPs represent some challenges for those wishing to use them in musical contexts. This is because one of the core features of ERPs, including the P300, is that when mixed with spontaneous EEG signals, or what we might call background EEG, they are not very easy to spot. In most cases, in order to detect ERPs we must remove the background noise that masks them. The problem is that ERPs themselves look quite a lot like the spontaneous potentials which make up the background noise, and occur in the same frequency bands. Therefore, simply filtering the signals using standard methods will not work. For these reasons, to remove the possibly random noise present in a time series of EEG data, a series of stimuli are usually presented, and the responses are averaged. This process reduces the amplitude of random signal components, and increases the amplitude of similar signal components—the ERPs. To test this approach we can do an oddball task, the basic process for which is detailed below (Fig. 3.2).

The Oddball task is a well-known P300 ERP paradigm (Polikoff et al. 1995). It is useful for testing if an EEG system with tagged stimuli is able to elicit and detect P300s. For the oddball task, two types of stimuli are required, for example, blue circles and red squares. The stimuli are flashed randomly on the screen at set intervals. If we decide that the red square is to be the less common and therefore more unexpected stimuli, the red squares are flashed less often—for example at a ratio of 10:2 for blue circles versus red squares. Each time a flash is triggered, a 400 ms chunk of EEG data is stored and tagged to the stimulus. At the end of each run, results for each stimulus type are averaged together. Both the blue circles and the red squares will then have only one single averaged EEG chunk that represents the average response of the brain for when there was either a blue circle, or a red square. So in total, there are only two averaged chunks at the end of the oddball test.

If the averaged signal contains an amplitude peak between 200 and 600 ms after the onset of the stimulus, and the average peak is greater than that of the averaged peak in the other signal, it is judged to be a possible P300 target signal, as the target would be the averaged signal with the highest amplitude. In the case of the oddball task, if we know that the red square appears less often, we'd expect it to have a higher average peak amplitude at that point. If this doesn't happen, it basically means either the equipment failed, or the user blinked/moved too much, or was thinking about something else. This sort of thing happens quite a lot, so we need to

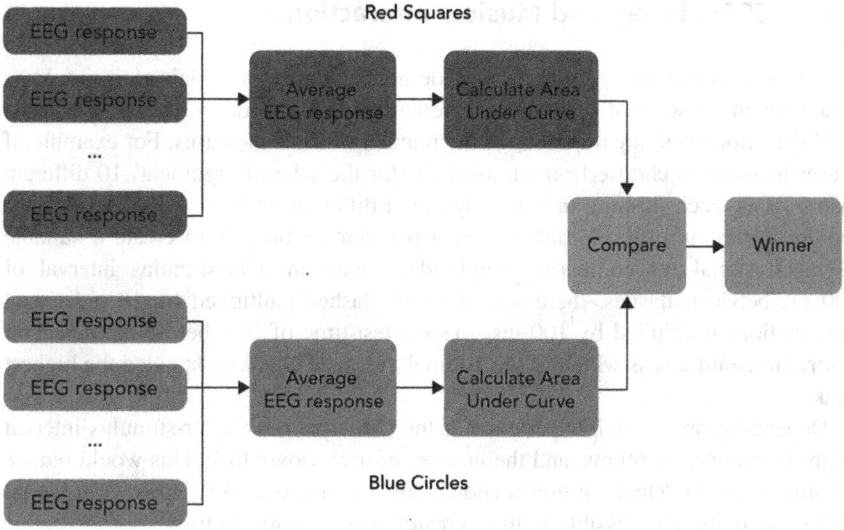

Fig. 3.2 The oddball paradigm. The approach represented *above* is slightly more efficient than the standard approach as instead of simply calculating which averaged chunk has the highest amplitude, here we calculate the area underneath the curve containing the highest positive amplitude peak (for details see Fig. 3.1), which is more robust as it helps to reduce inaccuracies introduced by high frequency amplitude spikes and other noise/unwanted signals

have ways to account for it, such as discounting epochs of data with extraordinarily large peaks consistent with head movement and blinking.

Of course, if we want to be able to choose between three or more different possible choices, we can have a grid containing all possible choices, and flash each choice randomly an equal number of times, whilst we ask the user to stare at their chosen target. Success rates for this method vary wildly depending on a range of factors, including the skill of the user, how much they are concentrating, the quality and/or position of the electrodes, the ease with which the stimuli can be seen, the number of times each stimulus is flashed, which algorithms are used for detection, how the signal is analysed prior to averaging, the accuracy of the signal baseline, and any number of less important but still occasionally significant variables. However, almost all of these variables also apply to all other forms of EEG BCI.

Despite these disadvantages, ERP detection is a surprisingly accurate method for BCI, with detection rates sometimes reaching over 90 %, meaning that 90 % of the time, the system is able to detect which stimuli is triggering the P300. As the method favours actual conscious attention measured with reference to the time when stimuli flashed, it is one of the best approaches for EEG-based BCMI in terms of accuracy and usability. Later in this chapter we will discuss a number of possible applications for using P300 BCI in music contexts. In addition we will also detail ways we can significantly improve both the speed and usability of P300 ERPs for BCMIs.

3.3 ERPs, Delay and Musical Interaction

Now that the averaging method for ERP de-noising has been explained, it should be clear that these sorts of techniques inherently involve a delay. Averaging-based ERP detection accuracy improves as the number of trials increases. For example, if a user is asked to choose from an array of (for the sake of argument) 10 different visual items, each flashing individually but at different times, it would be usual to require a minimum of 10 trials (flashes) per item to be able to create a suitable averaged signal for comparing amplitude. Given an inter-stimulus interval of 100 ms between flashes, there would be 10 flashes multiplied by 10 individual presentations multiplied by 100 ms—a total test-time of 10 s before the detection algorithm could assess which of the 10 final averaged chunks contained the highest peak.

Depending on the skill/experience of the participant, the inter-stimulus interval might be reduced to 60 ms, and the number of trials down to 7. This would reduce the time taken to detect the user's choice down to just over 4 s. However it would be less accurate and possibly require greater concentration to use.

Importantly, types of time delays are not incompatible with certain types of musical tasks, more specifically those types of tasks that are common to creating music with technology. It is only in the last 25 years that electronic and computer music has become a predominantly real-time activity, and many high quality processing techniques are still time consuming (time domain convolution for example).

Furthermore, many crucial aspects of musical interaction do not require real-time control. For example, composition can often be an 'offline' activity, requiring time for consideration and planning. In these cases, P300 ERP detection delay times are not a significant issue.

In addition, as P300 ERP approaches provide direct, time-tagged information about cognition of sensory events, it can be used as a passive aid to the composition process. That is to say, the oddball response might feasibly be used as a measure of novelty given the right presentation paradigm. For example, it is common for composers and recording artists to listen to complete performances/potential versions multiple times during production. If EEG recordings could be easily taken during these listening sessions, EEG responses to musical events could be used to determine how unexpected and/or attention grabbing each musical event was at the level of milliseconds. This approach was used in my Audiovisual Composition *Braindrop*, mentioned later in this chapter.

The holy grail of ERP detection is what is referred to as 'Single Trial'. This means that the ERP signal can be effectively detected and separated from the background EEG noise immediately, pointing directly to the element in the auditory or visual stream which caused the attentional shift without the need for averaging multiple tests/trials. These approaches are becoming more possible through machine learning, and we report results of our research in this area at the end of the chapter.

There are still a number of ways in which the averaging technique can be improved and refined for successful and fulfilling BCMI using the P300. What is unique and exciting about this approach is that as it indicates the brain's cognitive processes, it allows composers and performers to use knowledge of their own unconscious in creative decisions, whilst simultaneously giving very accurate results when users wish to indicate specific actions.

What follows are descriptions of P300 based BCMI systems that have been prototyped as part of my continuing research at Goldsmiths. In almost all cases the systems are created using bespoke software written in C++ for P300 ERP detection. In addition, the systems described use raw EEG data only, from a range of electrode locations. The type of EEG equipment used ranges from research grade to low-cost consumer EEG. All the approaches here are reproducible, but it should be stressed that the cheaper the equipment, the more challenging and creative the signal processing approach needs to be in order to get reliable output. If in doubt, one should use a professional research grade EEG system. Although multiple electrodes can lead to better results, a single electrode placed at Cz is sufficient for reasonable results for the majority of the use cases detailed below.

3.4 The P300 Composer

As has already been discussed, P300 approaches can be very accurate even with a high number of possible targets as long as the electrode quality, positioning and signal conditioning is adequate. Providing these caveats have been met, the P300 Composer is the simplest form of P300-based BCMI to create, and adequately demonstrates the power and effectiveness of the approach. This method of creating a successful visually controlled P300 BCI for music was first described in the 2008 paper "Composing With Brainwaves: Minimal Trial P300b Recognition as an Indication of Subjective Preference for the Control of a Musical Instrument", (Grierson 2008). The method used (which is detailed below) was also speculated upon by David Rosenboom as a potential future method for BCMI (Rosenboom 1976).

The P300 composer is capable of allowing a motionless user the ability to create a single monophonic pitched line from a range of possible pitches. The line itself can have unlimited total duration, and can be composed of notes of finite length, and also gaps/musical rests. The system is now much simpler to build than in the past as it can be produced by modifying the more-or-less standard P300 spellers (see below) available in the public domain. Such paradigms are available in a number of low-cost and free-to-use software libraries such as openVIBE[3] and BCI2000.[4] Both these systems can be made to interoperate with OSC[5] making the creation of P300 speller-based composer systems within reach of most researchers.

[3] http://openvibe.inria.fr/openvibe-p300-speller/.

[4] http://www.bci2000.org/wiki/index.php/User_Reference:P3SpellerTask.

[5] http://www.opensoundcontrol.org/.

The P300 Composer is like the P300 speller, but instead of displaying a grid of letters, it displays a grid of possible music note-names from A1 to G5, and additionally numbers and/or spaces that can indicate rests or other actions. Each grid element flashes an equal number of times, but in a random order, and the P300 averaging technique described above is used to detect the grid element having the highest average amplitude signal when compared against the others. As previously described, the user must attend to a specific grid element, actively noticing when it flashes. The user can do this either by mentally counting the number of flashes, or by some other method they find works for them.

In order to determine which note-name or rest a user is attending to (i.e. looking at), each position on the grid needs to flash enough times for the system to be able to compute a usable averaged signal representing each possible choice. As already mentioned, this depends on a number of factors, including the signal condition, the user's skill and experience, how tired the user is, how much they blink or move etc. The more time the user is able to spend, the more likely they are to be able to indicate their choice accurately, with the knock-on effect that with 42 different elements in the grid, the system can be slow to use. For example, if each note flashes 20 times, for a combined on/off duration of 100 ms each time, each note-name will take 84 s to be detected. However, it will be very accurate.

There are a number of methods for decreasing the time taken to detect ERPs. For example, each column and row can be flashed in sequence, which dramatically speeds up the process. If such a method is used, each XY position on the grid is found by flashing each row and column once in random order. The currently attended area of the screen is detected by comparing the average of all rows (X) and all columns (Y) to find the highest average peaks in each, such that they correspond to the target row and column. With this approach, the time taken to perform each test drops to around 26 s.

One can again speed up the process by reducing the flashing speed, and the inter-stimulus interval. For example, each flash might be fast as 50 ms (one-twentieth of a second), with a gap of 20 ms between flashes. This would reduce the time taken to detect the correct note-name choice to around 18 s.

Furthermore, the number of trials (flashes) can also be reduced. It is possible for an experienced BCMI user to work with a system which uses 7 trials per grid element or row/column in order to achieve success between 70 and 80 %. This can lead to a further improvement in speed of approximately 300 %, bringing the total test-time per note-name choice to just over 6 s if all speed improvements are applied.

Given the large number of potential interaction choices available to the user with this method (42 in the matrix presented in Fig. 3.3), some of the grid elements can be used to increase the user interaction. For example, these effectively 'empty slots' can be used for setting the current note duration, for setting rests of different durations, and also for indicating that a previous note choice should be removed. Additionally, play/stop functionality can be offered.

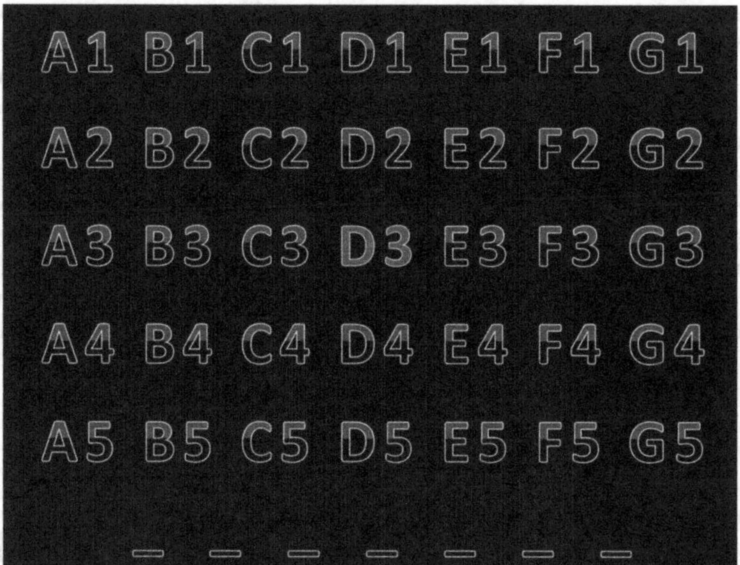

Fig. 3.3 A basic P300 composer interface. In the *above* image, the note-letter D3 is flashing. If the user is attending to D3 at that moment, there should be an increase in amplitude detectable in the averaged signal after each complete trial run. Note that although this display shows 5 octaves in the key of C, there are no restrictions in terms of which note-pitch values and controls can be represented

In this way, different types of user interactions can be explored quite easily without making changes to the underlying speller paradigm. All that is required is that each individual element in the grid be connected up to a synthesiser note or rest in a sequence, or to a user-interaction event, for example: play, stop, delete etc. Through these types of approaches, the standard P300 speller approach can be used to make a reasonably featured composition tool.

3.4.1 MusEEGk—The P300 Sequencer

An expansion of this approach is detailed in Yee Chieh (Denise) Chew's paper "MusEEGk: A Brain Computer Musical Interface", presented as work in progress at CHI 2011 (Chew and Caspary 2011). She describes a similar system but where P300 detection is used to indicate notes in a continuously looping sequence. This is particularly usable in the context of electronic music, and not unlike conventional approaches to loop-based electronic music composition and performance.

One of the useful adaptations apparent with this technique is the way in which the system becomes programmable whilst producing sound—that is to say, it allows a user to specify and play back a looping sequence, and then to iterate through it indicating any changes they wish to make whilst it continues to play. This mitigates against one of the central design problems of the P300 speller paradigm—that at some point, detection must momentarily stop and a decision must be reached, before beginning again. We discuss a separate solution to this problem later in this chapter.

The P300 composer is a good first step in the creation of P300 BCMIs. However, there are still a range of problems with the approach that users need to be made aware of. Significantly, if the user moves, blinks, loses attention or otherwise disturbs the signal, this will lead to inaccurate results. However, it is very challenging for users to avoid this without practice, and can occur even with experienced users. Various methods can be deployed to prevent this. For example, if the signal peak is very high, this may indicate facial or bodily movement. These trials can and should be discarded as and when they occur, as they will otherwise bias the signal average, creating inaccurate results.

3.5 P300 Scale Player

The P300 Scale Player is a simplification of the P300 Composer intended to be used as a close-to-real-time improvisation and performance system. I first used the system in 2008 as part of a demonstration to international media, and have used it on stage on a few occasions. The basic premise is that the system has only three visual targets—arrows. The left arrow indicates a drop in note pitch, the right arrow indicates an increase in note pitch, and the middle arrow indicates that the note pitch should remain the same. The system decides which note to play based on the outcome of a given number of trials, and testing is more or less continuous, with a new test beginning as soon as the previous test has completed.

The advantage of this approach is that with an experienced user, the time it takes for the system to make a decision is greatly reduced. For example, with a total inter-stimulus interval of 70 ms, and a total number of trials being no more than 5, decisions on note direction are reached in just over a second. It's of course accepted that this is slow when compared to playing an instrument, but it does allow for the control of slow moving melody lines with some reliability.

The primary disadvantages of this approach are that it only allows for adjacent notes to be played, and only within a particular predefined scale. So, for example, if the user is on D3, they can only choose to move to E3 or C3, or to stay on D3. However, it is precisely these restrictions that give the system its speed.

Although the scale player is certainly not an instrument with the capacity for varied melodic output, it is at least a BCMI that can be used in real-world scenarios, played as an instrument with an ensemble, and even to play the occasional well-known tune (for example, the main theme from Beethoven's Ode to Joy).

3.5.1 Using the Scale Player as a Game Controller

This approach can be modified to allow its use as a game controller to play simple 3D games for example. In this context, with a reaction time of around a second, it is possible to play a 3D driving game in a way that is not entirely unlike using other kinds of controllers, depending on the user's skill level. Although it is of course accepted that it is not an interface that in any way competes with a conventional game controller, it is an approach that with machine-learning based single trial detection might begin to be usable in such scenarios in the near future. We present results of trials carried out using user-derived machine learning classifiers later in this chapter.

3.6 P300 DJ

Another simple modification of the speller-based approach is the P300 DJ. In many ways, this system is the most usable of the ones described here as it assumes a specific interaction paradigm that requires little input from the user, and wherein musical data—the sounds used by the system—have specific restrictions.

In the use case presented here, the P300 DJ can use any audio file providing it is trimmed to a loop. The loop can be of any length or numbers of beats as long as they are multiples of 4. The system has the capacity to estimate the number of beats based on basic onset detection compared to the sample length, so long as it can assume that the audio file starts and stops on a bar division. It can then ensure that loops of different tempi can be synchronised.

The BCMI system is straightforward. The user has the opportunity to queue tracks into a playlist using the P300 Composer interface, but instead of selecting individual notes, the user is presented with a finite list of available tracks. The user can move back and forth, selecting or removing tracks from the playlist, and when each track is about to complete, the next playlist selection is beat-matched and automatically mixed in.

A different iteration of this system can be built using the P300 scale player interface, allowing the user to move within a long list of songs, then deliberately selecting a specific track.

This system was created to work with my commercial audiovisual mash-up tool, Mabuse, which has a number of features that can be used for creating beat-aligned interactive music and visuals. Again, as with the design of any P300 BCMI, the crucial approach is to find an interaction that is possible with the P300 averaging technique whilst remaining suitable for the user's needs. I have performed on a few occasions with this system, and it can produce as satisfying a set as with conventional methods, as the user need only specify something new very rarely when compared to other forms of music and sound performance.

3.7 P300 Influenced Algorithmic Improviser

Another approach that can provide fertile ground for experimentation is to combine the P300 selection system with an algorithmic improvisation tool, such as a Markov model-based melody and chord generator. In systems such as this, specific harmonic and melodic behaviour is encoded in a Markov model either by creating the model manually, or by analysis of a performed monophonic melodic line.

A similar system was used to create the piece *Braindrop*, whereby structural elements of an algorithmic piece are presented to the user through the use of accompanying visual cues. The P300 response to the visual cues causes the structure of the music to change. In this particular piece, models of attention are also used to gauge the performer's interest in the current structure, leading to a transition to the next section of music.

A similar approach was used as part of the Finn Peters project *Music of the Mind*, in collaboration with Dr Matthew YeeKing. Further details of kinds of approaches are detailed in our paper "Progress Report on the EAVI BCI Toolkit for Music: Musical Applications of Algorithms for Use with Consumer Brain Computer Interfaces", presented at ICMC in 2011 (Grierson et al. 2011).

3.8 Developing Novel Methods for P300 Detection

Significant aspects of our research involve attempts to create further refinements to P300 detection methods for the purposes of both increased speed and usability of generalised P300 BCI. As part of this research we have created novel signal processing methods that have as yet not been applied in any other BCI context. Two examples are presented here: the moving average method, and template matching through machine learning of Repetitive Serial Visual Presentation (RSVP) derived P300 features. In addition we have attempted to reproduce these results using the lowest cost BCI hardware available, including single dry electrode devices such as the NeuroSky MindSet. Importantly, good results on the MindSet are difficult to achieve with these methods, but we demonstrate these approaches can work on this hardware. Furthermore, these methods are equally as applicable to P300 BCIs that use any hardware that can provide access to a raw EEG signal given the caveats mentioned at the beginning of this chapter.

3.8.1 Collecting and Conditioning Raw EEG Signals from BCI Hardware

Almost all EEG devices are capable of providing raw untreated signals. When creating new approaches to BCI, it's vital we work with these as much as possible. This is usually fine, particularly when deploying custom software with research-grade devices such as g.Tech's mobiLab.

A crucial problem for consumer devices such as the NeuroSky MindSet and the Emotiv Epoch is lack of flexibility concerning electrode placement. With respect to the Neurosky, the forehead electrode is not well placed for P300 detection. Specifically, it is not considered possible to detect the P3b from the forehead, especially using the Mindset or similar devices. However, we have had success in detecting the P3a with such devices using the average technique. Although the P3a is less useful for creating speller-type applications, it can be used with certain approaches where only the oddball response is required. In addition, we have had some success in using machine learning to build a P300 classifier using the NeuroSky, and these results are presented at the end of this chapter.

Our system is based on our generalised C++ ERP detection algorithm previously described in [x]. This algorithm is agnostic to hardware, providing the same code interface for any device for which raw EEG data is available. There are some important aspects to the design of the algorithm which are worth noting if you are considering engineering your own P300 solution.

When producing custom signal processing techniques for ERP detection, it is vital that the signal is conditioned in the correct manner. Failure to do this will result in false and/or confusing detection rates. Providing that your hardware is correctly configured and operating normally, the derivation of a proper baseline for the raw EEG signal is the next priority. A good baseline is essential before any signal processing can begin—fundamentally if there is any positive or negative going offset in the signal, this will cause biased results, especially when averaging signals.

For example, if the baseline signal is positively biased, this will introduce a higher average peak in results that have a smaller number of signal blocks from which to derive an average. The offset will be reduced as more signal blocks are used to derive an average, but in the case of an oddball test, where the less common signal should contain a higher average than the more common signal, the positive offset biases the entire result.

In order to avoid these sorts of statistical anomalies, one can either use a high pass IIR filter, or subtract a continuous average signal made up of the last n samples. Neither approach is without flaws. An IIR filter may well introduce phase shift in significant areas of the signal, whereas subtracting the average will remove more than just the offset. This may well help rather than hinder your ERP detection —but either way it is a choice the reader must make for themselves.

3.8.2 The P300 Moving Average Method

Building on our baseline method, we created a modification of the standard ERP paradigm to allow control of directional movement within a 3D virtual world through continuous control. In order to achieve this, we created a windowed moving average algorithm that reliably detects ERP signals from raw EEG data in a continuous fashion, eliminating the need to stop the test in order to reach a decision. This has many applications in BCMI, for example, where continuous control is required to adjust parameter values.

In cases where the hardware or signal quality is poor, for example when using a low-cost consumer EEG system such as the NeuroSky MindSet, this approach can help to improve signal conditioning in a way that is more satisfying and less time consuming for the user. It does this by building up a history of the last n responses for each available choice during interaction.

Importantly, in our user test, each choice was flashed on the screen one at a time, instead of as a grid. Currently winning choices would flash red, providing feedback to the user. Stimuli are flashed randomly on the screen at regular intervals. For this test, we used the P300 scale player interface method detailed above to navigate a 3D world. Three arrows indicate left, right and straight ahead. These are arranged parallel to the floor of the 3D virtual environment (see Fig. 3.4). Each time a flash is triggered, a 400 ms chunk of EEG data is stored and tagged to the stimulus. At the end of any given number of windows (*n* windows), results for each stimulus are averaged together. Following this, each time a flash is triggered, the previous n results are averaged for each stimulus. Each stimulus therefore always has an averaged EEG signal associated with it representing the EEG response to the previous n windows. If at any time the averaged signal contains an amplitude peak between 200 and 600 ms after the onset of the stimulus, and the average area is greater than that of every other averaged peak area, it is judged to be a possible P300 target signal, as the target would be the peak with the highest average area under the peak. This target signal is judged to be the winner, and the stimulus changes from white to red to reflect the change in direction.

The movement system functions by increasing the amount of force in the direction of the current winner. In cases where subjects wish to move in a new direction, the system automatically cancels all force in the current direction in favour of generating force in the direction of the new winner.

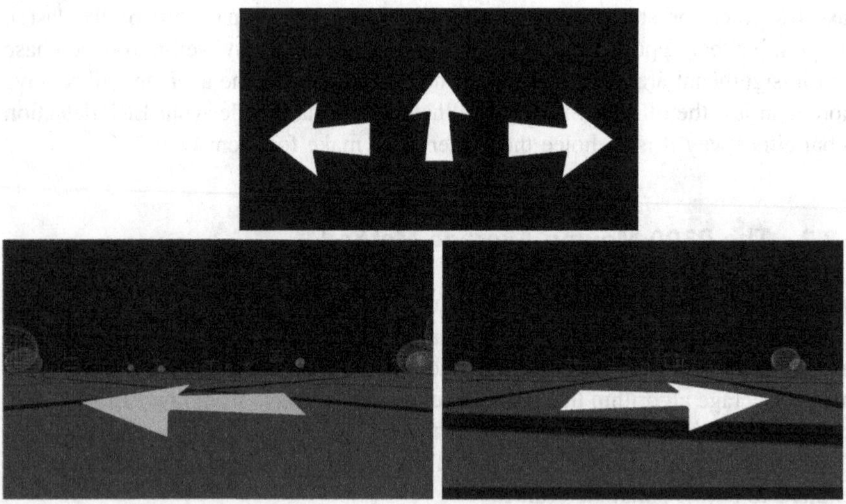

Fig. 3.4 The visual stimulus interface for the scale player (*above*). This interface has the advantage of being re-usable for P300 controlled game interactions (*below*). In the above example, to mitigate against distractions, each *arrow* is presented one at a time. The currently selected direction appears a different colour to provide visual feedback to the user (see 3.8)

The speed of the system is to some extent reliant on the number of windows used to create the moving average (n). Obviously, the greater the value of n, the more reliable the system becomes. However, as the system uses a first in first out (FIFO) queue, once the initial averages have been computed, it can take significantly less time for the change in average amplitude to bring about a change in direction.

We carried out tests with five participants in controlled conditions. Participants were tested in a sound insulated environment with low light-levels. We asked participants to attempt to concentrate on the forward moving arrow. We used this measure in order to judge how effective the system was, based on the amount of time that moving-average based EEG epochs relating to the forward arrow stimulus contained the greatest average area.

Results from controlled experiments demonstrated that this test performs as well as discrete P300 averaging approaches commonly used to create all BCIs, including BCMIs. This is not controversial as the main difference with our method is that a FIFO queue is used, and decisions are made by the system continually as the user interacts. This approach has some clear advantages for creating real-time continuous controllers for BCI generally, and can be used in BCMI for the control of mixers and crossfaders. For example, with two flashing arrows, one leftgoing, one rightgoing, the user can attend to either arrow in order to improve a continuous confidence level in any particular direction, with a maximum confidence level being equal to n averages. This scale can be used to assign a value to the user control, as opposed to representing a specific decision.

3.8.3 Template Matching Through Machine Learning of Repetitive Serial Visual Presentation (RSVP) Derived P300 Features

RSVP (Craston et al. 2006) is a variant on the oddball test. It offers the potential for collecting much higher quality datasets for P300 classification than the conventional oddball paradigm. In an RSVP test, the participant is presented with a series of symbols at high rate (in this case, 10 Hz) (Bowman et al. 2014). They must try to spot a symbol of one class in a stream of symbols from another. When the symbol is recognised by the participant, an ERP will occur. At the end of the test, the participant can be asked to identify the oddball symbol they spotted. The nature of this answer (either correct, incorrect or unsure) tells the experimenter whether the participant was attending to the task, giving a strong indication of whether a P300 is present in the EEG signal in the time window following the oddball presentation. This ground truth gives a significant improvement in data quality compared to data from the classic paradigm where this distinction cannot confidently be made.

An online RSVP experiment was run. Participants were asked to spot a single letter in a series of 30 numbers presented at 10 Hz, in repeated trials, while wearing a NeuroSky headset. 412 trials were collected from 19 separate user sessions. The data was preprocessed as follows: trials with bad signal quality (as reported by the headset) were rejected. The rest were DC filtered, and then the high frequencies

Block Size	Correctly Classified Instances	Incorrectly Classified Instances
10	70.3%	29.7%
3	60.6%	39.4%

Fig. 3.5 The results show that RSVP can be used successfully to build a separable data set for training a P300 classifier. The single trial results (Block Size 1) show 70 % accuracy

were removed using a cascading FIR filter, which was chosen to avoid phase distortion. From successful trials, the windows from 220 ms to 440 ms after the oddball presentation were collected (set A). Trials where the participant was unsure or gave an incorrect answer were collected in set B. Outliers were rejected from these sets, by removing windows that were more than three standard deviations in Euclidean distance from the mean, leaving 180 trails in set A and 40 trials in set B.

Data sets were created for classification, by creating each training example as an average over a number of windows. Drawing on set A, training sets were creating of 180 examples, using average sizes of 10, 3 and 1 (single trial). The corresponding sets of negative examples were creating using averages of windows from random time points in set B. These sets were used to train a bagging classifier, using a random forest as a sub-classifier. The results from tenfold validation tests were as follows (Fig. 3.5).

These results demonstrate that using machine learning, it is possible to create a P300 classifier using consumer hardware. This has exciting implications for the future usability of BCI systems. Given the interaction designs detailed in this chapter, it should be possible in the near future to create highly accessible, low-cost BCMI systems, and it is this goal that we continue to pursue.

3.9 Questions

1. What is an ERP?
2. How are ERPs different to spontaneous brainwave potentials?
3. How are ERPs different to SSVEPs?
4. What are the main drawbacks of the ERP technique for musical interaction?
5. What is the purpose of the P300 averaging technique?
6. Describe one way that ERP techniques could be used to get information about how listeners experience music.

7. What are the differences between conventional averaging methods and the continuous moving average method?
8. What is the difference between the P3a and P3b response?
9. What does "Single Trial" mean in the context of ERP-based BCI?
10. What is the simplest, most effective electrode placement for P300 detection using research-grade BCI?

References

Bowman H, Filetti M, Alsufyani A, Janssen DP, Su Li (2014) Countering countermeasures: detecting identity lies by detecting conscious breakthrough. PLoS ONE 9(3):1–17. ISSN 1932-6203

Craston P, Wyble B, Bowman H (2006) An EEG study of masking effects in RSVP [abstract]. J Vis 6(6):1016–1016. ISSN 1534-7362

Chew YC, Caspary E (2011) MusEEGk: a brain computer musical interface. In: Extended abstract of acm conference on human factors in computing systems (CHI)

Grierson M, Kiefer C, Yee-King M (2011) Progress report on the EAVI BCI toolkit for music: musical applications of algorithms for use with consumer brain computer interfaces. Proc ICMC, Huddersfield, UK

Grierson M (2008) Composing with brainwaves: minimal trial P300b recognition as an indication of subjective preference for the control of a musical instrument. Proc ICMC, Belfast

Miranda ER, Roberts S, Stokes M (2004) On generating EEG for controlling musical systems. Biomedizinische Technik 49:75–76

Nunez PL, Srinivasan R (2006) Electric fields of the brain: the neurophysics of EEG. Oxford University Press, Oxford

Polikoff JB, Bunnell HT, Borkowski Jr WJ (1995) Toward a P300-based computer interface. In: Proceedings of the RESNA '95 annual conference, RESNAPRESS, Arlington Va

Rosenboom D (1976) Biofeedback and the arts: results of early experiments. Aesthetic Research Centre of Canada, Vancouver

Prospective View on Sound Synthesis BCI Control in Light of Two Paradigms of Cognitive Neuroscience

4

Mitsuko Aramaki, Richard Kronland-Martinet, Sølvi Ystad, Jean-Arthur Micoulaud-Franchi and Jean Vion-Dury

Abstract

Different trends and perspectives on sound synthesis control issues within a cognitive neuroscience framework are addressed in this article. Two approaches for sound synthesis based on the modelling of physical sources and on the modelling of perceptual effects involving the identification of invariant sound morphologies (linked to sound semiotics) are exposed. Depending on the chosen approach, we assume that the resulting synthesis models can fall under either one of the theoretical frameworks inspired by the representational-computational or enactive paradigms. In particular, a change of viewpoint on the epistemological position of the end-user from a third to a first person inherently involves different conceptualizations of the interaction between the listener and the sounding object. This differentiation also influences the design of the control strategy

M. Aramaki (✉) · R. Kronland-Martinet · S. Ystad
Laboratoire de Mécanique et d'Acoustique (LMA), CNRS UPR 7051, Aix-Marseille University, Centrale Marseille, 31, Chemin Joseph Aiguier, 13402 Marseille Cedex 20, France
e-mail: aramaki@lma.cnrs-mrs.fr

R. Kronland-Martinet
e-mail: kronland@lma.cnrs-mrs.fr

S. Ystad
e-mail: ystad@lma.cnrs-mrs.fr

J.-A. Micoulaud-Franchi · J. Vion-Dury
Laboratoire de Neurosciences Cognitives (LNC), CNRS UMR 7291, Aix-Marseille University, Site St Charles, 3, Place Victor Hugo, 13331 Marseille Cedex 3, France
e-mail: jarthur.micoulaud@gmail.com

J. Vion-Dury
e-mail: jean.vion-dury@ap.hm.fr

E.R. Miranda and J. Castet (eds.), *Guide to Brain-Computer Music Interfacing*,
DOI 10.1007/978-1-4471-6584-2_4

enabling an expert or an intuitive sound manipulation. Finally, as a perspective to this survey, explicit and implicit brain-computer interfaces (BCI) are described with respect to the previous theoretical frameworks, and a semiotic-based BCI aiming at increasing the intuitiveness of synthesis control processes is envisaged. These interfaces may open for new applications adapted to either handicapped or healthy subjects.

4.1 Introduction

In this article, we present different approaches to sound synthesis and control issues and describe how these procedures can be conceptualized and related to different paradigms within the domain of cognitive neuroscience. A special emphasis is put on the notion of intuitive control and how such a control can be defined from the identification of signal invariants obtained both from the considerations of the physical or signal behaviour of the sound-generating sources and the perceptual impact of the sounds on the listeners.

Since the first sounds were produced by a computer in the late 1950s, computer-based (or synthesized) sounds have become subject to an increasing attention for everyday use. In early years of sound synthesis, the majority of applications were dedicated to musicians who learned to play new instruments that generally offered a lot of control possibilities, but required high skills to operate. Due to increasingly powerful computers, new applications linked to communication, virtual reality and sound design have made sound synthesis available for a broader community. This means that synthesis tools need to be adapted to non-expert users and should offer intuitive control interfaces that do not require specific training. The construction of such intuitive synthesis tools requires knowledge about human perception and cognition in general and how a person attributes sense to sounds. Why are we for instance able to recognize the material of falling objects simply from the sounds they produce, or why do we easily accept the ersatz of horse hooves made by the noise produced when somebody is knocking coconuts together? Is the recognition of sound events linked to the presence of specific acoustic morphologies that can be identified by signal analysis? In the approach presented here, we hypothesize that this is the case and that perception emerges from such invariant sound structures, so-called *invariant sound morphologies*, in line with the ecological approach of visual perception introduced by (Gibson 1986). From a synthesis point of view, this theoretical framework is of great interest, since if enables the conception of perceptually optimized synthesis strategies with intuitive control parameters.

Sound synthesis based on the modelling of physical sources is generally divided in two main classes, i.e. physical models and signal models. Physical models aim at simulating the physical behaviour of sound sources (i.e. the physical origin of sounds), while signal models imitate the recorded signal using mathematical representations without considering the physical phenomena behind the sound production. In the case of physical models, an accurate synthesis can only be achieved when physical phenomena linked to the sound production are well described by physics. This is not the case for complex sources (e.g. natural phenomena such as wind, rain, fire, etc.). In the case of signal models, any sound can generally be perfectly resynthesized for instance from the analysis of real sounds, independently of the complexity of the underlying physical phenomena of the sound source. However, the control of such sounds is a difficult issue due to the large number of synthesis parameters that generally are implied in such models and to the impossibility to physically interpret these parameters. The physical and signal models can also be combined to form so-called hybrid models (e.g. Ystad and Voinier 2001). The control of these models requires an expertise and the quality judgment of the control is based on an error function linked to the physical or signal precision between the model and the real vibration. Such controls necessitate a scientific expertise apart from certain cases such as musical applications where the control parameters correspond to physical values controlled by the musician (e.g. pressure, force, frequency, etc.). In this latter case, the musical expertise enables the control.

To propose efficient synthesis models that enable intuitive control possibilities, synthesis models combined with perceptual considerations have been developed lately. Perceptual correlates have been sought by testing the perceptual relevance of physical and/or signal parameters through listening tests (cf. Sect. 4.3.2). In the case of environmental sounds, we have identified such perceptually relevant sound morphologies through several experiments. These experiments have made it possible to identify sound elements, also described as sound "atoms", specific to given sound categories that enable definition of high-level control parameters for real-time synthesis applications. Such synthesis tools allow users to synthesize auditory scenes using intuitive rather than reflective processes. Intuitive processes appeal on intuition which is a kind of immediate knowledge, which does not require reasoning, or reflective thought. Intuition can also be defined as the knowledge of an evident truth, a direct and immediate seeing of a thought object (Lalande 1926). The quality of the control strategy is in this case based on perceptual judgments and on easily understandable control parameters on the user interface. Therefore, we call this synthesis control, *intuitive control*.

When searching for perceptually relevant sound morphologies, the understanding of attribution of sense of sounds becomes essential. This issue is a natural part of a more general research field called semiotics that consists in studying the general theory of signs. The notion of signs has been addressed since antiquity by the stoic philosophers (Nadeau 1999). Classically, semiotics is divided in syntax, semantics and pragmatics. Semiology is a part of semiotics, which concerns the social life, and dynamic impact of signs, as language (Nadeau 1999). For de Saussure, language constitutes a special system among all semiological facts. In linguistics, for de Saussure, a sign is the association of a signifier (acoustic image) and a signified (the correlated concept) linked together in a consubstantial way (de Saussure 1955). This consubstantial relationship is often difficult to understand. Semiotics span over both linguistic and non-linguistic domains such as music, vision, biology, etc. This means that it is possible to propose a semiotic approach of sounds, without referring to linguistic semiology. Like in de Saussure construction of signs, one can postulate that every natural (environmental) or social sound is linked to the afferent concept in the same consubstantial way. For example, if I hear a bell, I immediately know that it is a bell, and perhaps, but not always, I even manage to imagine the size of the bell, depending on its spectral contents. Except for "abstract sounds", i.e., sounds for which the sources cannot be easily identified, one can say that each sound can be considered as a non-linguistic sign whose origin can be described using language, in a reflective thought. Previous studies have shown that the processing of both linguistic and non-linguistic target sounds in conceptual priming tests elicited similar relationships in the congruity processing (cf. Sect. 4.5). These results indicate that it should be possible to draw up a real semiotic system of sounds, which is not the linguistic semiology, because phonemes can be considered only as particular cases of sounds.

So far, the identification of signal invariants has made it possible to propose an intuitive control of environmental sounds from verbal labels or gestures. An interesting challenge in future studies would be to propose an even more intuitive control of sound synthesis processes that bypasses words and gestures and directly uses a BCI that records electroencephalographic signals in a BCI/synthesizer loop. This idea is not new and several attempts have already been made to pilot sounds directly from the brain activity. In (Väljamäe et al. 2013), the authors made an exhaustive review in the field of EEG sonification in various applications (medical, neurofeedback, music, etc.) and concluded that the type of mapping strategy strongly depends on the applications. For instance, in the case of musical applications, the mapping is generally determined by artistic choices and does not necessarily mirror a strict semiotic relation. The intuitive BCI-controlled synthesizer that we aim at is intended for a generic context and should enable the identification of brain activity linked to specific signal morphologies that reflect the attribution of sense to a sound.

This paper is organized as follows. In Sect. 4.2, the methodology that leads to intuitive sound synthesis is viewed in the light of representational-computational and enactive perspectives. Then, in Sect. 4.3, two sound synthesis approaches are described and related to the previously presented perspectives. In Sect. 4.4, different control strategies emanating from the different synthesis approaches are described. In Sect. 4.5, some results from experiments supporting the existence of semiotics for non-linguistic sounds are presented. Finally, in Sect. 4.6, a prospective view on a control strategy for synthesis processes based on a BCI is proposed.

4.2 Two Conceptions on the Way We Interact with the Surrounding World

Sound synthesis that integrates perceptual effects from the morphology of their signal in order to enable intuitive control to the end-user brings forward the following questions: How do I attribute a meaning to a perceived sound (related to the semiotics)? What effect does this sound have on me? These questions induce a change in our position with respect to the sound from a third-person position (observer) in more traditional synthesis approaches where only acoustic considerations are taken into account, to a first-person position (implied) in the perceptual synthesis processes. This corresponds to a change from a representational to a neurophenomenological point of view in the field of cognitive neuroscience (Varela 1996). We here adopt a similar change of viewpoint to investigate the phenomenon of sound perception as it was seminally studied in (Petitmengin et al. 2009).

Classically, in the standard paradigm of cognitive neuroscience, there is, on one hand, the physical object and on the other hand, the subject that perceives this object according to his/her mental representation of the physical reality. From this conception of representation proposed by Descartes, a representational-computational paradigm has been developed. This paradigm involves the existence of a *correct* representation of the physical world and assumes that the perception of the object is all the more adequate when the subject's mental representation matches the physical reality, considered as the reference (Varela 1989). Less classically, in the neurophenomenological paradigm of cognitive sciences, it is the interaction between the subject and the object, which enables the subject to perceive an object. F. Varela called this interaction: *enaction* (Varela 1989; Varela et al. 1991). In the enactive paradigm, the mind and the surrounding world are mutually imbricated. This conception is inspired from the phenomenological philosophy of Husserl, who called this interaction a noetic–noematic correlation (Husserl 1950). He posited that there was a link between intentional content on the one hand, and extra-mental reality on the other, such that the structure of intentionality of the consciousness informs us about how we perceive the world as containing particular objects. In a certain manner, and quite caricatured, the physical reality is no more the reference, and the subject becomes the reference. The perception of the object is all the more adequate when the subject's perception makes it possible to efficiently conduct an action to respond to a task. As Varela puts it (Varela et al. 1991):

> The enactive approach underscores the importance of two interrelated points: 1) perception consists of perceptually guided action and 2) cognitive structures emerge from the recurrent sensorimotor patterns that enable action to be perceptually guided.

and concludes:

> We found a world enacted by our history of structural coupling.

In 1966, P. Schaeffer, who was both a musician and a researcher, published the "Traité des objets musicaux" (Schaeffer 1966), in which he reported more than ten years of research on electroacoustic music. He conducted a substantial work that was of importance for electroacoustic musicians. With a multidisciplinary approach, he intended to carry out fundamental music research that included both Concrete[1] and traditional music. Interestingly, he naturally and implicitly adopted a phenomenological approach to investigate the sound perception in listening

---

[1] The term "concrete" is related to a compositional method which is based on concrete material, i.e., recorded or synthesized sounds, in opposition with "abstract" music which is composed in an abstract manner, i.e., from ideas written on a score, and becomes "concrete" afterwards.

experiences. In particular, he introduced the notion of *sound object*. The proposition of P. Schaeffer naturally conducts the acoustician from the representational-computational paradigm to the enactive paradigm, since P. Schaeffer in line with the phenomenological viewpoint stresses the fact that sound perception is not only related to a correct representation of the acoustic signal. This is also coherent with later works of Varela and the conception of perception as an enactive process, where the sound and the listener constitute a unique imbricated system. The perception of the sound is modified by the intentionality of the subject directed towards the sound, which can induce an everyday listening, which is a source-oriented kind of listening, or musical (or acousmatic) listening, which involves the perception of the quality of the sound (Gaver 1993a, b). Thus, sound synthesis should not be limited to the simulation of the physical behaviour of the sound source. In other words, it is the sound object given in the process of perception that determines the signal to be studied, meaning that perception has to be taken into account during the signal reconstruction process.

In the work of P. Schaeffer, morphology and typology have been introduced as analysis and creation tools for composers as an attempt to construct a music notation that includes electroacoustic music and therefore any sound. This typological classification is based on a characterization of spectral (mass) and dynamical (facture) profiles with respect to their complexity and consists of 28 categories. There are nine central categories of "balanced" sounds for which the variations are neither too rapid and random nor too slow or non existent. Those nine categories include three facture profiles (sustained, impulsive or iterative) and three mass profiles (tonic, complex and varying). On both sides of the balanced objects in the table, there are 19 additional categories for which mass and facture profiles are very simple/repetitive or vary a lot. This classification reveals perceptually relevant sound morphologies and constitute a foundation for studies on intuitive sound synthesis.

Based on these previous theoretical frameworks from cognitive neuroscience, we suggest that the control of sound synthesis can be discussed in the framework of the representational-computational and the enactive points of view. In the approach inspired by the representational-computational framework, we consider that the user controls physical or signal parameters of the sound with the idea that the more actual (with respect to the physical reality) the parameter control, the better the perception. The physical or signal properties of sounds are considered as the reference for the sound control. In the approach inspired by the enactive framework, we consider that the user is involved in an interactive process where he/she controls the sound guided by the perceptual effect of his/her action. The idea is that the more recurrent (and intuitive) the sensorimotor manipulation, the better the perception. The sound control enables the perception to become a perceptually guided action. This is an enactive process because the sound influences the control effectuated by the subject and the control action modifies the sound perception. The sound as perceived by the subject is thus the reference for the sound control. Such enactive framework formed a theoretical basis for a recent research community centred on the conception of new human–computer interfaces (Enactive Network) and in a natural way, led to numerous interactive applications in musical contexts (*Journal of New Music Research, special issue "Enaction and Music"* 2009). A general review on fundamental research in the field of enactive music cognition can be found in (Matyja and Schiavio 2013).

4.3 Sound Synthesis Processes

To date, two approaches to synthesize sounds could be highlighted: sound synthesis based on the modelling of physical sources (from either physical or signal perspectives) and sound synthesis based on the modelling of perceptual effects. Interestingly, these synthesis approaches could be linked to the two paradigms related to our perception of the surrounding world (i.e. approaches inspired by the representational-computational and the enactive paradigms, cf. Fig. 4.2) described in the previous section.

4.3.1 Two Approaches for Sound Synthesis

4.3.1.1 Modelling the Physical Sources

In the case of sound synthesis based on the modelling of physical/vibrating sources, either the mechanical behaviour or the resulting vibration of the sound source is simulated.

Physical synthesis models that simulate the physical behaviour of sound sources can either be constructed from the equations describing the behaviour of the waves propagating in the structure and their radiation in air (Chaigne 1995) or from the behaviour of the solution of the same equations (Karjalainen et al. 1991; Cook 1992; Smith 1992; Bilbao 2009). Physical models have been used to simulate a large number of sound sources from voice signals to musical instruments. Several synthesis platforms based on physical modelling are now available, such as Modalys that is based on modal theory of vibrating structures that enable the simulation of elementary physical objects such as strings, plates, tubes, etc. These structures can further be combined to create more complex virtual instruments (http://forumnet.ircam.fr/product/modalys/?lang=en)n.d). Cordis-Anima is a modelling language that enables the conception and description of the dynamic behaviour of physical objects based on mass-spring-damper networks (http://www-acroe.imag.fr/produits/logiciel/cordis/cordis_en.htmln.d). Synthesis models for continuous interaction sounds (rolling, scratching, rubbing, etc.) were proposed in previous studies. In particular, models based on physical modelling or physically informed considerations of such sounds can be found (Gaver 1993a; Hermes 1998; van den Doel et al. 2001; Pai et al. 2001; Rath and Rocchesso 2004; Stoelinga and Chaigne 2007). In particular, Avanzini et al. (2005) developed a physically based synthesis model for friction sounds. This model generates realistic sounds of continuous contact between rubbed surfaces (friction, squeaks, squeals, etc.). The parameters of the model are the tribological properties of the contact condition (stiffness, dynamic or static friction coefficients, etc.) and the dynamic parameters of the interaction (mainly the velocity and the normal force). Also, a synthesis technique based on the modal analysis of physical objects (finite element modelling of each object for precomputation of shapes and frequencies of the modes) was proposed by (O'Brien et al. 2002) in the context of interactive applications. Note that this approach presents a limitation when the physical considerations involve complex modelling and can less easily be taken into account for synthesis perspectives especially with interactive constraints.

Signal synthesis models that simulate the resulting vibration of the sound source are based on a mathematical modelling of the signal. They are numerically easy to implement and can be classified in three groups as follows:

- Additive synthesis: The sound is constructed as a superposition of elementary sounds, generally sinusoidal signals modulated in amplitude and frequency (Risset 1965). For periodic or quasi-periodic sounds, these components have average frequencies that are multiples of one fundamental frequency and are called harmonics. The amplitude and frequency modulation (FM) laws should be precise when one reproduces a real sound. The advantage of these methods is the potential for intimate and dynamic modifications of the sound. Granular synthesis can be considered as a special kind of additive synthesis, since it also consists in summing elementary signals (grains) localized in both the time and the frequency domains (Roads 1978).
- Subtractive synthesis: The sound is generated by removing undesired components from a complex sound such as noise. This technique is linked to the theory of digital filtering (Rabiner and Gold 1975) and can be related to some physical sound generation systems such as speech (Flanagan et al. 1970; Atal and Hanauer 1971). The advantage of this approach is the possibility of uncoupling the excitation source and the resonance system. The sound transformations related to these methods often use this property to make hybrid sounds or crossed synthesis of two different sounds by combining the excitation source of a sound and the resonant system of another (Makhoul 1975; Kronland-Martinet 1989).
- Global (or non-linear) synthesis: The most well-known example of such methods is audio FM. This technique updated by Chowning (1973) revolutionized commercial synthesizers. The advantages of this method are that it calls for very few parameters, and that a small number of numerical operations can generate complex spectra. They are, however, not adapted to precise signal control, since slight parameter changes induce radical signal transformations. Other related methods such as waveshaping techniques (Arfib 1979; Le Brun 1979) have also been developed.

In some cases, both approaches (physical and signal) can be combined to propose hybrid models, which have shown to be very useful when simulating certain musical instruments (Ystad and Voinier 2001; Bensa et al. 2004).

4.3.1.2 Modelling the Perceptual Effects

In the case of sound synthesis based on the modelling of perceptual effects, the sound generation is not merely based on the simulation of the physical or signal phenomena. This approach enables the synthesis of any kind of sounds, but it necessitates the understanding of the perceptual relevance of the sound attributes that characterize the sound category in question. Concerning environmental sounds, several studies have dealt with the identification and classification of such sounds (Ballas 1993; Gygi and Shafiro 2007; Gygi et al. 2007; Vanderveer 1979). A hierarchical taxonomy of everyday sounds was proposed by Gaver (1993b) and is based on three main categories: sounds produced by vibrating solids (impacts, deformation, etc.), aerodynamic sounds (wind, fire, etc.) and liquid sounds (drops, splashes, etc.). This classification related with the physics of sound events and has shown to be perceptually relevant. Hence, the perceptual relevance of these categories encourages the search for invariant sound morphologies specific to each category. This notion is developed in the next section.

4.3.2 Invariant Sound Morphologies

The invariant sound morphologies associated with the evocation of sound attributes can either be linked to the physical behaviour of the source (Giordano and McAdams 2006), to the signal parameters (Kronland-Martinet et al. 1997) or to timbre qualities based on perceptual considerations (McAdams 1999). This means that different synthesis approaches can be closely related, since in some cases, physical considerations and in other cases, signal variations might reveal important properties to identify the perceived effects of the generated sounds. In particular for environmental sounds, several links between the physical characteristics of actions (impact, bouncing, etc.), objects (material, shape, size, cavity, etc.) and their perceptual correlates were established in previous studies (see Aramaki et al. 2009; Aramaki et al. 2011 for a review). In summary, the question of sound event recognition was subject to several inquiries (e.g. Warren and Verbrugge 1984; Gaver 1993a, b) inspired by Gibson's ecological approach (Gibson 1986) and latter formalized by McAdams and Bigand (1993). This led to the definition of structural and transformational invariants linked to the recognition of the object's properties and its interaction with the environment, respectively.

Sounds from impacted objects: Impact sounds have been largely investigated in the literature from both physical and perceptual points of view. Several studies revealed relationships between perceptual attributes of sound sources and acoustic characteristics of the produced sound. For instance, the attack time has been related to the perception of the hardness of the mallet that was used to impact the resonant object, while the distribution of the spectral components (described by inharmonicity or roughness) of the produced sound has been related to the perceived shape of the object. The perceived size of the object is mainly based on the pitch. A physical explanation can be found in the fact that large objects vibrate at lower eigenfrequencies than small ones. Finally, the perception of material seems to be linked to the damping of the sound that is generally frequency-dependent: high frequency components are damped more heavily than low frequency components. In addition to the damping, the density of spectral components, which is directly linked to the perceived roughness, was also shown to be relevant for the distinction between metal versus glass and wood categories (Aramaki et al. 2009, 2011).

Sounds from continuous interactions: Based on previous works described in Sect. 4.3, invariant sound morphologies related to the perception of interactions such as rubbing, scratching and rolling were investigated (Conan et al. 2013a, b; Thoret et al. 2013). An efficient synthesis model, init. An efficient synthesis model, initially proposed by (Gaver 1993a) and improved by (van den Doel et al. 2001), consists in synthesizing the interaction sounds by a series of impacts that simulates the successive micro-impacts between a plectrum and the asperities of the object's surface. Therefore, it has been highlighted that a relevant sound invariant morphology allowing the discrimination between rubbing and scratching interactions was the temporal density of these impacts, i.e., the more (respectively, the less) impacts that occur, the more the sound is associated to rubbing (respectively, to scratching) (Conan et al. 2012). For the rolling interaction, it has been observed, from numerical simulations based on a physical model, that the temporal structure of the generated impact series follows a specific pattern. In particular, the time intervals between impacts and associated amplitudes are strongly correlated. Another fundamental aspect supported by physical considerations is the fact that the contact time of the impact depends on the impact velocity. This dependency also seems to be an important auditory cue responsible for the evocation of a rolling interaction (Conan et al. 2013).

These studies related to such interaction sounds led us to address the perceptual relation between the sound and the underlying gesture that was made to produce the sound. Many works highlighted the importance of the velocity profile in the production of a movement and its processing may be involved at different levels of perception of a biological movement both in the visual and in the kinaesthetic system ((Viviani and Stucchi 1992; Viviani et al. 1997; Viviani 2002) for a review). Based on these findings, we investigated whether the velocity profile, in the case of graphical movements, was also a relevant cue to identify a human gesture (and beyond the gesture, the drawn shape) from a friction sound. Results from a series of perceptual experiments revealed that the velocity profile transmits relevant information about the gesture and the geometry of the drawn shape to a certain extent. Results also indicated the relevance of the so-called 1/3-power law, defined from seminal works by Viviani and his colleagues and translating a biomechanics constraint between the velocity of a gesture and the local curvature of the drawn shape, to evoke a fluid and natural human gesture through a friction sound (cf. Thoret et al. 2013, 2014 for details and review).

Other environmental sounds: For other classes of environmental sounds such as wave or aerodynamic sounds, physical considerations generally involve complex modelling and signal models are then useful. From a perceptual point of view, these sounds evoke a wide range of different physical sources, but interestingly, from a signal point of view, common acoustic morphologies can be highlighted across these sounds. We analysed several signals representative of the main categories of environmental sounds as defined by Gaver and we identified a certain number of perceptually relevant signal morphologies linked to these categories (Gaver 1993a, b). To date, we concluded on five elementary sound morphologies based on impacts, chirps and noise structures (Verron et al. 2009). This finding is based on a heuristic approach that has been verified on a large set of environmental sounds. Granular synthesis processes based on these five sound "atoms" then enabled the generation of various environmental sounds (i.e. solid interactions, aerodynamic or liquid sounds). Note that this atom dictionary may be completed or refined in the future without compromising the proposed methodology.

A first type of grain is the "tonal solid grain" that is defined by a sum of exponentially damped sinusoids. Such a grain is well adapted to simulate sounds produced by solid interactions. Nevertheless, this type of grain cannot alone account for any kind of solid impact sounds. Actually, impact sounds characterized by a strong density of modes or by a heavy damping may rather be modelled as an exponentially damped noise. This sound characterization stands for both perceptual and signal points of view, since no obvious pitch can be extracted from such sounds. Exponentially damped noise constitutes the second type of grain, the so-called "noisy impact grain". Such a grain is well adapted to simulate crackling sounds. The third type of grain concerns liquid sounds. From an acoustic point of view, cavitation phenomena (e.g. a bubble in a liquid) lead to local pressure variations that generate time-varying frequency components such as exponentially damped linear chirps. Hence, the so-called "liquid grain" consists of an exponentially damped chirp signal. Finally, aerodynamic sounds generally result from complicated interactions between solids and gases and it is therefore difficult to extract useful information from corresponding physical models. A heuristic approach allowed us to define two kinds of aerodynamic grains: the "whistling grain" (slowly varying narrow band noise) and the "background aerodynamic grain" (broadband filtered noise). Such grains are well adapted to simulate wind and waves.

By combining these five grains using an accurate statistics of appearance, various environmental auditory scenes can be designed such as rainy ambiances, seacoast ambiances, windy environments, fire noises, or solid interactions simulating solid impacts or footstep noises. We currently aim at extracting the parameters corresponding to these grains from the analysis of natural sounds, using matching pursuit like methods. Perceptual evaluations of these grains will further allow us to identify or validate signal morphologies conveying relevant information on the perceived properties of the sound source.

4.4 Control Strategies for Synthesis Processes

The choice of synthesis model highly influences the control strategy. Physical synthesis models have physically meaningful parameters, which might facilitate the interpretation of the consequence of the control on the resulting sound. This is less so for signal models obtained from mathematical representations of sounds.

Perceptual considerations might, however, be combined with these models to propose intuitive control strategies as described in the following sections.

4.4.1 Control of Synthesis Parameters

Although physical models can produce high-quality sounds that are useful for instance for musical purposes, this approach is less adapted to environmental sounds, when the physics of such sound sources is not sufficiently well understood or the existing models are not real-time compatible. In such cases, signal models that enable the simulation of the sound vibrations through mathematical models are useful. The control of these models consists in manipulating physical or signal parameters. Practically, these approaches might involve the control of physical variables (for instance, characterizing the tribological or mechanical properties of the source) or a high number of synthesis parameters (up to a hundred) that are generally not intuitive for a non-expert user. This means that a certain scientific (or musical) expertise is needed to use such models (expert control). In fact, the calibration of the control of these models is based on an *error function* that reveals the difference between the model and the actual physical sound vibration (cf. Fig. 4.2).

4.4.2 Control of Perceptual Effects

Common to all the previous approaches described in Sect. 4.4.1 is the lack of perceptual criteria. Actually, since the timbre of the resulting sound is generally related to the synthesis parameters in a non-linear way, the control process can quickly become complicated and non-intuitive. The design of a control of perceptual effects may lead to the definition of an intuitive control strategy. In particular, based on the identification of invariant sound morphologies (cf. Sect. 4.3.2), control processes mediating various perceptual evocations could be designed. In line with the previous definitions of structural and transformational invariants, the framework of our control strategy is based on the so-called {*action/object*} paradigm, assuming that the produced sound can be defined as the consequence of an action on an object. This approach supports the determination of sound morphologies that carry information about the action and the object, respectively.

Here we present several synthesis tools that we have developed for generating and intuitively controlling sounds. These synthesis models make it possible to relevantly resynthesize natural sounds. In practice, we adopted hierarchical levels of control to route and dispatch the parameters from an intuitive to the algorithmic level. As these parameters are not independent and might be linked to several signal properties at a time, the mapping between levels is far from being straightforward.

Sounds from impacted objects: We have developed an impact sound synthesizer offering an intuitive control strategy based on a three-level architecture (Aramaki et al. 2010a) (cf. Fig. 4.1). The top layer gives the user the possibility to define the impacted object using verbal descriptions of the object (nature of the

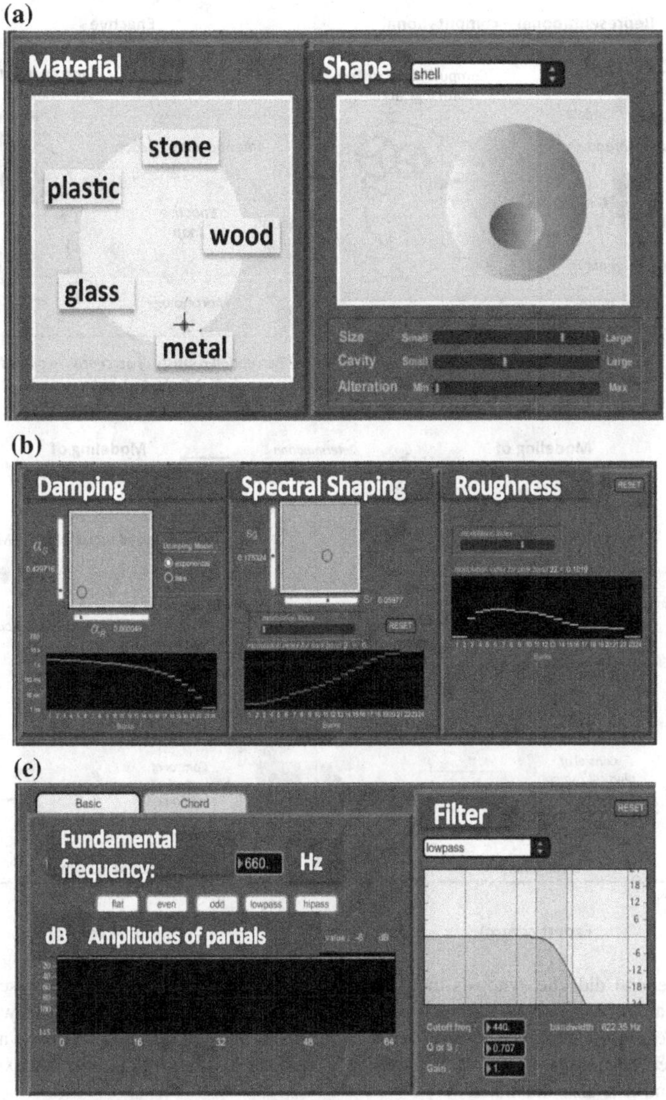

Fig. 4.1 **a** *Top layer* (semantic labels describing the perceived material and shape of the object), **b** *middle layer* (acoustic descriptors) and **c** *bottom layer* (synthesis parameters of the signal model) designed for the control of the impact sound synthesizer

perceived material, size and shape, etc.) and the excitation (force, hardness of the mallet, impact position, etc.). The middle layer is based on perceptually relevant acoustic descriptors linked to the invariant sound morphologies (cf. Sect. 4.3.2). The bottom layer consists of the set of synthesis parameters (for expert users). Two mapping strategies are implemented between the layers (we refer to (Aramaki et al. 2010a) for more details). The first one focuses on the relationships between verbal

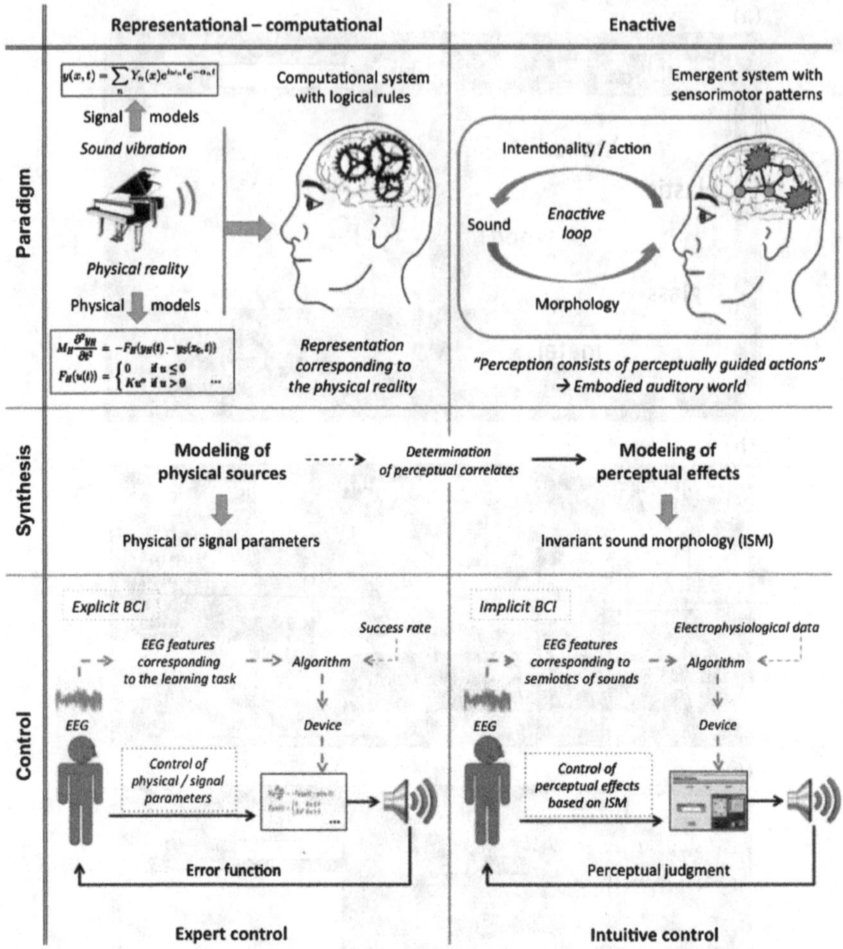

Fig. 4.2 General didactic synopsis including two approaches inspired by the representational-computational and enactive paradigms from cognitive neuroscience, the associated viewpoints for sound synthesis (modelling of physical sources and modelling of perceptual effects) and sound control (expert and intuitive control). A prospective view on the use of BCI in the context of sound synthesis control is also illustrated

descriptions of the sound source and the sound descriptors (damping, inharmonicity, roughness, etc.) characterizing perceptually relevant sound morphologies. The second one focuses on the relationships between sound descriptors and synthesis parameters (damping coefficient, amplitude and frequency of the components).

Sounds from continuous interactions: Control strategies for the synthesis processes of such sounds have recently been developed. In particular, an intuitive control strategy adapted to a non-linear friction sound model (producing phenomena such a creaky door, a singing glass or a squeaking wet plate) has been proposed. Inspired from Schelleng's diagrams, the proposed control is defined from a flexible physically informed mapping between a dynamic descriptor (velocity, pressure), and the synthesis parameters, and allows coherent transitions between the different non-linear friction situations (Thoret et al. 2013). Another intuitive control strategy dedicated to rolling sound synthesis has also been proposed (Conan et al. 2013). This strategy is based on a hierarchical architecture similar to that of the impacted object sounds (cf. previous paragraph). The high-level controls that can be manipulated by the end-user are the characteristics of the rolling ball (i.e. size, asymmetry and speed) and the irregularity of the surface. The low-level parameters (e.g. impacts' statistics, modulation frequency and modulation depth) are modified accordingly with respect to the defined mapping. Recently, a control strategy enabling to perceptually morph between the three continuous interactions, i.e. rubbing, scratching and rolling, was designed. For that purpose, we developed a synthesis process that is generic enough to simulate these different interactions and based on the related invariant sound morphologies (cf. Sect. 4.3.2). Then, a perceptual "interaction space" and the associated intuitive navigation strategy were defined with given sound prototypes considered as anchors in this space (Conan et al. 2013).

Finally, in line with the *action/object* paradigm, the complete synthesis process has been implemented as a source-filter model. The resulting sound is then obtained by convolving the excitation signal (related to the nature of the interaction) with the impulse response of the resonating object. The impulse response is implemented as a resonant filter bank, which central frequencies correspond to the modal frequencies of the object.

Immersive auditory scenes: An intuitive control of the sound synthesizer dedicated to environmental auditory scenes was defined. The control enables the design of complex auditory scenes and included the location and the spatial extension of each sound source in a 3D space so as to increase the realism and the feeling of being immersed in virtual scenes. This control is particularly relevant to simulate sound sources such as wind or rain that are naturally diffuse and wide. In contrast with the classical two-stage approach, which consists in first synthesizing a monophonic sound (timbre properties) and then spatializing the sound (spatial position and extension in a 3D space), the architecture of the proposed synthesizer yielded control strategies based on the overall manipulation of timbre and spatial attributes of sound sources at the same level of sound generation (Verron et al. 2010).

The overall control of the environmental scene synthesizer can be effectuated through a graphical interface where the sound sources (selected among a set of available sources: fire, wind, rain, wave, chimes, footsteps, etc.) can be placed around the listener (positioned in the centre of the scene) by defining the distance and the spatial width of each source. The sources are built from the elementary grains defined previously in Sect. 4.3.2. A fire scene is for instance built from a combination of a whistling grain (simulating the hissing), a background aerodynamic grain (simulating the background combustion) and noisy impact grains (simulating the cracklings). The latter grains are generated and launched randomly with respect to time using an accurate statistic law that can be controlled. A global control of the fire intensity, mapped with the control of the grain generation (amplitude and statistic law), is then designed. A rainy weather sound ambiance can be designed with a rain shower, water flow and drops, each of these environmental sounds being independently spatialized and constructed from a combination of the previous grains (see Verron et al. 2009 for more details). In case of interactive uses, controls can be achieved using either MIDI interfaces, from data obtained from a graphical engine or other external data sources.

4.5 Evidence of Semiotics for Non-linguistic Sounds

To propose an even more intuitive control of sound synthesis that directly uses a BCI, a relationship between the electroencephalogram (EEG) and the nature of the underlying cerebral processes has to be investigated. We here present results of several experimental studies aiming at supporting the existence of semiotics for non-linguistic sounds. In these studies, we used either synthetic stimuli using analysis/transformation/synthesis processes or sounds of a specific kind called "abstract" sounds promoting acousmatic listening (cf. Sect. 4.2). The participants' responses and reaction times (RTs) provided objective measurements to the processing of stimulus complexity.

Electrophysiological data: When appropriate, we also investigated the neural bases of the involved brain processes by analysing the EEG with the method of event-related potentials (ERP) time-locked to the stimulus onset during the various information processing stages. The ERP elicited by a stimulus (a sound, a light, etc.) are characterized by a series of positive (P) and negative (N) deflections relative to a baseline. These deflections (called components) are defined in terms of their polarity, their maximum latency (relative to the stimulus onset), their distribution among several electrodes placed in standard positions on the scalp and by their functional significance. Components P100, N100 and P200 are consistently activated in response to the auditory stimuli (Rugg and Coles 1995). Several late ERP components (N200, P300, N400, etc.) are subsequently elicited and associated with specific brain processes depending on the experimental design and the task in hand.

4.5.1 Perceptual Categorization of Sounds from Impacted Materials

In this experiment, we studied the perception of sounds obtained from impacted materials, in particular, wood, metal and glass (Aramaki et al. 2010a; Aramaki et al. 2010b; Aramaki et al. 2011). For this purpose, natural sounds were recorded, analysed, resynthesized and tuned to the same chroma to obtain sets of synthetic sounds representative of each category of the selected material. A sound-morphing process (based on an interpolation method) was further applied to obtain sound continua simulating progressive transitions between materials. Although sounds located at the extreme positions on the continua were indeed perceived as typical exemplars of their respective material categories, sounds in intermediate positions, which were synthesized by interpolating the acoustic parameters characterizing sounds at extreme positions, were consequently expected to be perceived as ambiguous (e.g. to be neither wood nor metal). Participants were asked to categorize each of the randomly presented sounds as wood, metal or glass.

Based on the classification rates, we defined "typical" sounds as sounds that were classified by more than 70 % of the participants in the right material category and "ambiguous" sounds, those that were classified by less than 70 % of the participants in a given category. Ambiguous sounds were associated with slower RTs than typical sounds. As might be expected, ambiguous sounds are therefore more difficult to categorize than typical sounds. This result is in line with previous findings in the literature showing that non-meaningful sounds were associated with longer RTs than meaningful sounds. Electrophysiological data showed that ambiguous sounds elicited more negative ERP (a negative component, N280, followed by a negative slow wave, NSW) in fronto-central brain regions and less positive ERP (P300 component) in parietal regions than typical sounds. This difference may reflect the difficulty to access information from long-term memory. In addition, electrophysiological data showed that the processing of typical metal sounds differed significantly from those of typical glass and wood sounds as early as 150 ms after the sound onset. The results of the acoustic and electrophysiological analyses suggested that spectral complexity and sound duration are relevant cues explaining this early differentiation. Lastly, it is worth noting that no significant differences were observed on the P100 and N100 components. These components are known to be sensitive to sound onset and temporal envelope, reflecting the fact that the categorization process occurs in later sound-processing stages.

4.5.2 Conceptual Priming for Non-linguistic Sounds

In language, a comprehensible linguistic message is for instance conveyed by associating words while respecting the rules of syntax and grammar. Can similar links be generated between non-linguistic sounds so that any variation will change the global information conveyed? From the cognitive neuroscience point of view, one of the major issues that arises from this question is whether similar neural networks are involved in the allocation of meaning in the case of language and that of sounds of other kinds. In a seminal study using a priming procedure, Kutas and Hillyard (Kutas and Hillyard 1980) established that the amplitude of a negative ERP component, the N400 component, increases when final sentence words are incongruous (e.g. *The fish is swimming in the river/carpet*). Since then, the N400 has been widely used to study semantic processing in language. In recent studies, priming procedures with non-linguistic stimuli such as pictures, odours, music and environmental sounds have been used (e.g. Holcomb and McPherson 1994; Castle et al. 2000; Koelsch et al. 2004; Daltrozzo and Schön 2009; Van Petten and Rheinfelder 1995; Orgs et al. 2006). Although the results of these experiments mostly have been interpreted as reflecting some kind of conceptual priming between words and non-linguistic stimuli, they may also reflect linguistically mediated effects. For instance, watching a picture of a bird or listening to a birdsong might automatically activate the verbal label "bird". Therefore, the conceptual priming cannot be taken to be purely non-linguistic because of the implicit naming induced by the processing of the stimulus. Such conceptual priming might imply at least language, generation of auditory scenes, and mental imaging, at various associative (non specific) cortex area levels. This might probably activate large neural/glial networks using long-distance synchronies, which could be investigated by a synchronous EEG activity measurement (Lachaux et al. 1999).

The aim of our first conceptual priming study (Schön et al. 2010) was to attempt to reduce as far as possible the likelihood that a labelling process of this kind takes place. To this end, we worked with a specific class of sounds called "abstract sounds", which physical sources cannot be easily recognized, meaning that verbal labelling is less likely to take place (Merer et al. 2011). We then conducted conceptual priming tests using word/sound pairs with different levels of congruence between the prime and the target. Subjects had to decide whether or not the prime and the target matched. In the first experiment, a written word was presented visually before the abstract sound, and in the second experiment, the order of presentation was reversed. Results showed that participants were able to assess the relationship between the prime and the target in both presentation orders (sound/word vs. word/sound), showing low inter-subject variability and good consistency. The presentation of a word reduced the variability of the interpretations of the abstract sound and led to a consensus between subjects in spite of the fact that the sound sources were not easily recognizable. Electrophysiological data showed the occurrence of an enhanced negativity in the 250–600-ms latency range in response to unrelated as compared to related targets in both experiments and the presence of a more fronto-central distribution in response to word targets and a more centro-parietal distribution in response to sound targets.

In a subsequent study (Aramaki et al. 2010b), we avoided the use of words as primes or targets. Conceptual priming was therefore studied using impact sounds (also used in the categorization experiment previously presented), as both primes and targets. As described in Sect. 4.5.1, these impact sounds were qualified as either typical or ambiguous with respect to a material category depending on their score in the categorization experiment. 3° of congruence were investigated through various combinations of typical and ambiguous sounds as prime and target: related, ambiguous and unrelated. The priming effects induced in these conditions were compared with those observed with linguistic sounds (spoken words) in the same group of participants. Results showed that N400-like components were also activated in a sound–sound design. This component may therefore reflect a search for meaning that is not restricted to linguistic meaning. Moreover, ambiguous targets also elicited larger N400-like components than related targets for both linguistic and non-linguistic sounds. These findings showed the existence of similar relationships in the processing of semiotics of both non-linguistic and linguistic target sounds. This study clearly means that it is possible to draw up a real language for non-linguistic sounds.

4.6 Towards a Semiotic-Based Brain Computer Interface (BCI)

BCIs provide a link between a user and an external electronic device through his or her brain activity, independently of the voluntary muscle activity of the subject. Most often BCIs are based on EEG recordings that allow for non-invasive measurements of electrical brain activity. As substitutional devices, BCIs open interesting perspectives for rehabilitation, reducing disability and improving the quality of life of patients with severe neuromuscular disorders such as amyotrophic lateral sclerosis or spinal cord injury (Wolpaw et al. 2002). Such interfaces, among many other possibilities, enable patients to control a cursor, to select a letter on a computer screen, or to drive a wheelchair. In addition to medical and substitutional applications, BCIs as enhancing devices can be used with healthy subjects. For example, in the field of video games, BCIs could capture the cognitive or emotional state of the user through the EEG to develop more adaptive games and to increase the realism of the gaming experience (Nijholt 2009). To date, two approaches to BCI could be highlighted: "explicit (or active) BCI" and "implicit (or passive) BCI" (George and Lécuyer 2010). These two classes of BCI could be linked with the two approaches inspired from the paradigms of cognitive science (described in Sect. 4.2) and the two approaches for sound synthesis (described in Sect. 4.3).

4.6.1 Explicit BCI

The explicit BCI is based on the principles of *operant conditioning*, the basic learning concept in experimental psychology, which assumes that the probability of occurrences of an animal or human behaviour is a function of a positive or negative reinforcement during the subject's learning process (Micoulaud-Franchi et al. 2013). Thus, the explicit BCI requires a learning period (George and Lécuyer 2010). In practice, the subject intentionally tries to control his/her cognitive activity to change his/her EEG activity and control an external electronic device. The EEG signal is recorded, processed in real time to extract the information of interest (e.g. spectral power EEG, slow cortical potential or ERP). This information is related to a cognitive activity that the subject intentionally produces. This information is further transmitted to the external electronic device using specific mapping that leads to the control of the device in the desired direction. The positive reinforcement (and the success rate) is determined by the capacity of controlling the external electronic device to achieve a given task.

This configuration fits with traditional neurofeedback therapeutics where the subject learns to intentionally control EEG through visual or auditory positive reinforcement, without any control of external device (Micoulaud-Franchi et al. 2013). In this context, the positive reinforcement could be an increase of a number of points, an advance of an animation on a computer screen, or a modification of a sound. When the EEG is related to symptoms of a disease, it has been shown that neurofeedback techniques can have a therapeutic effect, as is the case with attention deficit disorder with hyperactivity (Micoulaud-Franchi et al. 2011) or epilepsy (Micoulaud-Franchi et al. 2014).

4.6.2 Implicit BCI

In contrast with explicit BCI, the implicit BCI is not based on the principle of operant conditioning. The feedback in implicit BCI is used to optimize the interaction with an external device by directly modulating the brain activity and the cognitive activity of the subject (George and Lécuyer 2010). Implicit BCI does not require a learning period. In practice, the subject does not have to try to control intentionally his EEG. The EEG signal is recorded, processed in real time to extract the information of interest (e.g. power spectral EEG or ERP) corresponding to the subject's cognitive activity, and transmitted to the external electronic device to modulate and optimize the interaction between the device and the user.

This configuration fits with some non-traditional neurofeedback therapeutics that do not require specific cognitive tasks and are supposed to directly modulate the brain activity of the subject in order to optimize brain dynamics, although this remains largely hypothetical. Thus, unlike traditional neurofeedback approaches presented in the previous section, these non-traditional neurofeedback approaches have a very low level of therapeutic and clinical evidence (Micoulaud-Franchi et al. 2013).

4.6.3 Towards an Intuitive Control Using Semiotic-Based BCI

From the two approaches inspired by previous theoretical frameworks from cognitive neuroscience (Sect. 4.2), we propose a prospective view on a sound synthesis control strategy based on BCI. We reflect on whether EEG BCI would be helpful to increase the intuitiveness of control with the sound synthesizer. For a didactic perspective, we suggest to describe explicit and implicit BCI, respectively, from the representational-computational and from the enactive points of view.

We stress that in the explicit BCI, the user controls the external electronic device (positive reinforcement) as if it was an external object. In some way, there is a gap between the information of interest extracted from the recorded EEG activity and the positive reinforcement. The information feedback could be given to the subject by any kind of signal. The positive reinforcement mainly is useful for the learning process and for determining a success rate and is close to an error function (Sect. 4.4.1). We think that in many cases, explicit BCI does not permit to create recurrent sensorimotor patterns (from the enactive point of view) that enable action to be guided by the direct perception of the stimulus, which could be a limitation in the intuitiveness of BCI controllability.

We stress that in the Implicit BCI, the user and his/her brain is involved in an enactive process. In some way, there is a direct link between the information of interest extracted from the recorded EEG and the feedback. This feedback is not a positive reinforcement as defined by the operant-conditioning model. In fact, the aim of the feedback is not to inform the subject about the cognitive strategies that he/she develops during the learning process, but to directly influence the brain activity (and thus the EEG). Any kind of feedback cannot be used, but only those with the desired effect on the brain and the cognitive activity in order to enhance the interaction and the intuitiveness of the system.

Therefore, in the context of sound synthesis, a control strategy involving the use of explicit or implicit BCI would necessitate different mapping strategies. From a conceptual point of view, we stress that explicit and implicit BCI involve different levels of semiotic relation, i.e., the relation between the feedback and the meaning that the subject attributes to a sound. These two scenarios are discussed in the following paragraphs.

In the case of explicit BCI as defined above, the subject would have to control his/her cognitive activity to change his/her EEG and thus to control a specific parameter of the sound synthesizer. No semiotic relation between the EEG, the effect of the synthesized sound on the EEG, and the sound perception is therefore needed. In other words, the subject has to do something that is not necessarily related to the semiotics of the perceived synthesized sound to control the synthesizer. More so, an external algorithm is used to interpret the information of interest extracted from the EEG and to control the electronic device. For example, paying attention to a target to produce a P300 component that will be processed by the BCI and arbitrarily associated with a control parameter according to the output of the algorithm and to a success rate (Fig. 4.2). This situation that necessitates a certain expertise acquired during a learning period seems to be quite close to sound synthesis based on the physical or signal modelling of sound vibrations (Sect. 4.3).

In the case of implicit BCI as defined above, the aim would be to enhance the quality and the intuitiveness of the sound synthesizer by taking into account the EEG induced by the sound. Thus, a strict semiotic relation between the EEG and the influence of sounds on the EEG should be known. In other words, we need to understand the neural bases of sound semiotics ("electrophysiological data" in Fig. 4.2) to implement this information in an implicit BCI process dedicated to the sound synthesizer. We propose to call it "semiotic-based BCI". In this context, the results obtained from previous EEG experiments presented in Sect. 4.5 constitute an interesting starting point for the design of such a mapping strategy. This approach seems to be quite close to sound synthesis based on the modelling of perceptual effects, which does not necessitate a learning period (Sect. 4.3). This intuitive control implies that perceptual and cognitive aspects are taken into account in order to understand how a sound is perceived and interpreted. As shown in Fig. 4.2, a loop is thus designed between perception and action through the intuitive control of the sound synthesizer (Sect. 4.2). Implicit BCI offers the possibility of a second loop, between the sound effect on the EEG and the sound synthesizer that is likely to optimize the sound effect on both the perceptual judgment and the Implicit BCI.

4.7 Conclusion

To date, the design of a control strategy of sound synthesis processes that uses a BCI is still a challenging perspective. As discussed in (Väljamäe et al. 2013), a synthesis control of sounds directly from the brain through the measurement of its cerebral activity is still in its early stages. In particular, the mapping between electrophysiological signal features and synthesis parameters is generally validated on the basis on different metrics depending on applications. However, the definition of such metrics implies a given conception on the way we interact with the surrounding world.

To broach this issue, we introduced two conceptual approaches inspired from the representational-computational and the enactive paradigms from cognitive neuroscience. In light of these paradigms, we revisited the existing main approaches for synthesis and control of sounds. In fact, the viewpoints adopted to synthesize sounds are intricately underpinned by paradigms that differ in the epistemological positions of the observer (from a third or a first-person position) and have a substantial consequence on the design of a control strategy (cf. Figure 4.2). On one hand, synthesis processes based on the modelling of physical sources (from either the mechanical behaviour or the resulting vibration) are controlled by physical or signal parameters. This approach is based on the existence of a correct representation of the physical world and introduces the notion of an error function between the model and the physical reality as a quality criterion. Therefore, it requires a certain expertise from the end-user. On the other hand, synthesis processes based on the modelling of perceptual effects involve the identification of invariant sound morphologies specific to given perceptual attributes of the sound source. This approach assumes the emergence of an embodied auditory world from an enactive process. The perceptual judgments are considered as a quality criterion for the model, leading to the design of a more intuitive control.

By associating these conceptual and pragmatic considerations, we proposed a prospective view on the methodology to be used to design a BCI control. For the sake of illustration, we treated limited aspects of BCIs by addressing explicit BCI from the representational-computational point of view and implicit BCI from the enactive point of view. Actually, we are aware that the frontier between explicit and implicit BCI might be difficult to establish and less didactic than what this article presents. Indeed, the implicit communication channel might sometimes be used in an explicit way (George and Lécuyer 2010), and inversely brain plasticity can enable the participant to make use of the training experienced from the explicit BCI to generate implicit recurrent sensorimotor patterns (Bach-y-Rita and Kercel 2003). With current apparatus performances, the rate of transfer information between the BCI and the device is quite limited and the final task has to be defined accordingly. While this technique may represent a restricted interest for healthy users (in some cases, it would be easier to directly control the device manually), it constitutes a relevant medium for medical applications and can be used as a substitutional device for diseases. In the implicit BCI, the control is included in an optimization system in which the electrophysiological data supplies further information about the way the user perceives the sound (beyond verbal labels or gestures for instance). In contrast with the explicit BCI, this configuration is well adapted to intuitive synthesis control. Therefore, we suggested a "semiotic-based BCI" founded on identified links between the brain activity and invariant signal morphologies reflecting the attribution of sense to a sound that may enhance the interactivity and the intuitiveness of the system.

4.8 Questions

1. What are the characteristics of the representational-computational paradigm of perception?
2. What are the characteristics of the enactive paradigm of perception?
3. What is the difference between physical and signal sound synthesis models?
4. What are the main limitations of the use of physical models for sound synthesis?
5. How can the invariant sound morphologies be determined?
6. Which invariant sound morphologies are related to the perception of material in an impact sound?
7. Which aspects should be taken into account in the design of a control strategy based on a representational-computational or an enactive paradigm?
8. What are the characteristics of explicit (or active) BCI?
9. What are the characteristics of implicit (or passive) BCI?
10. What is the purpose of offering intuitive control of sound synthesis processes using BCI?

References

Aramaki M, Besson M, Kronland-Martinet R, Ystad S (2009) Timbre perception of sounds from impacted materials: behavioral, electrophysiological and acoustic approaches. In: Ystad S, Kronland-Martinet R, Jensen K (eds) Computer music modeling and retrieval—genesis of meaning of sound and music, vol 5493., LNCSSpringer, Berlin, Heidelberg, pp 1–17

Aramaki M, Besson M, Kronland-Martinet R, Ystad S (2011) Controlling the perceived material in an impact sound synthesizer. IEEE Trans Audio Speech Lang Process 19(2):301–314

Aramaki M, Gondre C, Kronland-Martinet R, Voinier T, Ystad S (2010a) Imagine the sounds: an intuitive control of an impact sound synthesizer. In: Ystad S, Aramaki M, Kronland-Martinet R, Jensen K (eds) Auditory display, vol 5954., Lecture notes in computer scienceSpringer, Berlin, Heidelberg, pp 408–421

Aramaki M, Marie C, Kronland-Martinet R, Ystad S, Besson M (2010b) Sound categorization and conceptual priming for nonlinguistic and linguistic sounds. J Cogn Neurosci 22(11):2555–2569

Arfib D (1979) Digital synthesis of complex spectra by means of multiplication of non-linear distorted sine waves. J Audio Eng Soc 27:757–768

Atal BS, Hanauer SL (1971) Speech analysis and synthesis by linear prediction of the speech wave. J Acoust Soc Am 50(2B):637–655

Avanzini F, Serafin S, Rocchesso D (2005) Interactive simulation of rigid body interaction with friction-induced sound generation. IEEE Trans Speech Audio Process 13(5):1073–1081

Bach-y-Rita P, Kercel W (2003) Sensory substitution and the human-machine interface. Trends in Cogn Sci 7:541–546

Ballas JA (1993) Common factors in the identification of an assortment of brief everyday sounds. J Exp Psychol Hum Percept Perform 19(2):250–267

Bensa J, Jensen K, Kronland-Martinet R (2004) A hybrid resynthesis model for hammer-strings interaction of piano tones. EURASIP J Appl Sig Process 7:1021–1035

Bilbao S (2009) Numerical sound synthesis: finite difference schemes and simulation in musical acoustics. Wiley, Chichester, UK

Castle PC, van Toller S, Milligan G (2000) The effect of odour priming on cortical EEG and visual ERP responses. Int J Psychophysiol 36:123–131

Chaigne A (1995) Trends and challenges in physical modeling of musical instruments, In: Proceedings of the international congress on acoustics', Trondheim, Norway

Chowning J (1973) The synthesis of complex audio spectra by means of frequency modulation. J Audio Eng Soc 21:526–534

Conan S, Aramaki M, Kronland-Martinet R, Thoret E, Ystad S (2012) Perceptual differences between sounds produced by different continuous interactions. Proceedings of the 11th Congrès Français d'Acoustique. Nantes, France, pp 409–414

Conan S, Aramaki M, Kronland-Martinet R, Ystad S (2013) Post-proceedings 9th International Symposium on Computer Music Modeling and Retrieval (CMMR 2012). Lecture notes in computer science, vol 7900. Springer, Berlin, Heidelberg, chapter Intuitive Control of Rolling Sound Synthesis

Conan S, Thoret E, Aramaki M, Derrien O, Gondre C, Kronland-Martinet R, Ystad S (2013) Navigating in a space of synthesized interaction-sounds: rubbing, scratching and rolling sounds. In: Proceedings of the 16th international conference on digital audio effects (DAFx-13), Maynooth, Ireland

Cook PR (1992) A meta-wind-instrument physical model, and a meta-controller for real-time performance control. In: Proceedings of the international computer music conference, pp 273–276

Daltrozzo J, Schön D (2009) Conceptual processing in music as revealed by N400 effects on words and musical targets. J Cogn Neurosci 21:1882–1892

de Saussure F (1955) Cours de linguistique générale. Payot, Paris

Flanagan JL, Coker CH, Rabiner LR, Schafer RW, Umeda N (1970) Synthetic voices for computer. IEEE Spectr 7:22–45

Gaver WW (1993a) How do we hear in the world? Explorations of ecological acoustics. Ecol Psychol 5(4):285–313

Gaver WW (1993b) What in the world do we hear? An ecological approach to auditory source perception. Ecol Psychol 5(1):1–29

George L, Lécuyer A (2010) An overview of research on "passive" brain-computer interfaces for implicit human-computer interaction. In: International conference on applied bionics and biomechanics ICABB 2010—workshop W1 Brain-Computer Interfacing and Virtual Reality, Venezia, Italy

Gibson JJ (1986) The ecological approach to visual perception, Lawrence Erlbaum Associates

Giordano BL, McAdams S (2006) Material identification of real impact sounds: effects of size variation in steel, wood, and plexiglass plates. J Acoust Soc Am 119(2):1171–1181

Gygi B, Kidd GR, Watson CS (2007) Similarity and categorization of environmental sounds. Percept Psychophys 69(6):839–855

Gygi B, Shafiro V (2007) General functions and specific applications of environmental sound research. Front Biosci 12:3152–3166

Hermes DJ (1998) Synthesis of the sounds produced by rolling balls. Internal IPO report no. 1226, IPO, Center for user-system interaction, Eindhoven, The Netherlands

Holcomb PJ, McPherson WB (1994) Event-related brain potentials reflect semantic priming in an object decision task. Brain and Cogn 24:259–276. http://forumnet.ircam.fr/product/modalys/?lang=en (n.d.). http://www-acroe.imag.fr/produits/logiciel/cordis/cordis_en.html (n.d.)
Husserl E (1950) Idées directrices pour une phénoménologie, Gallimard. J New Music Res, special issue “enaction and music” (2009), 38(3), Taylor and Francis, UK
Karjalainen M, Laine UK, Laakso T, Vilimtiki V (1991) Transmission-line modeling and real-time synthesis of string and wind instruments. In: I. C. M. Association (ed) Proceedings of the international computer music conference, Montreal, Canada, pp 293–296
Koelsch S, Kasper E, Sammler D, Schulze K, Gunter T, Friederici A (2004) Music, language and meaning: brain signatures of semantic processing. Nat Neurosci 7(3):302–307
Kronland-Martinet R (1989) Digital subtractive synthesis of signals based on the analysis of natural sounds. A.R.C.A.M. (ed), Aix en Provence
Kronland-Martinet R, Guillemain P, Ystad S (1997) Modelling of natural sounds by time-frequency and wavelet representations. Organ Sound 2(3):179–191
Kutas M, Hillyard SA (1980) Reading senseless sentences: brain potentials reflect semantic incongruity. Science 207:203–204
Lachaux JP, Rodriguez E, Martinerie J, Varela F (1999) Measuring phase synchrony in brain signals. Hum Brain Mapp 8:194–208
Lalande A (1926) Vocabulaire technique et critique de la philosophie, Edition actuelle, PUF n quadrige z 2002
Le Brun M (1979) Digital waveshaping synthesis. J Audio Eng Soc 27:250–266
Makhoul J (1975) Linear prediction, a tutorial review. In: Proceedings of the IEEE, vol 63. pp 561–580
Matyja JR, Schiavio A (2013) Enactive music cognition: background and research themes. Constr Found 8(3):351–357. http://www.univie.ac.at/constructivism/journal/8/3/351.matyja
McAdams S (1999) Perspectives on the contribution of timbre to musical structure. Comput Music J 23(3):85–102
McAdams S, Bigand E (1993) Thinking in sound: the cognitive psychology of human audition, Oxford University Press, Oxford
Merer A, Ystad S, Kronland-Martinet R, Aramaki M (2011) Abstract sounds and their applications in audio and perception research. In: Ystad S, Aramaki M, Kronland-Martinet R, Jensen K (eds) Exploring music contents, vol 6684., Lecture notes in computer scienceSpringer, Berlin, Heidelberg, pp 176–187
Micoulaud-Franchi JA, Bat-Pitault F, Cermolacce M, Vion-Dury J (2011) Neurofeedback dans le trouble déficit de l’attention avec hyperactivité : de l’efficacité à la spécificité de l’effet neurophysiologique. Annales Médico-Psychologiques 169(3):200–208
Micoulaud-Franchi JA, Cermolacce M, Vion-Dury J, Naudin J (2013) Analyse critique et épistémologique du neurofeedback comme dispositif thérapeutique. le cas emblématique du trouble déficit de l’attention avec hyperactivité’, L’évolution psychiatrique
Micoulaud-Franchi J, Lanteaume L, Pallanca O, Vion-Dury J, Bartolomei F (2014) Biofeedback et épilepsie pharmacorésistante : le retour d’une thérapeutique ancienne ? Revue Neurologique 170(3):187–196
Nadeau R (1999) Vocabulaire technique et analytique de l’épistémologie, PUF
Nijholt A (2009) BCI for games: a ‘state of the art’ survey. In: Stevens SM, Saldamarco SJ (eds) LNCS, vol 5309. Springer, Berlin, pp 225–228
O’Brien JF, Shen C, Gatchalian CM (2002) Synthesizing sounds from rigid-body simulations. In: Press A (ed) The ACM SIGGRAPH 2002 symposium on computer animation, pp 175–181
Orgs G, Lange K, Dombrowski J, Heil M (2006) Conceptual priming for environmental sounds and words: An ERP study. Brain Cogn 62(3):267–272
Pai DK, van den Doel K, James DL, Lang J, Lloyd JE, Richmond JL, Yau SM (2001) Scanning physical interaction behavior of 3D objects. In: Proceedings of SIGGRAPH 2001, computer graphics proceedings, annual conference series, pp 87–96

Petitmengin C, Bitbol M, Nissou JM, Pachoud B, Curalucci H, Cermolacce M, Vion-Dury J (2009) Listening from within. J Conscious Stud 16:252–284
Rabiner LR, Gold B (1975) Theory and application of digital signal processing. Prentice Hall, Englewood Cliffs, NJ
Rath M, Rocchesso D (2004) Informative sonic feedback for continuous human–machine interaction—controlling a sound model of a rolling ball. IEEE Multimedia Spec Interact Sonification 12(2):60–69
Risset JC (1965) Computer study of trumpet tones. J Acoust Soc Am 33:912
Roads C (1978) Automated granular synthesis of sound. Comput Music J 2(2):61–62
Rugg MD, Coles MGH (1995) Electrophysiology of mind. Event-related brain potentials and cognition, number 25. In: 'Oxford Psychology', Oxford University Press, chapter The ERP and Cognitive Psychology: Conceptual issues, pp 27–39
Schaeffer P (1966) Traité des objets musicaux, du Seuil (ed)
Schön D, Ystad S, Kronland-Martinet R, Besson M (2010) The evocative power of sounds: conceptual priming between words and nonverbal sounds. J Cogn Neurosci 22(5):1026–1035
Smith JO (1992) Physical modeling using digital waveguides. Comput Music J 16(4):74–87
Stoelinga C, Chaigne A (2007) Time-domain modeling and simulation of rolling objects. Acta Acustica united Acustica 93(2):290–304
Thoret E, Aramaki M, Gondre C, Kronland-Martinet R, Ystad S (2013) Controlling a non linear friction model for evocative sound synthesis applications. In: Proceedings of the 16th international conference on digital audio effects (DAFx-13), Maynooth, Ireland
Thoret E, Aramaki M, Kronland-Martinet R, Velay J, Ystad S (2014) From sound to shape: auditory perception of drawing movements. J Exp Psychol Hum Percept Perform
Thoret E, Aramaki M, Kronland-Martinet R, Ystad S (2013) Post-proceedings 9th International Symposium on Computer Music Modeling and Retrieval (CMMR 2012), number 7900. In: Lecture notes in computer science, Springer, Berlin, Heidelberg, chapter reenacting sensori-motor features of drawing movements from friction sounds
Väljamäe A, Steffert T, Holland S, Marimon X, Benitez R, Mealla S, Oliveira A, Jordà S (2013) A review of real-time EEG sonification research. In: Proceedings of the 19th international conference on auditory display (ICAD 2013), Lodz, Poland, pp 85–93
van den Doel K, Kry PG, Pai DK (2001) Foleyautomatic: physically-based sound effects for interactive simulation and animation. In: Proceedings of SIGGRAPH 2001, computer graphics proceedings, annual conference series, pp 537–544
Van Petten C, Rheinfelder H (1995) Conceptual relationships between spoken words and environmental sounds: event-related brain potential measures. Neuropsychologia 33 (4):485–508
Vanderveer NJ (1979) Ecological acoustics: human perception of environmental sounds, PhD thesis, Georgia Inst. Technol
Varela F (1989) Invitation aux sciences cognitives. Seuil, Paris
Varela F (1996) Neurophenomenology: a methodological remedy for the hard problem. J Conscious Stud 3:330–335
Varela F, Thompson E, Rosch E (1991) The embodied mind: cognitive science and human experience. MIT Press, Cambridge, MA, USA
Verron C, Aramaki M, Kronland-Martinet R, Pallone G (2010) A 3D immersive synthesizer for environmental sounds. IEEE Trans Audio Speech Lang Process 18(6):1550–1561
Verron C, Pallone G, Aramaki M, Kronland-Martinet R (2009) Controlling a spatialized environmental sound synthesizer. Proceedings of the IEEE workshop on applications of signal processing to audio and acoustics (WASPAA). New Paltz, NY, pp 321–324
Viviani P (2002) Motor competence in the perception of dynamic events: a tutorial. In: Prinz W, Hommel B (eds) Common mechanisms in perception and action. Oxford University Press, New York, NY, pp 406–442
Viviani P, Redolfi M, Baud-Bovy G (1997) Perceiving and tracking kinaesthetic stimuli: further evidence for motor-perceptual interactions. J Exp Psychol Hum Percept Perform 23:1232–1252

Viviani P, Stucchi N (1992) Biological movements look uniform: evidence of motor-perceptual interactions. J Exp Psychol Hum Percept Perform 18:603–623
Warren WH, Verbrugge RR (1984) Auditory perception of breaking and bouncing events: a case study in ecological acoustics. J Exp Psychol Hum Percept Perform 10(5):704–712
Wolpaw JR, Birbaumer N, McFarland DJ, Pfurtscheller G, Vaughan TM (2002) Brain-computer interfaces for communication and control. Clin Neuro physiol 113:767–791
Ystad S, Voinier T (2001) A virtually-real flute. Comput Music J 25(2):13–24

Machine Learning to Identify Neural Correlates of Music and Emotions

5

Ian Daly, Etienne B. Roesch, James Weaver and Slawomir J. Nasuto

Abstract

While music is widely understood to induce an emotional response in the listener, the exact nature of that response and its neural correlates are not yet fully explored. Furthermore, the large number of features which may be extracted from, and used to describe, neurological data, music stimuli, and emotional responses, means that the relationships between these datasets produced during music listening tasks or the operation of a brain–computer music interface (BCMI) are likely to be complex and multidimensional. As such, they may not be apparent from simple visual inspection of the data alone. Machine learning, which is a field of computer science that aims at extracting information from data, provides an attractive framework for uncovering stable relationships between datasets and has been suggested as a tool by which neural correlates of music and emotion may be revealed. In this chapter, we provide an introduction to the use of machine learning methods for identifying neural correlates of musical perception and emotion. We then provide examples of machine learning methods used to study the complex relationships between neurological activity, musical stimuli, and/or emotional responses.

5.1 Introduction

It is widely understood that music is able to induce a wide range of emotions in the listener. What is not so well understood is the specific neurological mechanism by which this process takes place.

I. Daly (✉) · E.B. Roesch · J. Weaver · S.J. Nasuto
School of Systems Engineering, University of Reading, Reading RG6 6AY, UK
e-mail: i.daly@reading.ac.uk

E.R. Miranda and J. Castet (eds.), *Guide to Brain-Computer Music Interfacing*,
DOI 10.1007/978-1-4471-6584-2_5

In order to allow a brain-computer music interface (BCMI) to interact with, and allow control of musical stimuli, it is first necessary to understand the relationships between music, the emotion(s) induced or perceived, and neurological activity. For this purpose, it is necessary to identify both neural correlates of the emotions induced by particular musical stimuli and neural correlates of the perception of music.

However, the feature space that may be extracted from—and used to describe—the brain may be very large, complex, and high dimensional. For example, the electroencephalogram (EEG) provides a time series of discrete measures of electrical activity recorded from the surface of the scalp and, due to high sampling rates, may be very large and described by a multitude of features. Similarly, functional magnetic resonance imaging (fMRI) provides a detailed three-dimensional measure of oxygen consumption by neurons throughout the brain resulting in a much higher-dimensional time series, which may also be described by a multitude of features.

Therefore, there is a need for advanced analysis methods to uncover relationships between cognitive processes and their corresponding neural correlates, which may not be immediately apparent.

Machine learning describes a set of methods which attempt to learn from the data (Alpaydin 2004). For the purposes of brain–computer interfacing and the study of the brain, this commonly takes the form of learning in which brain activation patterns are associated with particular cognitive processes or differentiate groups of processes.

Machine learning provides a data-driven approach to understanding relationships between neurological datasets and their associated cognitive processes. For example, it may allow the uncovering of complex neural correlates of specific emotions or music perception, which are not immediately apparent via other analysis techniques.

This chapter provides an introduction into the use of machine learning methods in the context of identifying neural correlates of emotion and music perception. We first introduce models of emotion and empirical measures of emotional responses, which are required by machine learning methods to allow training on labelled data. We then go on to review features which may be extracted from neurological data and audiological signals to describe relevant or interesting properties. Finally, machine learning methods are described and examples are provided to illustrate their use in uncovering neural correlates of emotion and music perception.

5.2 Measuring Emotion

5.2.1 Models of Emotion

Models of emotion differ significantly in the way they conceptualise the determinants of the emotional experience, the way they envision the neural mechanisms that give rise to this experience, and the predictions they formulate for the accompanying symptoms. Particularly, the landscape of emotion theories spans so-called discrete emotion theories and dimensional and appraisal theories, among others, which constitute many theoretical lenses through which results from psychology and neuroscience can be interpreted (see Chap. 6 for an introduction, and Cornelius 1996 for an accessible and comprehensive review).

Discrete emotion theories or basic emotion theories originated after Charles Darwin's seminal work on emotional expression. This theoretical tradition conceptualises emotions as differentiated entities that possess unique properties, distinguishing one emotion from the others. In this lens, each emotion is mediated by separate and distinct neural substrates, which are the result of evolutionary pressure. A basic emotion occurs rapidly, and automatically, upon perception and classification of a stimulus as belonging to a specific class of events (Tomkins 1962; Ekman 2003; Izard 2007).

Dimensional theories and appraisal theories are cousin traditions, which conceptualise affective phenomena as the result of the integration of a number of separate processes. Unlike basic emotion theories, these theoretical traditions attempt the description of mechanisms common to most, if not all emotions. Whereas appraisal theories focus on the mechanisms antecedent to the emotional experience (Scherer et al. 2001; Roesch et al. 2006; Fontaine et al. 2007), dimensional theories place the emphasis on the correlates of this experience, in the brain and the behaviour, and represent emotion in a low-dimensional space.

In the context of our understanding of the neural correlates underlying the emotional experience of music, the theoretical tradition that one chooses to depart from will have strong consequences for the design of empirical work and the practical decisions that researchers have to make for the analysis of their data. In particular, this choice will have consequences when attempting to capture the subjective experience of a participant placed in a specific context, as well as when choosing neural features of interest and implementing analysis of the data.

5.2.2 Self-reporting Felt Emotion

There are a wide variety of methods available for individuals to report their felt emotional response, all of which aim to accurately identify the emotional state of an individual. Techniques commonly focus on identifying one property of the induced affective state such as the intensity of the response (affect intensity measure, AIM) (Larsen and Diener 1987), the valence (profile of mood states, POMS) (Albrecht and Ewing 1989), or how aroused the individual is (UWIST mood adjective checklist) (Matthews 1990). Most of these methods rely on participants accurately interpreting descriptive emotive terms as well as having an adequate level of emotional intelligence (Salovey and Pizarro 2003) to be able to assess their own affective state.

To avoid any semantic misunderstanding of the adjective descriptions as well as the difficulty of reflecting them to one's emotional state, many researchers use the self-assessment manikin (SAM) (Bradley and Lang 1994). SAM is comprised of descriptive visual images that range across a nine-point rating scale for pleasure, arousal, and intensity, respectively. This tool for reporting emotional response has been extensively used to label emotionally loaded stimuli for a range of databases such as the International Affective Picture System (IAPS) (Lang et al. 2008), International Affective Digital Sounds (IADS) (Bradley et al. 2007), and the dataset for emotional analysis using physiological signals (DEAP) (Soleymani et al. 2012).

However, one downside to the use of SAM and other self-assessment tools when exploring the influence of music is that they only provide information of the individuals' affective state at a discrete time, post-stimulus in most experimental paradigms. As such, there is a growing interest in emotional research for a tool which provides the participant the ability to continuously report how they feel in an easy, quick, and accurate manner. Current tools available to researchers include FEELTRACE (Cowie and Douglas-Cowie 2000) and EMuJoy (Hevner 2007), both of which allow the subject to position and navigate their affective state through two-dimensional emotional models.

5.3 Measuring Neurological Activity

There are three broad groups of feature types that may be extracted from neurological data: those based upon the amplitude of the recorded signals, those based in the frequency content, and those based upon the phase content (Lotte et al. 2007; Hwang et al. 2013). Additionally, combinatorial features, which combine two or more feature types, may also be considered. For example, time-frequency activity maps may be used to describe changes in the amplitude and frequency distribution of the EEG or magnetoencephalogram (MEG) over time.

Amplitude-based features may include measures such as event-related potential (ERP) amplitude, peak latency, statistical measures of the distribution of the data, measures of relationships within the data (e.g. correlation), voxel strength in specific regions of interest (ROIs) in a magnetic resonance imaging (MRI) map, etc. Frequency-domain features are used to describe how the frequency content of the data changes with specific control tasks or over time. This can include measures of the magnitude of specific frequencies, the distribution of frequencies across the power spectra (power spectral density; PSD), coherence measures, or the relative power of specific frequencies (Wang et al. 2011). Phase-domain features are used much less frequently in BCI research (Hwang et al. 2013), but nonetheless show some promise, particularly when used in the investigation of relationships between different spatial regions during specific tasks (Daly et al. 2012). Combinatorial features are being used in an increasing proportion of BCI research (Hwang et al. 2013). This is most likely due to the increasing interest in hybrid BCIs (hBCIs), in which paradigms or signal types are combined (Pfurtscheller et al. 2010; Müller-Putz et al. 2011).

Figure 5.1 provides a taxonomy of feature types which may be used to describe neurological signals, including examples of each type.

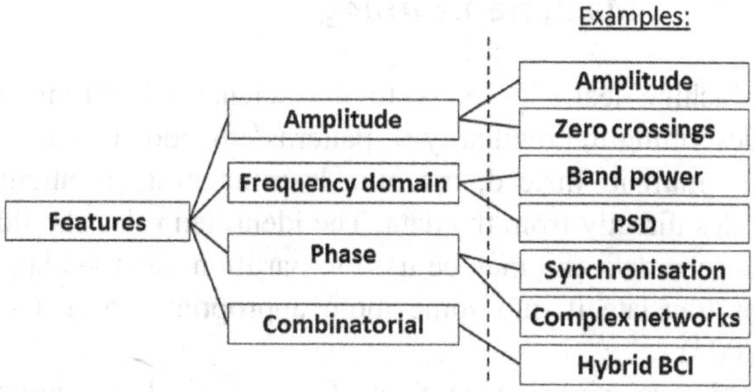

Fig. 5.1 Taxonomy of feature types used to describe neurological datasets

5.4 Measuring Music

Musical properties and structure may be described by a number of models or methodologies. However, many of these methods rely on specific descriptions of musical pieces, which are often grounded in specific musical styles or cultures. For example, descriptions of chord structures in a piece of music stem from European musical history (Christensen 2002) and may not apply equally well to music from other cultural backgrounds.

An alternative view of music is to treat it as a complex time-varying set of sounds. From this view, music is merely a label one may apply to a set of complex sounds with specific structural properties. Thus, one may take acoustic properties of a recording of a piece of music and use them as alternate descriptors of the music.

The advantage of such an approach is that it allows one to describe all sounds in the same manner. Thus, music from any cultural background, genre, or style may all be described in the same manner and via the same framework of features. In addition, non-musical sounds such as speech, environmental noise, animal cries, etc., may also be described under the same framework.

A very large number of feature types may be extracted from a piece of sound (Mitrovic et al. 2010). These may be broadly grouped into six types: temporal-domain features, frequency-domain features, cepstral features, modulation frequency-domain features, eigen-domain features, and phase space features.

Temporal- and frequency-domain audio features are analogous to EEG features. Cepstral-domain features are heavily used in speech analysis (Liu and Wan 2001) and attempt to capture timbre and pitch information by taking frequency-smoothed representations of the log magnitude spectrum of the signal. Modulation frequency features capture low-frequency modulation information; sounds induce different hearing sensations in human hearing (e.g. rhythm) (Tzanetakis and Cook 2002). Eigen-domain features describe long-term information in the audio signal, such as statistical markers of noise versus structured sound (Mitrovic et al. 2010). Finally, phase space features attempt to capture nonlinear properties of the auditory signal, such as turbulence introduced by the vocal tract (Kokkinos and Maragos 2005).

5.5 Machine Learning

Machine learning refers to the science of getting a computer to learn a rule describing the regularity of patterns embedded in data without having to explicitly program it. Instead, machine learning systems attempt to identify generalisable rules directly from the data. The identified rules should be applicable to previously unseen data and may be used to partition them via appropriate decision boundaries or translate it into some more appropriate space (Alpaydin 2004; Müller et al. 2004).

For example, given a set of neurological measurements, it may be desirable to identify a rule which relates them to measures of musical stimuli from a piece of music played to the participant. Alternatively, one may uncover neural correlates of emotion by identifying a rule which relates neurological activity to participants' self-reported emotions.

Rule identification often amounts to identification of decision boundaries which may be applied to the data. For example, given EEG recorded during two tasks (e.g. listening to music vs. listening to noise), rule identification may amount to identifying a rule for finding whether a new EEG segment corresponds to piece of music or a noisy auditory stimulus.

More formally, a two-class problem classification learning may be expressed as the process of identifying a function $f: R^N \rightarrow \{-1, +1\}$ from a function class F using a set of training data such that f will classify unseen examples with minimum error. For problems with more than two classes, f is modified appropriately.

Machine learning methods can be broadly described as either supervised or unsupervised. Supervised machine learning methods use labelled data as a part of the learning process, whereas unsupervised methods do not.

5.5.1 Unsupervised Machine Learning Methods

Unsupervised machine learning does not use labelled data and hence concentrates on removing redundancy in the dataset or on emphasising components of the data which may be of interest, for example, components of high variance. This means unsupervised machine learning methods are often used for dimensionality reduction, for example, principal component analysis (PCA) (Smith 2002; Lee and Seungjin 2003), for identifying advantageous translations that may be applied to the data, for example, independent component analysis (ICA) (Comon 1994; Qin et al. 2005), or for identifying hidden structure in the data, for example, clustering algorithms (Guyon and Elisseeff 2003; Dy 2004) or Markov modelling (which may be used for either clustering or classification) (Obermaier et al. 2001). They may use translations or transformations to reduce the dimensionality of the dataset and hence select subsets of the data that may better illustrate or highlight features or structures of interest.

To illustrate the ways in which unsupervised machine learning methods can be applied to uncover neural correlates of musical perception or emotions induced by listening to music, a set of case studies is described below.

5.5.1.1 Case Study 1: PCA for Uncovering EEG Dynamics During Music Appreciation

In work by Lin et al. (2009), PCA was used to help uncover neural correlates of emotions during a music listening task. EEG was recorded from 26 participants who listened to 16 different musical segments, selected to induce particular emotional responses. Power spectral densities (a frequency-domain feature) were then estimated from 1-s non-overlapping windows over 30 s of EEG recorded during each of the musical clips. This was done for each of the traditional EEG frequency bands (delta: 1–3 Hz, theta: 4–7 Hz, alpha: 8–13 Hz, beta: 14–30 Hz, and gamma: 31–50 Hz), forming an initial feature set containing 2,400 features per subject (5 frequency bands × 30 time windows × 16 musical clips) per channel. The final feature set was then produced by taking the spectral differences between left and right hemisphere channel pairs for all channels from the International 10/20 system (e.g. Fp1-Fp2 etc.).[1] The final feature set is denoted as X where each element $X_{i,k}$ denotes a feature k extracted from the EEG recorded while the participant listened to a piece of music i.

PCA attempts to identify an orthogonal transformation that translates a set of potentially correlated variables into a set of linearly uncorrelated variables. These new variables are referred to as the principal components (PCs) (Smith 2002).

PCA operates by first subtracting the mean from the data to centre it. Thus, in our case study, Lin and colleagues took their original feature set X and derived a new, zero-mean, feature set $\overline{X}$ by subtracting the mean from X.

The covariance matrix of the feature set $\overline{X}$ is then used to measure the strength of the relationships between all rows of $\overline{X}$. This is defined as the matrix C where each element $C_{i,j}$ denotes the covariance between rows i and j (corresponding to musical pieces i and j) in the feature set $\overline{X}$.

$$C_{i,j} = \frac{\sum_{k=1}^{n}(X_{i,k} - \overline{X}_{i,:})(X_{j,k} - \overline{X}_{j,:})}{(n-1)} \tag{5.1}$$

where $X_{i,k}$ and $X_{j,k}$ denote the kth features from different musical pieces i and j, and $\overline{X}_{i,:}$ denotes the mean over a feature vector for an individual piece of music i.

Eigen decomposition is then applied to analyse the structure of this covariance matrix. The covariance matrix is decomposed into a matrix of eigenvectors and a vector of eigenvalues. This may be defined as

$$Cu = \lambda u, \tag{5.2}$$

[1] Please refer to Chap. 2 for an introduction to EEG electrode placement systems.

where u denotes an eigenvector of the covariance matrix C and λ denotes the corresponding eigenvalue.

The eigenvalues identified for C may be ranked in increasing value with the corresponding eigenvectors containing projections of the feature set onto principal components, which are placed in the order of decreasing variance. Thus, the eigenvector corresponding to the largest eigenvalue contains a projection of the feature set X which has the greatest variance over the selection of musical pieces played. This eigenvector may then be used to classify musical pieces with high accuracy with respect to their emotional valence while using a subset of the data.

Lin et al. (2009) select a set of the first few eigenvectors calculated from the EEG such that they contain more than 80 % of the variance of the feature set. These eigenvectors are then used as features in a classification stage, which is reported to produce classification accuracies of up to 85 %.

A similar approach can also be seen in (Ogawa et al. 2005) in which PCA and canonical discriminant analysis (CDA) are each used to identify features that may be extracted from the EEG. These methods are applied to EEG to attempt to identify metrics for the identification of pieces of music for use in music therapy.

5.5.1.2 Case Study 2: ICA to Identify Neural Correlates of Music

ICA is used by Cong et al. (2013) to identify feature sets which are able to identify neural correlates of perception of long pieces of naturalistic music. EEG was recorded from 14 participants, while they listened to an 8.5-min-long piece of music (modern tango by Astor Piazzolla), via 64 electrodes.

ICA is based upon the assumption that the recorded neurological data are derived from recording a mixture of statistically independent sources. This may be explained by an analogy of a cocktail party. In a party, you may be able to hear several people talking at the same time. Specifically, although the sounds produced by each speaker may be independent of one another, the sound you hear contains a linear mixture of each of the speakers. Thus, in order to attend to what one specific speaker is saying, you must attempt to separate this linear mixture of sounds into their original sources.

ICA attempts to solve this problem by identifying a linear transformation from the recorded multichannel EEG data x to a set of independent neural sources s. It does so by imposing the assumption that the neural sources are statistically independent of one another and identifying a transformation matrix which, when applied to the EEG, results in maximally independent estimated sources. This may be defined as

$$s = A^{-1}x, \tag{5.3}$$

where A^{-1} denotes the transformation matrix, to translate the EEG into the estimated sources s and may be inverted to reconstruct the recorded EEG from the estimated sources,

$$x = As. \tag{5.4}$$

After application of ICA decomposition to the 64-channel EEG recorded from their participants, Cong and colleagues produced spatial maps of the projections of each independent component (IC) onto the scalp.

The musical piece played to the participants was then decomposed into a set of temporal and spectral features. Features were selected that attempted to describe the tonal and rhythmic features of the music. The ICs identified from the EEG were then clustered and used as the basis for a neural feature set, which was observed to significantly correlate with the acoustical features extracted from the musical piece.

5.5.2 Supervised Machine Learning Methods

In contrast to the unsupervised methods, supervised techniques need information about data class membership in order to estimate the function $f{:}R^N \to \{-1, +1\}$, effectively amounting to learning the class membership or inter-class decision boundary. This information comes in the form of a labelled training dataset where each datum, representing in the feature space the object of interest, is accompanied by the label denoting the class to which this object belongs. The class information can be used either explicitly or implicitly in construction of the class membership function.

The methods may use labels explicitly if the latter are represented by numeric values, typically −1 and 1 for a binary classification problem. In this case, the entire training set, data, and their labels are used in training the classifier. Typically, this is performed by feeding the data into the classifier with randomly initialised parameter and comparing the classifier output representing the proposed class membership with the data class labels. The discrepancies between obtained and the true labels are accumulated and form the basis of an error cost function. Thus, the classifier training is cast as an optimisation problem—traversing the classifier parameter space in order to find the optimal parameter configuration, such that the error cost function is minimised.

In classification methods based on implicit use of class membership in training, the class labels need not be numeric and are not used explicitly to adjust the classifier parameters during traversing the parameter space. Rather, the class information is used by grouping data belonging to an individual class and extracting relevant information groupwise in the process of constructing the classifier decision boundary.

There are different ways in which classifiers can be categorised. Discriminative classifiers need to use information from both classes in their training as they learn by differentiating properties of data examples belonging to different classes. In contrast, generative classifiers assume some functional form of the class models which are then estimated from the training set and subsequently used in order to perform the classification. The additional benefit of such an approach is that the models can also be used to construct new data with properties consistent with the class description. The discriminative methods do not afford such a use but construct decision boundary making somewhat less explicit assumptions.

Classifiers can also be differentiated with respect to the nature of the data. Most standard classifiers are applied to categorise objects. This typically amounts to representing objects as points in an appropriately dimensional feature space, which typically is a Euclidean vector space. However, it is also possible to use classification approach to analyse processes rather than objects. Data from such processes typically come in the form of time series.

Although it is possible to unfold time series in time and use the highly dimensional vector space representation and hence static classifiers, there have been also a number of dynamic classification techniques developed to deal with temporal dynamics of processes.

Another way to categorise classifiers considers the nature of the decision boundary they can construct. Linear classifiers construct hyperplanes as their decision boundaries, whereas nonlinear classifiers can generate more flexible hypersurfaces in order to achieve class separation.

Once the decision boundary is constructed, using explicit or implicit class membership information, the unseen data can be fed into classifier which will then suggest the most likely class for a given datum (i.e. on which side of decision boundary it falls).

The subsequent classification performance will depend on the nature of the decision boundary a given classifier architecture can realise, on the properties of the training set and how the training process was conducted. Any finite training set will by necessity contain only limited information about the data class structure. This is because of the inherent error and incompleteness of the representation of classified objects in the feature space, as well as finite data size. Moreover, the dataset may actually contain some peculiar characteristics which is specific to only this finite sample rather than representative of the entire class. Thus, the training of the classifier must be monitored in order to arrive at a decision boundary that represents true class information rather than overfitting to the specific dataset. At the same time, how well the true decision boundary can be captured depends on the classifier complexity. These conflicting constraints can be formalised within the framework of mean square cost error where the latter can be shown to decompose into three terms corresponding, respectively, to the noise, bias, and variance, where only the bias (the constraints implied by a particular type of the decision boundary class supported by the classifier) and variance (the dependence on the training set used) can be minimised. Unfortunately, the bias–variance trade-off implies it is actually typically not possible to minimise both simultaneously; the increase of classifier flexibility from enlarging the class of the boundary decision functions it can represent typically makes it more prone to overfitting. On the other hand, more robust classifiers that are not sensitive to the peculiarities of the data tend to have low complexity (highly constrained decision boundary form).

In order to strike the balance implied by the bias–variance trade-off dilemma, the preparation of the classifier is often performed in stages supported by splitting the data into several sets, with the training subset used to adjust the classifier parameters, validation set used to monitor classifier progress or complexity, and testing used for the finally selected classifier.

Many authors use machine learning techniques to analyse EEG for BCI applications, and the number of studies investigating automated emotion recognition is growing (Lin et al. 2010, 2011; Wang et al. 2013; Sohaib et al. 2013).

The following two cases studies will illustrate the use of the supervised machine learning methods used to study emotional responses to music using EEG.

5.5.2.1 Case Study 3: SVMs for Identifying Neural Correlates of Emotion

Sohaib et al. (2013) used the support vector machine as a method for automatic emotion recognition from the EEG. The increased use of computer technology by humans and hence raise in prominence of human computer interaction motivates incorporating automated emotion recognition into such technologies. The authors used the IAPS as emotional stimuli. Images in the IAPS database were tagged with their emotional content along dimensions of valence, arousal, and dominance, although the authors only considered valence and arousal in their study. EEG was recorded while the subjects viewed the images and assessed their emotion using a self-assessment manikin.

The EEG was recorded at 2,048 Hz from 6 channels—Fp1, Fp2, C3, C4, F3, and F4—and referenced to Cz and were preprocessed using ICA in order to remove artefacts. Four features for each channel (minimum, maximum, mean, and standard deviation) were obtained forming 24-dimensional feature vectors which were then used to train the classifiers.

The support vector machine belongs to the family of discriminative classifiers. It is also a member of the so-called kernel methods, because of its use of kernel functions which allow it to form a map of the input data space into a feature space where the classification is attempted. The mapping is only done implicitly, and hence, the computational costs are associated with the dimensionality of the input space, even though the classification is performed in the potentially highly dimensional feature space. Additionally, because of this feature mapping, although the decision boundary in the input data space is nonlinear, it is an image (via feature mapping) of the decision boundary which is a linear hyperplane in the feature space. The strength of the SVM comes from using this kernel trick—as the problem is formulated as linear classification yet whether the actual classification is linear or not and what kind of nonlinearities are involved depends on the choice of the kernel. Moreover, the problem of finding a hyperplane in the feature space is formulated as a constrained optimisation where a hyperplane is sought that maximises the separation margin between the classes. This also gives rise to the selection of the data which are important for ultimate construction of the optimal separating hyperplane, the support vectors, which lend the name to the entire classifier.

Sohaib et al. (2013) report that their SVM obtained classification rates of over 56 % on the binary classification problem (negative/positive arousal/valence), higher than four other classifiers used. The classification rate was raised to an average of over 66 % when the 15 subjects were split into 3 equal groups and each group classified separately.

5.5.2.2 Case Study 4: Classifying Discrete Emotional States

Murugappan et al. (2010) used two classifiers, K nearest neighbour (KNN) and linear discriminant analysis (LDA), for classification of discrete emotional states from audio-visual stimuli. They used video clips from the international standard emotional clips set (Yongjin and Ling 2005). They collected EEG from 62 channels and used a surface Laplacian (SL) filter to remove the artefacts. They then extracted both standard (signal power, standard deviation, and variance) and novel features from each channel based on a discrete wavelet decomposition into 3 frequency bands: alpha, beta, and gamma. The wavelet features were functions of the relative contribution of energy within a frequency band to the total energy across the three bands.

The KNN classifier takes a new data point and assigns it to the most frequent class among a group of the k labelled training examples. The LDA works by identifying a decision boundary hyperplane which maximises the inter-class distance, while simultaneously minimising within-class variance. The authors report the highest classification accuracy for discrete wavelet power-derived features (83 % for KNN and 75 % for LDA) on the entire set of 62 channels with the classification accuracy dropping to 72 % for KNN and 58 % for LDA on a subset of 8 channels. The traditional features provided consistently worse results.

Machine learning methods have been used in very diverse ways ranging from generic approaches to increase machine "intelligence" (Warwick and Nasuto 2006), to analysis of pictorial (Ruiz and Nasuto 2005) or numeric data, such as EEG time series for BCI applications (Aloise et al. 2012; Rezaei et al. 2006; Daly et al. 2011). Lotte et al. (2007) provide an extensive discussion of supervised approaches used in brain–computer interfaces.

5.6 Summary

Neurological data may be described in a large multitude of ways by a range of different feature types. Therefore, often relationships between neurological data and relevant measures of behaviour, stimuli, or responses may not be immediately apparent. Such relationships may in fact be complex and comprised of multiple, potentially weakly interacting components.

Machine learning provides a statistically sound framework for uncovering these relationships. It has, therefore, been proposed by a number of authors as a suitable mechanism for identifying neural correlates of music perception and emotional responses to music.

We suggest that in order to construct a brain-computer music interface (BCMI) based upon the interaction of the brain and a musical generator, an understanding of these relationships is required. Machine learning provides a suitable framework through which such an understanding may be acquired.

5.7 Questions

1. How can the theoretical landscape of emotion theories be described?
2. What are the predictions about the emotional experience evoked by music that one can formulate from a discrete emotion perspective? From a dimensional perspective? From an appraisal perspective?
3. What are the practical implications of favouring one theory over an other for the application of machine learning techniques to neural signals in the emotional experience evoked by music?
4. What are the advantages of using the self-assessment manikin for assessing emotional states of individuals? Are there any disadvantages?
5. Describe an experimental paradigm which would benefit more from a continuous self-assessment tool than a discrete approach, explain your reasons?
6. What information you would need to collect during experiments aimed at assessing EEG correlates of emotional states in order to use supervised techniques to learn to recognise the brain emotional states?
7. What class of the machine learning techniques is suitable for EEG analysis if one does not have objective information about the emotional states of the subject?
8. What are the advantages of generative classifiers over the discriminative ones? Can you list also some of their disadvantages?
9. An EEG experiment is conducted to measure neurological activity during a music listening task. The experimental hypothesis is that listening to music with a faster tempo may increase the power spectral density in the alpha frequency band recorded from the prefrontal cortex. Describe the types of features that may be extracted from (1) the EEG and (2) the music, to test this hypothesis.
10. How might ICA be applied to identify neural correlates of emotional responses to stimuli in the EEG?

References

Albrecht R, Ewing S (1989) Standardizing the administration of the profile of mood states (POMS): development of alternative word lists. J Pers Assess 53(1):31–39

Aloise F, Schettini F, Aricó P et al (2012) A comparison of classification techniques for a gaze-independent P300-based brain–computer interface. J Neural Eng 9(4):045012

Alpaydin E (2004) Introduction to machine learning. MIT Press, Cambridge

Bradley MM, Lang PJ (1994) Measuring emotion: the self-assessment manikin and the semantic differential. J Behav Ther Experim Psychiatry 25(1):49–59

Bradley MM, Lang PJ, Margaret M et al (2007) The international affective digitized sounds affective ratings of sounds and instruction manual. Technical report B-3, University of Florida, Gainesville, Fl

Christensen T (2002) The Cambridge history of western music theory. Cambridge University Press, Cambridge

Comon P (1994) Independent component analysis, a new concept? Sig Process 36(3):287–314
Cong F, Alluri V, Nandi AK et al (2013) Linking brain responses to naturalistic music through analysis of ongoing EEG and stimulus features. IEEE Trans Multimedia 15(5):1060–1069
Cornelius RR (1996) The science of emotion. Prentice Hall, Upper Saddle River
Cowie R, Douglas-Cowie E (2000) "FEELTRACE": an instrument for recording perceived emotion in real time. In: Proceedings of the ISCA workshop on speech and emotion: a conceptual framework for research, Belfast, pp 19–24
Daly I, Nasuto SJ, Warwick K (2011) Single tap identification for fast BCI control. Cogn Neurodyn 5(1):21–30
Daly I, Nasuto SJ, Warwick K (2012) Brain computer interface control via functional connectivity dynamics. Pattern Recogn 45(6):2123–2136
Dy J (2004) Feature selection for unsupervised learning. J Mach Learn Res 5:845–889
Ekman P (2003) Emotions revealed: recognizing faces and feelings to improve communication and emotional life. Weidenfeld & Nicolson, London
Fontaine JRJ, Scherer KR, Roesch EB et al (2007) The world of emotions is not two-dimensional. Psychol Sci 18(12):1050–1057
Guyon I, Elisseeff A (2003) An introduction to variable and feature selection. J Mach Learn Res 3:1157–1182
Hevner SK (2007) EMuJoy: software for continuous measurement. Behav Res Methods 39 (2):283–290
Hwang HJ, Kim S, Choi S et al (2013) EEG-based brain–computer interfaces (BCIs): a thorough literature survey. Int J Human Comput Interact 29(12):814–826
Izard CE (2007) Basic emotions, natural kinds, emotion schemas, and a new paradigm. Perspect Psychol Sci 2:260–280
Kokkinos I, Maragos P (2005) Nonlinear speech analysis using models for chaotic systems. IEEE Trans Speech Audio Process 13(6):1098–1109
Lang PJ, Bradley MM, Cuthbert BN (2008) International affective picture system (IAPS): affective ratings of pictures and instruction manual. Technical Report A-6, University of Florida, Gainesville, Fl
Larsen RJ, Diener E (1987) Affect intensity as an individual difference characteristic: a review. J Res Pers 21(1):1–39
Lee H, Seungjin C (2003) PCA + HMM + SVM for EEG pattern classification. In: Proceedings of signal processing and its applications, vol 1, pp 541–544
Lin Y, Juny T, Chen J (2009) EEG dynamics during music appreciation. In: Annual international conference of the IEEE Engineering in Medicine and Biology Society. IEEE Engineering in Medicine and Biology Society, pp 5316–5319
Lin Y-P, Wang C-H, Jung T-P et al (2010) EEG-based emotion recognition in music listening. IEEE Trans Biomed Eng 57(7):1798–1806
Lin Y-P, Chen J-H, Duann J-R et al (2011) Generalizations of the subject-independent feature set for music-induced emotion recognition. In: 33rd annual international conference of the IEEE, EMBS, Boston, Massachusetts, USA
Liu M, Wan C (2001) Feature selection for automatic classification of musical instrument sounds. In: Proceedings of the first ACM/IEEE-CS joint conference on digital libraries—JCDL '01. ACM Press, New York, pp 247–248
Lotte F, Congedo M, Lécuyer A et al (2007) A review of classification algorithms for EEG-based brain-computer interfaces. J Neural Eng 4(2):1–13
Matthews G (1990) Refining the measurement of mood: the UWIST mood adjective checklist. Br J Psychol 81(1):17–42
Mitrović D, Zeppelzauer M, Breiteneder C (2010) Features for content-based audio retrieval. Adv Comput 78:71–150
Müller KR, Krauledat M, Dornhege G et al (2004) Machine learning techniques for brain-computer interfaces. Biomed Eng 49(1):11–22

Müller-Putz GR, Breitwieser C, Cincotti F et al (2011) Tools for brain–computer interaction: a general concept for a hybrid BCI. Front Neuroinform 5:30
Murugappan M, Nagarajan R, Yaacob S (2010) Classification of human emotion from EEG using discrete wavelet transform. J Biomed Sci Eng 3:390–396
Obermaier B, Guger C, Neuper C et al (2001) Hidden Markov models for online classification of single trial EEG data. Pattern Recogn Lett 22(12):1299–1309
Ogawa S, Ota S, Ito S et al (2005) Influence of music listening on the cerebral activity by analyzing EEG. In: Proceedings of the 9th international conference on knowledge-based intelligent information and engineering systems (KES'05), pp 657–663
Pfurtscheller G, Allison BZ, Brunner C et al (2010) The hybrid BCI. Front Neuroprosthetics 4:30
Qin J, Li Y, Cichocki A (2005) ICA and committee machine-based algorithm for cursor control in a BCI system. In: Advances in neural networks—ISNN 2005. Springer, Berlin, pp 973–978
Rezaei S, Tavakolian K, Nasrabadi AM et al (2006) Different classification techniques considering brain computer interface applications. J Neural Eng 3(2006):139–144
Roesch EB, Fontaine JB, Scherer KR (2006) The world of emotion is two-dimensional or is it? Paper presented to the HUMAINE Summer School 2006, Genoa, Italy
Ruiz VF, Nasuto SJ (2005) Biomedical image classification methods and techniques. In: Costaridou L (ed) Medical image analysis methods. Taylor & Francis, New York, p 504
Salovey P, Pizarro DA (2003) The value of emotional intelligence. In: Sternberg RJ, Lautrey J, Lubart TI (eds) Models of intelligence: international perspectives. American Psychological Association, Washington, DC, pp 263–278
Scherer KR, Schorr A, Johnstone T (2001) Appraisal processes in emotion: theory, methods, research. Oxford University Press, Oxford
Smith LI (2002) A tutorial on principal components analysis. Technical report, Cornell University
Sohaib AT, Qureshi S, Hagelbäck J et al (2013) Evaluating classifiers for emotion recognition using EEG. In: Foundations of augmented cognition. Lecture notes in computer science, vol 8027, pp 492–501
Soleymani M, Member S, Lee J (2012) DEAP: a database for emotion analysis using physiological signals. IEEE Trans Affect Comput 3(1):18–31
Tomkins SS (1962) Affect, imagery, consciousness. Springer, New York
Tzanetakis G, Cook P (2002) Musical genre classification of audio signals. IEEE Trans Speech Audio Process 10(5):293–302
Wang XW, Nie D, Lu BL (2011) EEG-based emotion recognition using frequency domain features and support vector machines. In: Lu B-L, Zhang L, Kwok J (eds) Proceedings of the 18th international conference on neural information processing (ICONIP'11), vol 7062, pp 734–743
Wang X-W, Nie D, Lu B-L (2013) Emotional state classification from EEG data using machine learning approach. Neurocomputing 129:94
Warwick K, Nasuto SJ (2006) Historical and current machine intelligence. IEEE Instrum Meas Mag 9(6):20–26
Yongjin W, Ling G (2005) Recognizing human emotion from audiovisual information. In: IEEE international conference on acoustics, speech, and signal processing (ICASSP '05), vol 2, pp 1125–1128

Emotional Responses During Music Listening

6

Konstantinos Trochidis and Emmanuel Bigand

Abstract

The aim of this chapter is to summarize and present the current knowledge about music and emotion from a multi-disciplinary perspective. Existing emotional models and their adequacy in describing emotional responses to music are described and discussed in different applications. The underlying emotion induction mechanisms beside cognitive appraisal are presented, and their implications on the field are analyzed. Musical characteristics such as tempo, mode, loudness, and so on are inherent properties of the musical structure and have been shown to influence the emotional states during music listening. The role of each individual parameter on emotional responses as well as their interactions is reviewed and analyzed. Different ways of measuring emotional responses to music are described, and their adequacy in accounting for emotional responses to music is discussed. The main physiological responses to music listening are briefly discussed, and their application to emotion recognition and to emotion intelligence in human–machine interaction is described. Music processing in the brain involves different brain areas and several studies attempted to investigate brain activity in relation to emotion during music listening through EEG signals. The issues and challenges of assessing human emotion through EEG are presented and discussed. Finally, an overview of problems that remain to be addressed in future research is given.

K. Trochidis (✉) · E. Bigand
Department of Cognitive Psychology, University of Burgundy, Dijon, France
e-mail: konstantinos.trochidis@gmail.com

E. Bigand
e-mail: emmanuel.bigand@u-bourgogne.fr

E.R. Miranda and J. Castet (eds.), *Guide to Brain-Computer Music Interfacing*,
DOI 10.1007/978-1-4471-6584-2_6

6.1 Introduction

Music by its nature has the ability to communicate strong emotions in everyday life. Given the important role of emotion in music, the topic of music and emotion has recently become an expanding field of research with up-to-date developments.

The aim of this chapter is to summarize and present the current knowledge about music and emotion from a multi-disciplinary perspective.

We intended to make this chapter accessible to a wide range of readers from different disciplines including computer science, engineering, musicology, music information retrieval, and brain–computer interfaces.

The chapter is organized into eight sections. The first section consists of this introduction, which describes the objectives of the chapter and its structure.

The first issue raised in the research on music and emotion is the modeling and representation of emotions. Section 6.2 focuses on the existing emotional models and their adequacy in modeling emotional responses to music. The two main approaches, discrete and dimensional, and their variants are described and discussed in different applications. A comparison between the two approaches is presented, and the advantages and drawbacks of each approach are analyzed.

How music evokes emotions? A fascinating question that still remains open. To explain how music can induce strong emotions in listeners is of great importance for numerous disciplines including psychology, musicology, and neuroscience. All these years, however, it was assumed that musical emotion can be studied without necessarily knowing the underlying induction mechanisms. In Sect. 6.3, the existing approaches are reviewed and state-of-the-art proposed mechanisms beside cognitive appraisal are presented and discussed.

Musical characteristics such as tempo, mode, loudness, and so on are inherent properties of the musical structure and have been shown to influence the emotional states during music listening. Although affective associations of both tempo and mode are fairly well established, relatively little is known about how these musical parameters interact. In Sect. 6.4, the role of each individual parameter on emotional responses as well as their interactions is reviewed and analyzed.

Section 6.5 discusses and compares different ways of measuring emotional responses. Emotions are subjective phenomena, and therefore, their measurement is a difficult task. Measurement of emotional responses is closely related to the existing emotional models. When using discrete models, distinct labels are employed such as happiness, anger, etc. In contrast, when dimensional models are used, rating scales of valence and arousal are employed. Both approaches are mainly based on "subjective" self-reports. Physiological responses to music offer an "objective" alternative for measuring emotional responses to music.

Emotional responses during music listening are related to physiological responses. To understand the relationship between the two is not an easy task and has been the subject of intensive investigation. In Sect. 6.6, the main physiological responses to music listening are briefly discussed and their application to emotion recognition and to emotion intelligence in human–machine interaction is described.

Music processing in the brain involves different brain areas and several studies attempted to investigate brain activity in relation to emotion during music listening through EEG signals. Furthermore, a growing interest has been recently developed to build brain-computer music interfaces (BCMI) that use real-time brain signals to communicate with the environment. Recent research in BCMI aims at the ability to access the user's brain activity to gain insight into the user's emotional state. Deeper understanding of the influence of the emotional state on brain activity patterns can allow the BCMI to adapt its recognition algorithms, so that the intention of the user is correctly interpreted in spite of deviations caused by the subject's emotional state. Furthermore, the ability to recognize emotions can be used to provide the user with more ways of controlling the BCMI through affective modulation. In Sect. 6.7, the issues and challenges of assessing different human emotions through EEG are presented and discussed.

Finally, Sect. 6.8 provides an overview of the chapter and focuses on problems that remain to be addressed in future research.

6.2 Models of Musical Emotions

The first problem raised in the research of music emotion is a model of emotions. There are different approaches as to how emotions can be conceptualized and described. The two main approaches that have strongly influenced research in the area are the discrete or categorical approach and the dimensional approach.

6.2.1 Discrete Emotion Models

According to the discrete model, all emotions can be derived from a limited number of basic universal emotions such as fear, anger, disgust, sadness, and happiness (Ekman 1992a, b, 1999; Panksepp 1998). Each emotion is independent of the others in its behavioral, psychological, and physiological manifestation, and each arises from activation of independent neural systems. In studies investigating music and emotion, the discrete model has been modified to better represent the emotions induced by music. Emotions such as disgust are rarely expressed by music and therefore have been replaced by tenderness, which is more suitable in the context of music (Balkwill and Thompson 1999; Gabrielsson and Juslin 1996). Although the number of basic emotions has been a matter of debate, the discrete model has proven robust against cross-cultural, neural and physiological studies (Panksepp 1992). The discrete model has found so far applications in music psychological (Dalla Bella et al. 2001), physiological (Baumgartner et al. 2006), and neurological studies (Peretz et al. 1998).

Basic emotions have been investigated by exploring peripheral physiological responses, and it was assumed that each basic emotion is associated with a specific physiological pattern. It was found, however, that basic emotions are not associated with specific patterns of autonomic activation (Cacioppo et al. 2000). Moreover, listeners may experience both sad and happy feelings at the same time depending on the stimulus (Hunter et al. 2008). It was also proposed that each basic emotion is associated with a characteristic facial expression. This assumption, however, has not been confirmed. Certain facial expressions are often associated with more than one emotion (smile, for example, is associated with both happiness and pride).

It was argued that the basic emotions of the discrete model are not adequate to capture the richness of musical emotion (Zentner et al. 2008). Therefore, a new model was proposed based on a study that used self-reports of the listeners. The listeners were asked to list how frequently perceived a group of affective terms related to music. Principal component analysis of the results showed that affective responses can be grouped into nine categories. The resulted Geneva Emotion Music Scale model (GEMS) includes wonder, transcendence, tenderness, nostalgia, peacefulness, power, joyful activation, tension, and sadness. Zentner et al. (2008) compared GEMS model with the discrete and dimensional emotion models by asking listeners to rate music-induced emotions in a list containing all emotions. The results of this comparison showed that the listeners preferred to describe the emotions induced in terms of GEMS rather than the other two. Moreover, the most effective discrimination of musical excerpts was obtained using the terms provided by GEMS. Although it was reported that GEMS model outperformed both discrete and dimensional models, the results have to be further investigated and tested on larger collections including various music genres.

6.2.2 Dimensional Emotion Models

The alternative to discrete models is the dimensional approach (Fig. 6.1). While the discrete approach focuses on the distinct characteristics that distinguish emotions from each other, in the dimensional models, emotions are expressed on a plane along two axes such as valence and arousal. In contrast to basic emotion concept, dimensional models suggest that an interconnected neurophysiological system is responsible for all affective states.

The circumplex emotion model (Russel 1980) proposes that emotions can be expressed in terms of two dimensions, one related to arousal (activation-deactivation) and valence (pleasure-displeasure) that are orthogonally situated in the affective plane. Thus, all emotions can be considered as varying degrees of both valence and arousal. Although Russell's model has found wide application and is the dominant model in emotion research other potential variants of two-dimensional models have been proposed. Thayer (1989) proposed a different two-dimensional model. He suggested that the two affective dimensions are two separate arousal dimensions: energetic arousal and tension arousal. According to this model, valence can be explained as varying combination of energetic and tension arousal.

Another variant of the circumplex model is the Tellegen–Watson model (Watson and Tellegen 1985). This model extends the two-dimensional models by

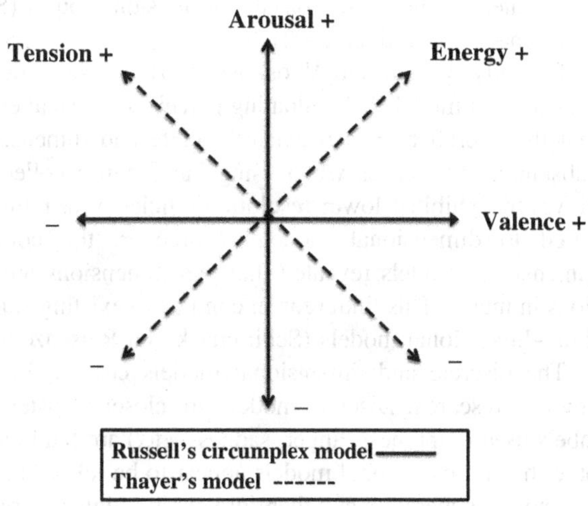

Fig. 6.1 Schematic diagram of the dimensional emotion model

emphasizing the value of a hierarchical perspective of emotional expressivity. It analyzes a three-level hierarchy incorporating at the highest level a general bipolar happiness versus unhappiness dimension, an independent positive affect versus negative affect dimension at the second-order level below it and discrete expressivity factors (joy, sadness hostility, fear) at the base. The key to this hierarchical structure is the recognition that the general bipolar factor of happiness and independent dimensions of positive affect (PA) and negative affect (NA) are better viewed as different levels of abstraction within a hierarchical model, rather than as competing models at the same level of abstraction. Thus, the hierarchical model of affect accounted for both bipolarity of pleasantness–unpleasantness and the independence of PA and NA effectively.

Both circumplex and Tellegen-Watson models have gained empirical support, and their advantages and drawbacks have been actively debated in the literature (Watson et al. 1999; Russel and Caroll 1999).

The dimensional models have been criticized in the past that fail to differentiate between emotions such as anger and fear that are very close on the affect plane. Many studies using valence and arousal showed that two-dimensional models cannot capture all the variance in music-induced emotions (Collier 2007). In a multi-dimensional scaling approach used to explore the underlying structure of emotional responses to music, it was found that in addition to the two dimensions, a third dimension related to kinetics is necessary (Bigand et al. 2005).

Over the years, three-dimensional models with different dimensions have been proposed. Wundt (1896) proposed a three-dimensional model with the three dimensions of pleasure–displeasure, arousal–calmness, and tension–relaxation. Schlossberg (1954) proposed a three-dimensional model with three main dimensions related to arousal, valence, and control. The most known three-dimensional model is a modification of Russell's model with a Thayer's variant having three axes, valence, energetic arousal, and tension arousal (Schimmack and Grob 2000; Schimmack and Rainer 2002).

Recently, Eerola and Vuoskoski (2011) systematically compared discrete and dimensional models by evaluating perceived musical emotions. The results showed that the overall ratings between discrete and dimensional models did not reveal substantial differences when using large music collections. The discrete models, however, exhibited lower resolution efficiency in rating ambiguous emotion compared to dimensional models. Moreover, the comparison between different dimensional models revealed that two dimensions are sufficient to describe emotions in music. This finding is in contrast to existing studies supporting the need for three-dimensional models (Schimmack and Reisenzein 2002).

The discrete and dimensional models coexist for a long time in music and emotion research. Discrete models are closer to listener's experience because the labels used (happiness, anger, sadness, etc.) are familiar from everyday life. On the other hand, dimensional models appear to be related to the underlying mechanisms of emotion generation and therefore exhibit higher resolution in cases of ambiguous emotions.

6.3 How Does Music Evoke Emotions?

Although the dominant view in the field of music and emotion is that music is able to induce real emotions, there were music philosophers who challenged the existence of music-evoked emotions (Kivy 1990; Konenci 2008). The so-called cognitivists argued that listeners refer to a music piece as happy or sad because the music piece expresses happiness or sadness and not because music actually makes them feel happy or sad. In contrast, "emotivists" argue that music evokes real emotions to the listeners (Davies 2001). There is growing experimental evidence that musical emotions can be reflected in physiological measures, supporting the emotivist position that musical emotions are felt emotions (Krumhansl 1997; Nyklicek et al. 1997; Witvliet and Vrana 2007; Lundqvist et al. 2009). Even if it is accepted that music induces emotions, the fascinating question of how music evoke emotions is still a matter of controversy. The answer to this question is a key issue with implications for future research on the field of music and emotion.

6.3.1 Appraisal Theory

The most common discussed mechanism of music emotion elicitation is cognitive appraisal (Ekman 1992a, b; Scherer 1999). Appraisal theory suggests that emotions result on the basis of a person's subjective evaluation or appraisal of an event. One person feels sad, for example, by hearing the news of death of a beloved person or feels happy by hearing the news of a great success. The result of the appraisal is an emotion, which is expressed or externalized in physiological response symptoms.

Appraisal theory has been criticized to have little relation to music-evoked emotions because most reactions to music do not involve implications of life goals (Juslin et al. 2010). In a recent study, listeners were asked to indicate what might have caused their emotion by choosing among different proposed mechanisms (Juslin et al. 2008). The above study investigated different mechanisms evoking musical emotions in daily life. The results show that cognitive appraisal was the least important.

Emotions are often considered as a multi-componential phenomenon including physiological, behavioral, subjective, expressive, and cognitive components related to different organismic subsystems. It is argued (Scherer and Zentner 2001) that the most salient criterion for an emotion event is the degree of synchronization of all organismic subsystems involved through rhythmic entrainment. If music can influence one of the components, peripheral mechanisms can be triggered to cause spread to other emotions components. Rhythmic entrainment is the process where an emotion is induced by a piece of music because the powerful external rhythm of music interacts and synchronizes to an internal biological rhythm of the listener such as heart rate or respiration. The synchronized heart rate may then spread to other emotional components through proprioceptive feedback, causing increased arousal (Juslin and Sloboda 2010). Existing research suggests that coupling of internal biological oscillators (heart rate, respiration) and external stimuli exists (Boiten et al. 1994). Such coupling could provide an explanation of emotion-inducing effect of music. Rhythmic entrainment, however, has not been so far systematically investigated with respect to musical emotion. In order to answer the question of whether music induces emotion, all pertinent indicators in the respective organismic subsystems need to be accurately measured and the degree of their synchronization assessed using reliable mathematical techniques. Recent research provides some evidence that the mechanism of rhythm entrainment causes increased arousal during visual stimulation (Valenza et al. 2012).

6.3.2 The BRECVEM Model

Research on the possible mechanisms for music emotion induction was mainly limited on one or a few mechanisms (Berlyne 1971; Meyer 1956; Scherer and Zentner 2001). There was no attempt, however, to develop a general framework including several induction mechanisms. Recently, a novel theoretical framework for music-induced emotions was proposed (Juslin and Västfjäll 2008; Juslin et al. 2010). This framework was based on both existing research (Berlyne 1971; Meyer 1956) as well as on recent research (Juslin et al. 2008). It is suggested that seven physiological mechanisms are involved in the induction of musical emotions, in addition to cognitive appraisal: (1) brain stem reflex, (2) rhythmic entrainment (3) evaluative conditioning, (4) emotional contagion, (5) visual imagery, (6) episodic memory, and (7) musical expectancy.

Brain stem reflex is the process of emotion induction by music because one or more acoustical characteristics of music are taken by the brain stem to signal an important event.

Rhythmic entrainment is the process whereby an emotion is induced by a piece of music because the external rhythm of music interacts with and synchronizes to an internal physical rhythm of the listener such as heart rate or respiration (Clayton et al. 2005). The synchronized heart rate may then spread to other components of emotion through proprioceptive feedback producing increased emotional arousal (Juslin and Sloboda 2010).

Evaluative conditioning is a mechanism of emotion induction by a piece of music because this piece has been repeatedly paired with other positive or negative stimuli.

Emotional contagion is a process whereby an emotion is induced because the listener perceives an emotion and then mimics it internally.

Visual imagery refers to the mechanism whereby images evoked by music act as cues to emotion.

Episodic memory is the mechanism whereby a piece of music is associated with a particular event of the listener's life, which in turn is associated with a particular emotion. When the memory is evoked, the emotion associated with the memory is also induced.

Finally, *musical expectancy* is an induction mechanism whereby the listener's expectations of music are confirmed, violated, or suspended.

The above-described mechanisms are considered to be distinct brain functions with different origins and different characteristics. The mechanisms are not mutually exclusive but rather complementary ways of music emotion induction. What mechanisms may be activated depends on several factors including music style, the listener and the circumstances of listening. The co-activation of different mechanisms is possible leading to complicated interactions among the mechanisms. On the other hand, it cannot be excluded that the mechanisms can be activated in isolation from each other since they are related to different brain regions and process different types of information. All these issues, however, have to be resolved by further research.

6.3.3 Implications

A deeper understanding of the mechanisms underlying music emotion induction has important implications on both theoretical research in the field of music and emotion as well as on applications including multimedia, health care, and music therapy. Further experimental studies are needed to test the mechanisms of both componential arousal and BRECVEM model.

In the first case, multivariate techniques have to be used in a consistent and accurate methodological way. Significant progress has been achieved using multivariate approaches. It seems, however, that multivariate approaches should be broadened to include subjective emotion ratings (self-reports), facial activity (EMG), autonomic responses (heart and respiration rate, skin conductivity), and brain activity (EEG). The observed responses should be linked to particular features of the musical structure to see which features are responsible for the observed phenomena. The tests should also include powerful mathematical modeling to asses the degree of synchronization between the biological subsystems involved.

As far as the BRECVEM model concerns, it could help resolve the debate between "cognitivists" and "emotivists," remove existing disagreements in the field and provide answers in a series of open issues. One issue is about which emotions music can induce. This is of importance for emotional models used for decades in music and emotion research. Some researchers argue that music can induce basic emotions, while others argue that can induce both basic and complex emotions. It seems that which emotions can be induced depends on the mechanism activated. Furthermore, the proposed framework allows the induction of mixed emotions when two or more mechanisms are activated simultaneously. Another issue is whether music emotions are different from other emotions in everyday life. It appears that the emotions evoked by music are similar to other emotions since the mechanisms involved are to some extend common. Another implication concerns physiological responses to music. Most studies investigated musical emotions by trying to establish links between music and physiological measures without considering the underlying mechanisms. Better understanding of the relationship between physiological data and emotions, by considering the mechanisms involved, will help the interpretation of physiological and brain imaging data.

It is widely accepted that music has positive health effects and is used to regulate emotions and mood which in turn positively influences reactions and stress. The understanding of the underlying mechanisms will contribute to music therapy practice by highlighting the processes involved in different therapy techniques.

6.4 The Role of Musical Structure on Emotion

Musical characteristics, such as tempo, mode, loudness, pitch, timbre, and so on, are inherent properties of the structure of music, and it has been shown to influence emotional responses to music (Juslin and Sloboda 2010). The relation between characteristics of musical structure and emotional responses during music listening has been the subject of investigation for decades (see Gabrielson and Lindstroem 2010 for a review).

6.4.1 Effect of Mode and Tempo

The most widely investigated musical characteristics are mode and tempo. The pioneering work of Hevner (1935, 1937) was the first to demonstrate that both tempo and mode affect the emotional response to music. Short pieces of tonal music were selected, and variations of the original pieces by mode (a piece in major was also played in minor) and tempo (fast versus slow) were constructed. Clusters of adjectives were given, and the listeners were instructed to mark adjectives they found appropriate for each piece of music. Hevner concluded that faster tempi were associated with happiness, whereas sadness was associated to slow tempi. Rigg (1940a, b) studied the effect of music structure including mode and tempo on emotional responses. Music phrases supposed to express pleasant/happy and sad/serious emotional states were composed, and they were then systematically modified regarding tempo and mode. The listeners rated their perceived emotion by choosing between these two categories. The results showed that shifts an octave upward makes the phrase happier and faster tempo results in happier ratings. Since these pioneer works, numerous studies demonstrated that mode manipulations are strongly associated with happiness and sadness, indicating that mode is a reliable indicator of mood (Peretz et al. 1998). Even among 8-years-old children, the major mode is associated with happiness and joy, whereas minor mode is associated with sadness (Dalla Bella et al. 2001).

It is generally agreed that music stimulates wide networks across the brain and that specific areas of the brain appear to be involved for the perception of different aspects of music such as melody, rhythm, and timbre (Zatorre and Samson 1991, see also Sect. 6.6 in this chapter). In that vein, Tsang et al. (2001) investigated the effect of mode and tempo separately on music emotion using EEG recordings in the frontal region. They reported that both tempo and mode in the happier direction resulted in greater relative left frontal activity, whereas changes in both tempo and mode in the sadder direction resulted in greater relative right frontal activation, in agreement with the hemispheric specialization of emotional valence (Davidson 1988). There are few studies that investigated the effect of mode and tempo on brain activity using fMRI. Khalfa et al. (2005) used the manipulation of mode and tempo in musical excerpts to test the lateralization of brain regions involved in the recognition of negatively and positively valenced musical emotion. They found that the minor mode (sad excerpts) involved the left frontal cortex, which does confirm the valence lateralization model. In the same line, Green et al. (2008) investigated the effect of mode on emotional responses to music. Although the reported results are in some cases contradictive, minor mode melodies were evaluated as sadder than major melodies and caused increased activity in limbic structure.

6.4.2 Interactive Effects

All the above-described studies investigated the affect of tempo and mode on musical emotion separately without considering possible interactive effects between these parameters. Therefore, little is known about how tempo and mode interact. Scherer and Oshinsky (1977) studied emotional responses to musical excerpts with varying tempo and mode of Beethoven melodies but no interaction effects between tempo and mode were reported. Husain et al. (2000) studied the effect of tempo and mode on arousal and mood. A Mozart sonata was manipulated to produce four versions that varied both in tempo and mode. The results show that tempo manipulation affected arousal but not mood, whereas mode manipulation affected mood but not arousal. Furthermore, musical excerpts that have fast or slow tempo were judged to be happy or sad, respectively. Cagnon and Peretz (2003), although they were not interested in the interaction between mode and tempo, reported that happy–sad conditions were influenced more strongly by tempo than they were by mode. The results confirm that both mode and tempo determine the "happy–sad" judgments with tempo being more salient. Webster and Weir (2005) investigated systematically the effect of tempo, mode, and texture on emotional responses to music. They concluded that the effects of mode, tempo, and texture were interactive in nature. Major modes presented at fast tempi were positively valenced, whereas minor modes presented at slow tempi were negatively valenced. Recently, the combined influence of mode and tempo on emotional responses to music was studied (Ramos et al. 2011). Three musical pieces composed in the Ionian mode and then played in the remaining six Greek modes without affecting the melodic contour were used. The resulted musical excerpts were then played at three different tempi. The reported results showed some interactive effects between tempo and mode but the effect of the two parameters was mainly additive. The research reported so far in the interaction between musical characteristics of mode and tempo provided contradictory results. On the one hand, it has been shown that fast tempo music excerpts increase valence and arousal up to a certain degree of happiness, and on the other hand, it has been found that tempo increases decrease the effect of happiness on major mode and support happiness appraisal on minor mode.

The main body of the research on the effect of mode and tempo on emotional responses to music was primarily based on self-reports instead of physiological responses or a combination of both. Physiological responses, compared to self-reports, provide unbiased responses and are able to capture changes in emotions that would be undetected in self-reports. Using physiological responses to music stimuli including heart and respiration rate and skin conductance, Van der Zwaag et al. (2011) studied the effect of tempo, mode, and percussiveness on emotion. Percussiveness can be considered as a descriptor of timbre (Skowronek and Mc-Kinney 2007). They found, in agreement with previous research, that fast tempo increases arousal and tension. Minor mode, however, evoked higher arousal compared to major mode. This is in contradiction with existing research. They also found interdependencies of musical characteristics in affecting emotion. Percussiveness is strengthening the influence of either mode or tempo on the intensity of positive feelings. Fast tempo and major mode music are both experienced more positively in combination with high percussiveness compared to low percussiveness. The combined interactions of mode and tempo on emotional responses to music were recently investigated using both self-reports and EEG activity (Trochidis and Bigand 2013). It was reported that musical modes influence the valence of emotion with major mode being evaluated happier and more serene, than minor and locrian modes. In EEG frontal activity, major mode was associated with an increased alpha activation in the left hemisphere compared to minor and locrian modes, which, in turn, induced increased activation in the right hemisphere. The tempo modulates the arousal value of emotion with faster tempi associated with stronger feeling of happiness and anger, and this effect is associated in EEG with an increase of frontal activation in the left hemisphere. By contrast, slow tempo induced decreased frontal activation in the left hemisphere.

6.4.3 Effect of Pitch, Timbre, and Loudness

The effect of pitch and rhythm on the perceived emotional content of musical melodies was also examined (Schellenberg et al. 2000). The pitch and rhythm parameters were manipulated to obtain altered versions with different pitch and rhythm. It was found that ratings were influenced more by differences in pitch rather than differences in rhythm. Whenever rhythm affected ratings, there was an interaction between pitch and rhythm.

Few studies investigated the effect of loudness and timbre on musical emotion. Loudness causes higher levels of activation and tension (Ilie and Thompson 2006) and negative feelings (Kellaris and Rice 1993). It was shown that with increasing loudness, arousal increases, whereas with decreasing loudness, arousal decreases (Schubert 2004). Timbre is considered to play a less important role on emotional affect (Blackwill et al. 2004). There is some evidence that soft timbres are associated with sadness, whereas sharp timbres are associated with anger (Juslin 1997).

6.5 Measurement of Musical Emotions

Music has the ability to induce strong emotions to the listeners. Emotions, however, are by their nature subjective phenomena, and therefore, measuring a person's emotional state is a quite difficult task. The measures commonly used to assess emotional responses to music fall in three main categories: self-reports, physiological measures, and behavioral measures. In what follows, self-reports and physiological measures are described and their adequacy in accounting for emotional responses to music is analyzed and discussed. Behavioral measures are not discussed because their use is rare in emotional responses to music.

6.5.1 Self-reports

Music studies based on self-reports use either a discrete or a dimensional approach. The discrete emotion perspective is based on the assumption that there is a universal set of basic emotions including fear, anger, disgust, sadness, and happiness (for details see Sect. 6.1 of this chapter). The ratings of emotion categories are gathered by asking participants to rate how much a music piece expresses each emotion category. Usually, a list of emotion terms is provided to the listener, and the latter is asked to check terms that describe the emotion experienced or to rate the intensity of the emotion experienced on a certain scale. A standardized emotion scale of this kind is the Differential Emotion Scale (Izard et al. 2003). It contains 30 expressions to characterize 10 basic emotions. Most of the researchers, however, prefer to create their own self-reports, which can serve better the needs of a specific research. About one-third of the studies on musical emotion research used self-reports based on the discrete emotion model. This is due to the fact that discrete emotions are easily used in recognition paradigms and in physiological and neurological studies. Furthermore, discrete emotions can provide insight into mixed emotions.

The main alternative to discrete approach is the dimensional approach. According to the dimensional approach, there are two fundamental dimensions that describe emotional responses to music. The most common assumed dimensions are valence and arousal. The valence dimension contrasts states of pleasure with states of displeasure (positive–negative), whereas the arousal dimension contrasts states of low arousal with states of high arousal (calm–excited). In using the dimensional approach, Russell's circumplex model has dominated. The listeners are asked to rate valence (how positive or negative they feel) and arousal (low or high excitation) independently in bipolar scales. Thus, the emotional state of the listener is described as a point in the arousal–valence affective space. The results obtained using the dimensional approach are reliable, easy to analyze, and admit advanced statistical processing. A standardized scale based on the dimensional approach is the Positive and Negative Affect Schedule (Watson et al. 1988). In some cases, pictorial versions of rating scales are used. The Self-assessment Manikin scale, for example, rates pleasure and arousal by using images of human characters with different facial expressions. Moreover, photographs or drawings of various facial expressions are used when basic emotions are studied. The adequacy of two-dimensional models has been questioned, and three-dimensional models have been proposed. The question that arises is whether dimensional or discrete approaches are better to capture emotional responses to music. In a recent study, discrete and dimensional models were systematically compared by evaluating perceived musical emotions (Eerola and Vuoskoski 2011). The results showed that the overall ratings between discrete and dimensional models did not reveal substantial differences when using large music collections. Moreover, the comparison between different dimensional models revealed that two dimensions are sufficient to describe emotions in music.

Many studies rely on self-reports because they are easy and cheap to use and interpret. Self-reports, however, have serious drawbacks. One of the main drawbacks is demand characteristics. It refers to the possibility of transferring the experimental hypothesis to the listener and consequently cause hypothesis influenced response. Another serious drawback is self-presentation bias. It refers to the difficulty of a person to report and describe emotional states that can be considered undesirable.

6.5.2 Physiological Measures

An alternative to self-reports is physiological measures. Physiological changes during music listening are related to the activation of the autonomic nervous system (ANS), which is responsible for regulating a variety of peripheral functions. There is a large amount of studies establishing the relationship between physiological responses and musical emotion during music listening (see Sect. 6.6 of this chapter). Physiological measures are considered to be unbiased, more objective measures of emotional responses to music. They can be easily and noninvasively recorded and analyzed. They are also able to capture changes in emotional responses that would remain unnoticed in self-reports (Cacioppo et al. 2000).

The most commonly assessed physiological measures are electrodermal, cardiovascular, and respiratory responses. Electrodermal activity is quantified in terms of skin conductance (SC), which is considered to reflect arousal (Boucsein 1992). It has been shown that SC increases linearly with arousal of emotional stimuli, whereas no differences in valence were found. In general, emotionally powerful music tends to increase SC more than less emotional music (Rickard 2004).

The most frequently used cardiovascular measures include heart rate (HR), blood pressure (BP), and heart rate variability (HRV). Most of the existing studies on the effect of music on heart rate indicate that music listening can cause changes in heart rate (Nyklicek et al. 1997; Bernardi et al. 2006). These changes can be easily measured through ECG (Electrocardiogram). It was shown that high arousal music tends to increase the heart rate, whereas sedative music tends to decrease it. On the other hand, HRV is associated with valence and has been found to be higher during high positive valence (Cacioppo et al. 2000). Krumhansl (1997) reported increases in HRV during sad and happy music.

Respiration is also strongly linked to emotional responses to music. Most of the existing studies show an increase in respiration or breathing rate during music listening (Krumhansl 1997; Gomez and Danuser 2004; Nyklicek et al. 1997). Breathing rate is closely related to heart rate changes during music listening since the two systems are considered as two weakly coupled oscillators and through this coupling respiration regulates heart rate. Recent experiments provide evidence that both respiration rate and heart rate entrain musical rhythm.

Muscular tension and particularly facial expressions are among the potential measures of emotional states. Facial expressions are measured through electromyography (EMG) by placing electrodes on zygomaticus (associated with furrowing of the eyebrows) and corrugator (associated with rising of the corners of the lips) muscles. The results of existing measurements on zygomaticus and corrugator showed increased zygomatic muscle activity during high arousal and positive valence music, whereas greater corrugator activity was reported for musical excerpts of negative valence (Larsen et al. 2003; Witvliet and Vrana 2007). Thus, EMG activity can be considered as a promising measure of valence. An additional important result is that using facial EMG, discrete emotions can be recognized (Thayer and Faith 2001; Khalfa et al. 2008). Facial expressions have been so far used mainly in video applications and to a lesser degree in measuring emotional responses to music.

It is well established that emotional processing involves activation of wide networks of central nervous system (Blood et al. 1999; Blood and Zatore 2001). In that vein, several studies have used brain activity measures to explore emotional responses to music (Koelsch 2005; Koelsch et al. 2006). An approach taken to examine emotional processing in brain are EEG experiments during music listening (see Sect. 6.6 of this chapter). The most commonly used measure is alpha power, which is considered to be inversely related to cortical activation. When alpha power in the left frontal hemisphere is contrasted with alpha power in the right frontal hemisphere, an asymmetry is found depending on the stimulus (Davidson 1988). This frontal asymmetry is linked to emotional valence. Pleasant music induces greater left frontal activity, whereas unpleasant music leads to greater right frontal activation.

There are, however, results providing evidence that frontal asymmetry is related to motivational direction rather than emotional valence. Using EEG measurements, Davidson et al. (1990) found substantial evidence for the asymmetric frontal brain activation. Since then, several EEG studies using various sets of musical stimuli provided support for the hemispheric specialization hypothesis for emotional valence (Schmidt and Trainor 2001).

In summary, it appears that different measures of emotions are sensitive to different aspects of emotional states, and therefore, emotion cannot be captured by a single measure. Emotions are often considered to have different components (cognitive, behavioral, and physiological). Therefore, multivariate approaches involving the investigation of various physiological responses could differentiate among different emotions. Significant progress has been achieved using multivariate approaches. It seems, however, that multivariate approaches should be broadened to include subjective emotion ratings (self-reports), facial activity (EMG), autonomic responses (heart and respiration rate, skin conductivity), and brain activity (EEG). Moreover, models that might explain how various response systems are coordinated (synchronized) should be also included.

6.6 Physiological Responses to Music

There is a large amount of studies establishing the relationship between physiological responses and musical emotion during music listening (see Hodges 2010 for a review). Physiological changes during music listening are related to the activation of the ANS, which regulates a variety of organs and controls somatic processes. Research on physiological effects of music includes mainly changes in heart rate (HR), respiration rate (RR), blood pressure (BP), skin conductance (SC), finger temperature and muscle tension (EMG). Ongoing brain activation (EEG responses) is an important physiological response to music related to central nervous system

(CNS) and should be included. This topic, however, will be discussed in a separate section of this chapter (see Sect. 6.6).

Physiological responses to music are important in many aspects. From the theoretical perspective, a key issue is to demonstrate whether basic emotions induced by music are related to specific physiological patterns (Nyklicek et al. 1997; Lundqvist et al. 2009; Krumhansl 1997; Khalfa et al. 2008). The relation between discrete emotions and emotion-specific physiological response patterns predicted by theorists, however, still remains an open problem (Scherer 2004). As far as practical applications concerns, physiological responses are important for many applications including emotion recognition, health care, and human–computer interfaces. They can be continuously monitored and used as robust measures (descriptors) of emotional states. In what follows, the main physiological responses to music listening will be briefly discussed and their application to emotion recognition and consequently to emotion intelligence in human–machine interaction will be described.

6.6.1 Effect of Music on Different Physiological Processes

6.6.1.1 Heart Rate

Most of the existing studies on the effect of music on heart rate indicate that music listening can cause changes in heart rate (Nyklicek et al. 1997; Lundqvist et al. 2000; Krumhansl 1997; Baumgartner et al. 2006; Samler et al. 2007; Blood and Zatorre 2001; Bernardi et al. 2006). These changes can be easily measured through ECG (Electrocardiogram). It was shown that high arousal music tends to increase the heart rate, whereas sedative music tends to decrease it. There are also studies reported that music caused no changes in heart rate (Iganawa et al. 1996; Gomez and Danuser 2004). The style of music used in existing studies was not systematically investigated and could be one cause of the existing inconsistencies, the other being the methodology employed. Recent studies provide evidence that tempo is the most influential factor on heart rate. This is due to the fact that rhythm entrainment between tempo and heartbeat seems to be the mechanism through which the changes are caused.

6.6.1.2 Respiration

Breathing rate is the second physiological response linked to musical emotion. Most of the existing studies show an increase in respiration or breathing rate during music listening (Krumhansl 1997; Gomez and Danuser 2004; Nyklicek et al. 1997; Thayer and Faith 2001). Few studies reported no substantial changes in respiration during music listening (Davis 1992; Davis-Rollan and Cunningham 1987; Iganawa et al. 1996). Breathing rate is closely related to heart rate changes during music listening. The two systems are considered as two weakly coupled oscillators and through this coupling respiration regulates heart rate. Recent experiments provide evidence that both respiration rate and heart rate entrain musical rhythm.

6.6.1.3 Skin Conductance

Skin conductance (SC) is a measure of the electrical resistance of the skin and is frequently used as a physiological effect of music. Skin conductance is related to emotional arousal and has been proved to be a reliable measure of emotional response in domains other than music. For high arousal conditions, skin conductance increases. Most of the existing studies show a significant increase in skin conductance during music listening (Khalfa et al. 2002; Gomez and Danuser 2007; Lindqvist et al. 2009). There are, however, studies indicating no substantial changes in skin conductance during music listening (Blood and Zatorre 2001).

6.6.1.4 Blood Pressure

Blood pressure (BP) has been also used as a measure of physiological effect of music. It can be easily measured by a sphygmomanometer. The existing results, however, are contradictive. Most of the studies show an increase of blood pressure to stimulative music and a decrease to sedative music (Baumgartner et al. 2006; Gomez and Danuser 2004; Krumhansl 1997; Thayer and Faith 2001). There are, however, studies where blood pressure decreased (Yamamoto et al. 2007; Iwanaga et al. 1996) or did not change during music listening (Davis 1992; Davis-Rollans and Cunningham 1987).

6.6.1.5 Muscular Tension

Muscular tension and particularly facial expressions are among the potential measures of emotional states. Facial expressions are measured through EMG (Electromyography) by placing electrodes on zygomaticus, corrugator, and orbicularis oculi muscles. The results of existing measurements on zygomaticus and corrugator showed increased zygomatic muscle activity during high arousal and positive valence music, whereas greater corrugator activity was reported for musical excerpts of negative valence (Witvliet and Vrana 2007). Thus, EMG activity can be considered as a promising measure of valence. An additional important result is that using facial EMG discrete emotions can be recognized (Thayer and Faith 2001; Khalfa 2008). Facial expressions have been so far used mainly in video applications and to a lesser degree in musical applications (music listening).

6.6.1.6 Finger or Body Temperature

Skin temperature changes have been linked to music listening in several studies. Most of the existing studies show an increase of skin temperature during listening to stimulating music (Baumgartner et al. 2006; Lundqvist et al. 2009). There are, however, cases where a decrease (Krumnhansl 1997; Nater et al. 2006) or no changes at all (Blood and Zatorre 2001) were observed resulting in inconsistent results.

6.6.2 Physiological Measures and Emotion Recognition

This section will emphasize the significance of physiological responses during music listening to emotion recognition and its applications. Music emotion recognition is of considerable importance for many research fields including music retrieval, health applications, and human–machine interfaces. Music collections are increasing rapidly, and there is a need to intelligently classify and retrieve music based on emotion. Training computers to recognize human emotional states, on the other hand, is a key issue toward successful realization of advanced computer–human interaction systems. The goal is to develop computational models that are able to link a given physiological pattern to an emotional state.

Relatively little attention has been so far paid to physiological responses compared to other modalities (audio–visual for example) for emotion recognition.

A significant amount of work has been conducted showing that musical emotions can be successfully recognized based on physiological measures such as heart rate, respiration, skin conductance, and facial expressions. Picard et al. (2001) were the first who showed that certain affective states can be recognized by using physiological signals including heart rate, respiration, skin conductivity, and muscle activity. Nasoz et al. (2003) used movie clips to induce emotions in 29 subjects and combining physiological measures and subjective components achieved 83 % recognition accuracy. Wagner et al. (2005) recorded four biosignals from subjects listening to music songs and reached a recognition accuracy of 92 %. Kim and Andre (2008) used music excerpts to spontaneously induce emotions. Four biosensors were used during the experiments to measure electromyogram, electrocardiogram, skin conductivity, and respiration changes. The best features were extracted, and their effectiveness for emotion recognition was tested. A classification accuracy of 70–90 % for subject-independent and subject-dependent classification respectively was achieved. Koelstra et al. (2011) used a multimodal approach based on physiological signals for emotion recognition. They used music video clips as stimuli. During the experiments EEG signals, peripheral physiological signals and frontal video were recorded. A variety of features was extracted and used for emotion recognition by using different fusion techniques. The results show a modest increase in the recognition performance, indicating limited complementarity of the different modalities used. Recently, a combination of acoustic features and physiological responses was used for emotion recognition during music listening (Trochidis et al. 2012). The reported results indicate that by merging acoustic and physiological modalities substantially improves participant's ratings of felt emotion recognition rate compared to the results using single modalities.

One of the main problems toward assessing musical emotions using physiological measures is to extract features that are relevant. In the current state, most studies try to extract features by simply removing non-relevant and keeping relevant based on statistical measures. It seems that by equally weighting features of different modalities does not lead to improved recognition accuracy. Alternative approaches should be developed treating valence arousal separately. To combine the two modalities, one has to decide at which level the individual modalities should be fused. A straightforward approach is to simply merge the features from each modality (feature-level fusion). The alternative is to fuse the features at the decision level based on the outputs of separate single classifiers (decision-level fusion) or to use a hybrid method. This issue needs further investigation and new effective fusion strategies should be developed.

6.7 Brain Responses to Music

It is generally agreed that emotional processing involves activation of wide networks of central nervous system (Blood et al. 1999; Blood and Zatore 2001). In that vein, several studies have used brain activity measures to explore emotional responses during music listening (Koelsch 2005; Koelsch et al. 2006). An approach taken to examine emotional processing in brain are EEG experiments during music listening. Davidson (1988) suggested that the left frontal area is involved in the experience of positive emotions such as joy and happiness. In contrast, the right frontal region is involved in the experience of negative emotions such as fear, angry, and sadness. There are, however, results providing evidence that frontal asymmetry is related to motivational direction rather than emotional valence (Harmon-Jones and Allen 1998). Using EEG measurements, Davidson et al. (1990) found substantial evidence for the asymmetric frontal brain activation. Since then, several EEG studies using various sets of musical stimuli provided support for the hemispheric specialization hypothesis for emotional valence. That is, musical stimuli which are considered positive or negative in valence, elicited asymmetric frontal EEG activity. Schmidt and Trainor (2001) investigated patterns of EEG activity induced by musical excerpts in a group of undergraduates. They found greater left and right frontal activity during music listening to pleasant and unpleasant music. Furthermore, they were the first to show that the overall power of frontal activity distinguishes the intensity of musical emotion. Moreover, faster tempi and the major mode produced greater responses in the left hemisphere, whereas slower tempi and minor mode were associated with greater responses in the right hemisphere (Tsang et al. 2001). Sammler et al. (2007) investigated electrophysiological correlates during the processing of pleasant (consonant) and unpleasant (dissonant) music using both heart rate and EEG measurements. In the EEG, they found an increase of frontal midline theta power for pleasant music in contrast to unpleasant music. Altenmueller et al. (2002) presented musical excerpts from four different genres to students who provided judgments for each excerpt. Positively valenced stimuli elicited bilateral fronto–temporal activations predominantly of the left hemisphere, whereas negatively valenced stimuli elicited bilateral activations predominantly of the right hemisphere. Females showed greater valence-related differences than males did. In consequence, the frontal temporal lobes seem to be involved in emotional evaluation and judgment rather than the perceptual analysis of emotional information (Heilman 1997). Flores-Gutierez et al.

(2007) studied emotional reactions to pleasant and unpleasant music induced by dissimilar piano excerpts. They employed Principal Component Analysis on fMRI and EEG, and they reported that a left cortical network involved with pleasant feelings. In contrast, unpleasant emotions involved the activation of the right frontal and limbic brain areas.

The frontal activation emotion hypothesis has been also tested across different modalities, age groups, and measures. Davidson and Fox (1989) found that asymmetrical frontal brain activity discriminated sweet and sour tastes in newborns. Schmidt et al. (2003), tested 3-month-old to 12-month-old infants using musical excerpts of varying valence and arousal (happy, sad, and fear). The authors suggest that taken together their findings, which showed an emerging asymmetry of activation in the presence of an overall decrease of EEG power, indicate maturation of cortical music processing as well as a “calming” influence of music by the end of the first year of life. Baumgartner et al. (2006) investigated neural correlates of sadness, fear, and joy. They observed that auditory information interacts with visual information in several limbic and paralimbic structures. Activity changes in these structures were stronger during combined presentation of fearful and joy photographs with fearful and joy music, compared to when only visual information was present.

The research on asymmetries of EEG activity mainly focused on the analysis of alpha band power. Relatively few studies have examined frequency bands other than alpha including theta, beta, and gamma (Aftanas and Golocheikine 2001; Sammler et al. 2007; Pizzagalli et al. 2002, 2003; Flores-Gutierez et al. 2007). The results of these studies provide evidence that theta band plays a more important role in emotion processing that previously believed. Therefore, it is important to examine other frequency bands than alpha carefully as these may provide additional information not reflected in alpha.

6.7.1 EEG and Emotion Recognition

In addition to peripheral physiological responses to music, EEGs from the brain gained recently great attention for emotion recognition. Estimating the emotion from EEG is important because brain waves are generated by the brain and are deeply related to cognition processes. Furthermore, the ongoing brain activity provides noninvasive measurement with high resolution. It appears that EEGs provide more insight into emotional processes compared to peripheral CAN signals.

There are an increasing number of studies on EEG-based emotion recognition. In these studies, different approaches with respect to both feature extraction and classification algorithms were investigated. Most of the early studies on EEG-based emotion recognition focused on spectral power changes in few bands and specific brain areas. Power spectra of EEG signals in different frequency bands were used to examine the relationship between brain activity and emotional states. A common indicator of musical emotion is the alpha-power asymmetry at the anterior region of the brain (Schmidt and Trainor 2001; Trochidis and Bigand 2013). There is strong evidence that other spectral changes and brain regions are involved in emotional responses. These include frontal midline theta power (Samler et al. 2007), beta-power asymmetry (Schutter et al. 2008) and gamma spectral changes at right parietal areas (Balconi and Lucchiari 2008; Li and Lu 2009).

A variety of research studies on EEG-based emotion recognition and classification has been reported. These studies used different features and classification algorithms. Ishino and Hagiwara (2003) proposed a system based on neural networks. They applied FFT, WT, and PCA to extract features from EEG signals. Consequently, neural networks were applied for classification of four emotions (joy, relax, sorrow, and anger) achieving accuracy of 67 %. Murugappan et al. (2008) used a lifting-based wavelet transform for feature extraction from measured EEG signals. Next, Fuzzy C-Means clustering was employed for classification of four emotions (disgust, happy, surprise, and fear).

Ko et al. (2009) reported an EEG-based emotion recognition system. They divided measured EEG signals into five frequency ranges on the basis of power spectral density and employed Bayesian network to predict the user's emotional states. Lim et al. (2010) proposed an EEG-based emotion recognition system. Using measured EEG responses from 26 subjects during music listening, they extracted features related to power spectral density, to power asymmetry of 12 electrode pairs across different frequency bands and to the corresponding rational asymmetry. They employed SVM classifiers and reported a recognition accuracy of 82 %. The reported results showed that features of spectral power asymmetry across different frequency bands were the most sensitive parameter characterizing emotional states. Petrantonakis and Hadjileontiadis (2010) employed higher order crossings for feature extraction from EEG signals. Using the extracted features, four different classifiers (QDA, k-nearest neighbors, Mahalanobis distance, and SVM) were tested for the classification of six emotions (happiness, surprise, anger, fear, disgust, and sadness). Depending on the classifier, recognition accuracies from 63 to 83 % were reported. Sourina and Liu (2011) proposed a real-time emotion recognition and visualization system based on fractal dimension. They applied a fractal-based algorithm and a valence–arousal emotion model. They calculated FD values from the EEG signals and used a SVM classifier for arousal and valence prediction for six basic emotions.

Despite the substantial progress achieved in EEG-based emotion recognition many issues need to be further improved (resolved). Relatively limited number of emotional states can be recognized using EEG. The best performance reported so far involves only six different emotions (Petrantonakis and Hadjileontiadis 2010). Another important issue is the number of electrodes needed to extract an optimal number of features. In current research, studies a big number of electrodes are used resulting in complications both during the experiments and the processing of the data. Research on the best features is needed to reduce the number of electrodes. Solving the above constrains will allow real-time EEG-based emotion recognition and realization of BCMI applications.

6.8 Conclusion

In this chapter, we attempted to give an overview of the current knowledge in the topic of music and emotion from an interdisciplinary perspective. At the present state, the area of music and emotion has grown and became an important field of research with implications for a number of disciplines. Despite the progress achieved, several issues still remain open and need to be further explored. In what follows, a few of the main issues in the area, which deserves particular attention for future research, will be briefly described.

The underlying mechanisms through which music evokes emotions is one of the main issues that remain unresolved. Better understanding of the mechanisms underlying music emotion induction has important implications on both theoretical research in the field as well as on applications including multimedia, health care, and music therapy. Recently, a novel theoretical model of music-induced emotions was proposed (see Sect. 6.2). Further theoretical and experimental studies are needed to test the mechanisms featured in the model. First, the characteristics of each mechanism should be specified to allow distinguishing among different mechanisms. Second, well-controlled experiments should be designed to test the proposed mechanisms. The proposed framework allows the induction of mixed emotions when two or more mechanisms are activated simultaneously. Thus, stimuli are needed able not only to activate a certain mechanism but at the same time to isolate the effect of others. To this end, sophisticated acoustical techniques can be used which allows the manipulation of certain acoustic features, while living other intact. A different approach could be to design experiments in such a way to prevent the information processing required for a particular mechanism to be activated. For that purpose, new techniques such as transcranial magnetic stimulation (TMS) could be employed to disrupt brain activity at certain brain areas. This would prevent the activation of these brain areas by music stimuli.

Another issue is the temporal aspect of musical emotion. Music evolves in time, and therefore, emotional responses change in the course of time. There is need to explore the temporal dynamics of music-induced emotions in long pieces of music (a symphony, for example) during the course of which one may experience different emotions. To capture the temporal dynamics of musical emotions, continuous measurements are needed. Most of the studies on music-induced emotions use self-report measures. Self-reports, however, provide ratings for short music excerpts after a stimulus has been heard. On the other hand, physiological measures are by their nature temporal and therefore more efficient in capturing the temporal dynamics of music and of music-induced emotions. The use of continuous measures is more demanding concerning both the quantity and the complexity of the recorded data. Therefore, extra effort and more complicated techniques of processing and analysis of long time series should be used.

In addition to the above issues, the efficient coupling of psychological predictions with physiological and neuroimaging techniques is important. Most of the neuroimaging and physiological studies of music and emotion look for simple, direct relationships between music and physiological or brain responses without considering the underlying physiological processes. Deeper understanding of the relationship between physiological data and emotions will enhance the interpretation of physiological and brain imaging data and can constitute the psychophysiological foundation of musical emotion.

6.9 Questions

1. What is the difference between discrete and dimensional models of emotions?
2. How does music evoke emotions?
3. Explain the seven physiological mechanisms of the BRECVEM model involved in the induction of musical emotions
4. What is the role of mode and tempo on emotion induction?
5. What is the effect of pitch, timbre, and loudness on emotion induction?
6. How do we measure musical emotions?
7. What is the difference between behavioral and physiological measures of musical emotions?
8. What is the effect of music on heart and respiration rate?
9. Discuss the effect of music on brain responses
10. Explain an emotion recognition system based on brain responses

References

Aftanas LI, Golocheikine SA (2001) Human anterior and frontal midline theta and lower alpha activity reflect emotional positive state and internalized attention: high-resolution EEG investigation of meditation. Neuroscience 310(1):57–60

Altenmueller E, Schuermann K, Lim VK, Parlitz D (2002) Hits to the left, flops to the right: different emotion during music listening reflected in cortical lateralization patterns. Neuropsychologia 40(13):2242–2256

Balconi M, Lucchiari S (2008) Consciousness and arousal effects on emotional face processing as revealed by brain oscillations. A gamma band analysis. Int J Psychophysiol 67(1):41–46

Balkwill LL, Thompson WF (1999) A cross-cultural investigation of the perception of the emotion in music: psychophysical and cultural cues. Music Percept 17(1):43–64

Baumgartner T, Esslen M, Jaencke L (2006) From emotion perception to emotion experience: emotions evoked by pictures and classical music. Int J Psychophysiol 60(1):34–43

Berlyne DE (1971) Aesthetics and psychobiology. Appleton Century Crofts, New York

Bernardi L, Porta C, Sleight P (2006) Cardiovascular, cerebrovascular and respiratory changes induced by different types of music in musicians and non-musicians: the importance of silence. Heart 92(4):445–452

Bigand E, Vieillard S, Madurell F, Marozeau J, Dacquet A (2005) Multidimensional scaling of emotional responses to music: the effect of musical expertise and of the duration of the excerpts. Cogn Emot 19(8):1113–1139

Blackwill LL, Thompson WF, Matsounaga R (2004) Recognition of emotion in Japanese Western and Hindustani music by Japanese listeners. Jpn Psychol Res 46(4):337–349

Blood AJ, Zatorre RJ, Bermudez P, Evans AC (1999) Emotional responses to pleasant and unpleasant music correlate with activity in paralimbic brain regions. Nat Neurosci 2 (4):382–387
Blood AJ, Zatorre RJ (2001) Intensely pleasurable responses to music correlate with activity in brain regions implicated in reward and emotion. Proc Natl Acad Sci 98(20):11818–11832
Boiten FA, Frijda NH, Wientjes CJE (1994) Emotions and respiratory pattern: review and critical analysis. Int J Psychophysiol 17(2):103–128
Boucsein W (1992) Electrodermal activity. Plenum Press, New York
Cacioppo JT, Berntson GG, Larsen JT, Poehlmann KM, Ito TA (2000) The psychophysiology of emotion. In: Lewis M, Havilnd-Jones JM (eds) Handbook of emotions. Guilford Press, New York
Cagnon L, Peretz I (2003) Mode and tempo relative contribution to "happy-sad" judgments in equitone melodies. Cogn Emot 17(1):25–40
Clayton M, Sager R, Will U (2005) In time with the music: the concept of entrainment and its significance for ethnomusicology. Eur Meetings Ethnomusicology 11:3–142
Collier GL (2007) Beyond valence and activity in the emotional connotations of music. Psychol Music 35(1):110–131
Dalla Bella S, Peretz I, Rouseau L, Grosselin N (2001) A development study of the effective value of tempo and mode in music. Cognition 80(3):B1–B10
Davidson RJ (1988) EEG measures of cerebral asymmetry: conceptual and methodological issues. Int J Neurosci 39(1–2):71–78
Davidson RJ, Fox NA (1989) Frontal brain asymmetry predicts infant's response to maternal separation. J Abnorm Psychol 98(2):127–131
Davidson RJ, Eckman P, Saron CD, Senulis J, Friegenn WW (1990) Approach-withdrawal and cerebral asymmetry: emotional expression and brain physiology. J Pers Soc Psychol 158 (2):330–341
Davies S (2001) Philosophical perspectives on music's expressiveness. In: Juslin PA, Sloboda JA (eds) Music and emotion: theory and research. Oxford University Press, Oxford, pp 23–44
Davis-Rollans C, Cunningham S (1987) Physiologic responses of coronary care patients to selected music. Heart Lung 16(4):370–378
Davis C (1992) The effects of music and basic relaxation instruction on pain and anxiety of women undergoing in-office gynecological procedures. J Music Ther 29(4):202–216
Eerola T, Vuoskoski J (2011) A comparison of the discrete and dimensional models of emotion in music. Psychol Music 39(1):18–49
Ekman P (1992a) An argument for basic emotions. Cogn Emot 6(3–4):169–200
Ekman P (1999) Basic emotions. In: Dalgleish T, Power MJ (eds) Handbook of cognition and emotion. Willey, New York, pp 301–320
Ekman P (1992b) Are they basic emotions. Physiol Rev 99(3):550–553
Flores-Gutierez EO, Diaz JL, Barrios FA, Favila-Humara R, Guevara MA, del Rio-Portilla Y, Corsi-cabrera M (2007) Metabolic and electric brain patterns during pleasant and unpleasant emotions induced by music masterpieces. Int J Psychophysiol 65(1):69–84
Gabrielsson A, Juslin PN (1996) Emotional expression in music performance: between the performance intention and the listener's experience. Psychol Music 24(1):68–91
Gabrielsson A, Lindstroem E (2010) The role of structure in the musical expression of emotions. In: Juslin PN, Sloboda JA (eds) Handbook of music and emotion: theory, research, applications. Oxford University Press, New York, pp 367–400
Gomez P, Danuser B (2007) Relationships between musical structure and psychophysiological measures of emotion. Emotion 7(2):377–387
Green AC, Baerentsen KB, Stoedkilde-Joergensen H, Wallentin M, Roepstorff A, Vuust P (2008) Music in minor activates limbic structures: a relationship with dissonance? Neuroreport 19 (7):711–714
Heilman KM (1997) The neurobiology of emotional experience. J Neuropsychiatry Clin Neurosci 9:439–448

Hevner K (1937) The affective value of pitch and tempo in music. Am J Psychol 49(4):621–630
Hevner K (1935) The affective character of the major and minor modes in music. Am J Psychol 47 (1):103–118
Hodges DA (2010) Psychophysiological responses to music. In: Juslin PN, Sloboda JA (eds) Handbook of music and emotion: theory, research, applications. Oxford University Press, New York, pp 279–311
Hunter PG, Schellenberg EG, Schimmack U (2008) Mixed affective responses to music with conflicting cues. Cogn Emot 22(2):327–352
Husain G, Thompson WF, Schellenberg EG (2000) Effects of musical tempo and mode on mood, arousal and spatial abilities. Music Percept 20(2):151–171
Ilie G, Thompson WF (2006) A comparison of acoustic cues in music and speech for three dimensions of affect. Music Percept 23(4):319–330
Ishino K, Hagiwara M (2003) A feeling estimation system using a simple electroencephalograph. Proc IEEE Int Conf Syst Man Cybern 5:4204–4209
Iwanaga M, Ikeda M, Iwaki T (1996) The effects of repetitive exposure to music on subjective and physiological responses. J Music Ther 33(3):219–230
Izard CE, Libero D, Putnam P, Haynes O (2003) Stability of emotion experiences and their relations to personality traits. J Pers Soc Psychol 64(5):847–860
Juslin PN, Liljestroem S, Västfjäll D, Lundqvist LO (2010) How does music evoke emotions? Exploring the underlying mechanisms. In: Juslin PN, Sloboda JA (eds) Handbook of music and emotion: theory, research, applications. Oxford University Press, Oxford, pp 605–642
Juslin PN (1997) Perceived emotional expression in synthesized performances of a short melody: capturing the listener's judgment policy. Musicae Scientiae 1(2):225–256
Juslin PN, Sloboda JA (eds) (2010) Handbook of music and emotion: theory, research, applications. Oxford University Press, Oxford
Juslin PN, Liljestroem S, Västfäll D, Baradas G, Silva A (2008) An experience sampling study of emotional reactions to music: listener, music and situation. Emotion 8(5):668–683
Juslin PN, Västfjäll D (2008) Emotional responses to music: the need to consider underlying mechanisms. Behav Brain Sci 31(5):559–575
Kellaris JJ, Rice RC (1993) The influence of tempo, loudness and gender of listener on responses to music. Psychol Mark 10(1):15–19
Khalfa F, Schon D, Anton J, Liegois-Chauvel C (2005) Brain regions involved in the recognition of happiness and sadness in music. Neuroreport 16(18):1981–1984
Khalfa S, Roy M, Rainwille P, Dalla Bella S, Peretz I (2008) Role of tempo entrainment in psychophysiological differentiation of happy and sad music. Int J Psychophysiol 68(1):17–26
Kim J, Andre E (2008) Emotion recognition based on physiological changes in music listening. IEEE Trans Pattern Anal Mach Intell 30(12):2067–2083
Kivy P (1990) Music alone: Reflexions on a purely musical experience. Cornel University Press, Ithaca
Ko KE, Yang HC, Sim KB (2009) Emotion recognition using EEG signals with relative power values and Bayesian network. Int J Control Autom Syst 7(5):865–870
Koelsch S, Fritz T, Cramon DY, Mueller K, Friederici A (2006) Investigating emotion with music: an fMRI study. Hum Brain Mapp 27(3):239–250
Koelsch S (2005) Investigating emotion with music: Neuroscientific approaches. Ann N Y Acad Sci 1060:412–418
Koelstra S, Muehl C, Soleymani M, Lee JS, Yazdani A, Ebrahimi T, Pun T, Nijholt A, Patras I (2011) DEAP: a database for emotion analysis using physiological signals. IEEE Trans Affect Comput 3(1):18–31
Konenci VJ (2008) Does music induce emotion? A theoretical and methodological analysis. Psychol Aesthetics Creativity Arts 2(2):115–129
Krumnhansl C (1997) An exploratory study of musical emotions and psychophysiology. Can J Exp Physiol 51(4):336–352

Larsen JT, Norris CJ, Cacioppo JT (2003) Effects of positive and negative affect on electromyographic activity over zygomaticus major and corrugator supercilii. Psychophysiology 40(5):776–785
Li M, Lu BL (2009) Emotion classification on gamma-band EEG. In: Proceedings of the annual international conference of the IEEE engineering in medicine and biology society EMBS 2009, pp 1223–1226
Lim Y-P, Wang C-H, Jung T-P, Wu T-L, Jeng S-K, Duann J-R, Chen J-H (2010) EEG-Based Emotion Recognition in Music Listening. IEEE Trans Biomedical Eng 57(7):1798–1806
Lundquist L, Carlsson F, Hilmersson P, Juslin PN (2009) Emotional responses to music: experience, expression and physiology. Psychol Music 37(1):61–90
Meyer LB (1956) Emotion and meaning in music. Chicago University Press, Chicago
Murugappan M, Rizon M, Nagarajan R, Yaacob S, Zunaidi I, Hazry D (2008) Lifting scheme for human emotion recognition using EEG. In: International symposium on information technology, ITSim, vol 2
Nasoc F, Alvarez K, Lisetti CN, Finkelstein (2003) Emotion recognition from physiological signals for presence technologies. Int J Cogn Technol Work-Special Issue Presence 6(1):4–14
Nater U, Krebs M, Ehlert U (2006) Sensation seeking, music preference and physiological reactivity to music. Musicae Scientiae 9(2):239–254
Niklicek L, Thayer J, Van Doornen L (1997) Cardiorespiratory differentiation of musically-induced emotions. J Psychophysiol 11(4):304–321
Pankssep J (1992) A critical role for "affective neuroscience" in resolving what is basic emotions. Psychol Rev 99(3):554–560
Pankssep J (1998) Affective neuroscience: the foundations of human and animal emotions. Oxford University Press, Oxford
Peretz I, Gagnon L, Bouchard B (1998) Music and emotion: perceptual determinants, immediacy, and isolation after brain damage. Cognition 68(2):111–141
Petrantonakis PC, Hadjileontiadis LJ (2010) Emotion recognition from EEG using higher order crossings. IEEE Trans Inf Technol Biomed 14(2):186–197
Pickard RW, Vryzas E, Healey J (2001) Towards machine emotional intelligence: analysis of affective physiological state. IEEE Trans Pattern Anal Mach Intell 23(10):1175–1191
Pizzagalli DA, Greischar LL, Davidson RJ (2003) Spatio-temporal dynamics of brain mechanism in aversive classical conditioning: high-density event-related potential and brain electrical tomography analyses. Neurophysiologia 41(2):184–194
Pizzagalli DA, Nitschke JB, Oakes TR, Hendrick AM, Horras KA, Larson CL, Abercrombie HC, Scaefer SM, Keger GV, Benca RM, Pasqual-Marqi RD, Davidson RJ (2002) Brain electrical tomography in depression: the importance of symptom severity, anxiety and melancholic features. Biol Psychiatry 52(2):73–85
Ramos D, Bueno JLO, Bigand E (2011) Manipulating Greek musical modes and tempo affects perceived musical emotion in musicians and non-musicians. Braz J Med Biol Res 44 (2):165–172
Rickard N (2004) Intense emotional responses to music: a test of the physiological arousal hypotheses. Psychol Music 32(4):371–388
Rigg MG (1940a) Speed as a determiner of musical mood. J Exp Psychol 27(5):566–571
Rigg MG (1940b) The effect of register and tonality upon musical mood. J Musicology 2(2):49–61
Russel JA, Caroll JM (1999) On the bipolarity of positive and negative affect. Psychol Bull 125 (1):3–30
Russell JA (1980) A circumplex model of affect. J Pers Soc Psychol 39(6):1161–1178
Sammler D, Grigutsch M, Fritz T, Koelsch S (2007) Music and emotion: electrophysiological correlates of the processing of pleasant and unpleasant music. Psychophysiology 44 (2):293–304
Schellenberg EG, Krysciak A, Campbell RJ (2000) Perceiving emotion in melody: interactive effects of pitch and rhythm. Music Percept 18(2):155–171

Scherer KR, Zentner MR (2001) Emotional effects of music: production rules. In: Juslin PN, Sloboda JA (eds) Music and emotion: theory and research. Oxford University Press, Oxford, pp 361–392

Scherer KR (2004) Which emotions can be induced by music? What are the underlying mechanisms? And how can we measure them? J New Music Res 33(3):239–251

Scherer KR, Oshinski JS (1977) Cue utilization in emotion attribution from auditory stimuli. Motiv Emot 1(4):331–346

Scherer KR (1999) Appraisal theories. In: Dalgleish T, Power M (eds) Handbook of cognition and emotion. Willey, Chichester, pp 637–663

Schimmack U, Reisenzein R (2002) Experiencing activation: energetic arousal and tense arousal are not mixtures of valence and activation. Emotion 2(4):412–417

Schimmack U, Grob A (2000) Dimensional models of core affect: a quantitative comparison by means of structural equation modeling. Eur J Pers 14(4):325–345

Schimmack U, Reiner R (2002) Experiencing activation: energetic arousal and tense arousal are not mixtures of valence and activation. Emotion 2(4):412–417

Schlossberg H (1954) Three dimensions of emotion. Psychol Rev 61(2):81–88

Schmidt LA, Trainor LJ, Santesso DL (2003) Development of frontal encephalogram (EEG) and heart rate (ECG) responses to affective musical stimuli during the first 12 months of post-natal-life. Brain Cogn 52(1):27–32

Schmidt LA, Trainor LJ (2001) Frontal brain electrical activity (EEG) distinguishes valence and intensity of musical emotion. Cogn Emot 15(4):487–500

Schutter DJ, De Weijer AD, Meuwese JD, Morgan B, Van Honk J (2008) Interrelations between motivational stance, cortical excitability and the frontal electroencephalogram asymmetry of emotion: a transcranial magnetic stimulation study. Hum Brain Mapp 29(5):574–580

Skowronek J, McKinney M (2007) Features for audio classification: percussiveness of sounds. In: Vergaegh W, Aarts E, Korst J (eds) Intelligent algorithms in ambient and biomedical computing. Phillips research book series, vol 7. Springer, Dordrecht, pp 103–117

Sourina O, Liu Y (2011) A fractal-based algorithm of emotion recognition from EEG using arousal-valence model. In: Proceedings of the international conference on bio-inspired systems and signal processing. BIOSIGNALS 2011, Rome, pp 209–214

Thayer J, Faith M (2001) A dynamic systems model of musically induced emotions. Ann N Y Acad Sci 999(1):452–456

Thayer RE (1989) The biophysiology of mood and arousal. Oxford University Press, Oxford

Trochidis K, Bigand E (2013) Investigation of the effect of mode and tempo on emotional responses to music using EEG power asymmetry. J Psychophysiol 27(3):142–147

Trochidis K, Sears D, Trân L, McAdams S (2012) Psychophysiological measures of emotional response to romantic orchestral music and their musical and acoustic correlates. In: Aramaki M et al (eds) Proceedings of the 9th international symposium on computer music modeling and retrieval, 19–22 June, London. Lecture notes in computer science, vol 7900. Springer, Heidelberg, pp 44–57

Tsang CD, Trainor LJ, Santesso DL, Tasker CL, Schmidt LA (2001) Frontal EEG responses as a function of affective musical features. Ann N Y Acad Sci 930(1):439–442

Valenza G, Lanata A, Scilingo P (2012) Oscillations of heart rate and respiration synchronize during affective visual stimulation. IEEE Trans Inf Technol Biomed 16(4):683–690

Van der Zwaag MD, Westerink JHD, Van den Broek EL (2011) Emotional and psychophysiological responses to tempo, mode and percussiveness. Musicae Scientae 15(20):250–269

Wagner J, Kim J, Andre E (2005) From physiological signals to emotion. In: International conference on multimedia and expo, pp 940–943

Watson D, Wiese D, Vaidya J, Tellegen A (1999) The two general activation systems of affect: structural findings, evolutionary consideration, and psychophysiological evidence. J Pers Soc Psychol 76(5):820–838

Watson D, Tellegen A (1985) Toward a consensual structure of mood. Psychol Bull 98 (2):219–235

Watson D, Clark LA, Tellegen A (1988) Development and validation of brief measures of positive and negative affect: the PANAS scales. J Pers Soc Psychol 54(6):1063–1070

Webster G, Weir CG (2005) Emotional responses to music: interactive effects of mode, tempo and texture. Motiv Emot 29(1):19–39

Witvliet C, Vrana S (2007) Play it again Sam: repeated exposure to emotionally evocative music polarizes liking and smiling responses and influences other affects reports, facial EMG and heart rate. Cogn Emot 21(1):3–25

Wundt WM (1896) Outlines of psychology (Wilhelm Engelmann), Leipzig (translated by CH Judd)

Zattore RJ, Samson S (1991) Role of the right temporal neocortex in retention of pitch in auditory short memory. Brain 114(6):2403–2417

Zentner MR, Grandjean D, Scherer KR (2008) Emotions evoked by the sound of music: characterization, classification, and measurement. Emotion 8(4):494–521

A Tutorial on EEG Signal-processing Techniques for Mental-state Recognition in Brain–Computer Interfaces

7

Fabien Lotte

Abstract

This chapter presents an introductory overview and a tutorial of signal-processing techniques that can be used to recognize mental states from electroencephalographic (EEG) signals in brain–computer interfaces. More particularly, this chapter presents how to extract relevant and robust spectral, spatial, and temporal information from noisy EEG signals (e.g., band-power features, spatial filters such as common spatial patterns or xDAWN, etc.), as well as a few classification algorithms (e.g., linear discriminant analysis) used to classify this information into a class of mental state. It also briefly touches on alternative, but currently less used approaches. The overall objective of this chapter is to provide the reader with practical knowledge about how to analyze EEG signals as well as to stress the key points to understand when performing such an analysis.

7.1 Introduction

One of the critical steps in the design of brain–computer interface (BCI) applications based on electroencephalography (EEG) is to process and analyze such EEG signals in real time, in order to identify the mental state of the user. Musical EEG-based BCI applications are no exception. For instance, in (Miranda et al. 2011), the application had to recognize the visual target the user was attending to from his/her EEG signals, in order to execute the corresponding musical command. Unfortunately, identifying

F. Lotte (✉)
Inria Bordeaux Sud-Ouest/LaBRI, 200 Avenue de la Vieille Tour,
33405 Talence Cedex, France
e-mail: fabien.lotte@inria.fr

E.R. Miranda and J. Castet (eds.), *Guide to Brain-Computer Music Interfacing*,
DOI 10.1007/978-1-4471-6584-2_7

the user's mental state from EEG signals is no easy task, such signals being noisy, non-stationary, complex, and of high dimensionality (Lotte et al. 2007). Therefore, mental-state recognition from EEG signals requires specific signal-processing and machine-learning tools. This chapter aims at providing the reader with a basic knowledge about how to do EEG signal processing and the kind of algorithms to use to do so. This knowledge is—hopefully—presented in an accessible and intuitive way, by focusing more on the concepts and ideas than on the technical details.

This chapter is organized as follows: Sect. 7.2 presents the general architecture of an EEG signal-processing system for BCI. Then, Sect. 7.3 describes the specific signal-processing tools that can be used to design BCI based on oscillatory EEG activity while Sect. 7.4 describes those that can used for BCI based on event-related potentials (ERP), i.e., brain responses to stimulus and events. Section 7.5 presents some alternative tools, still not as popular as the one mentioned so far but promising, both for BCI based on oscillatory activity and those based on ERP. Finally, Sect. 7.6 proposes a discussion about all the tools covered and their perspectives while Sect. 7.7 concludes the paper.

7.2 General EEG Signal-processing Principle

In BCI design, EEG signal processing aims at translating raw EEG signals into the class of these signals, i.e., into the estimated mental state of the user. This translation is usually achieved using a pattern recognition approach, whose two main steps are the following:

- **Feature Extraction**: The first signal-processing step is known as "feature extraction" and aims at describing the EEG signals by (ideally) a few relevant values called "features" (Bashashati et al. 2007). Such features should capture the information embedded in EEG signals that is relevant to describe the mental states to identify, while rejecting the noise and other non-relevant information. All features extracted are usually arranged into a vector, known as a feature vector.
- **Classification**: The second step, denoted as "classification," assigns a class to a set of features (the feature vector) extracted from the signals (Lotte et al. 2007). This class corresponds to the kind of mental state identified. This step can also be denoted as "feature translation" (Mason and Birch 2003). Classification algorithms are known as "classifiers."

As an example, let us consider a motor imagery (MI)-based BCI, i.e., a BCI that can recognize imagined movements such left hand or right hand imagined movements (see Fig. 7.1). In this case, the two mental states to identify are imagined left hand movement on one side and imagined right hand movement on the other side. To identify them from EEG signals, typical features are band-power features, i.e., the power of the EEG signal in a specific frequency band. For MI, band-power features are usually extracted in the μ (about 8–12 Hz) and β (about 16–24 Hz)

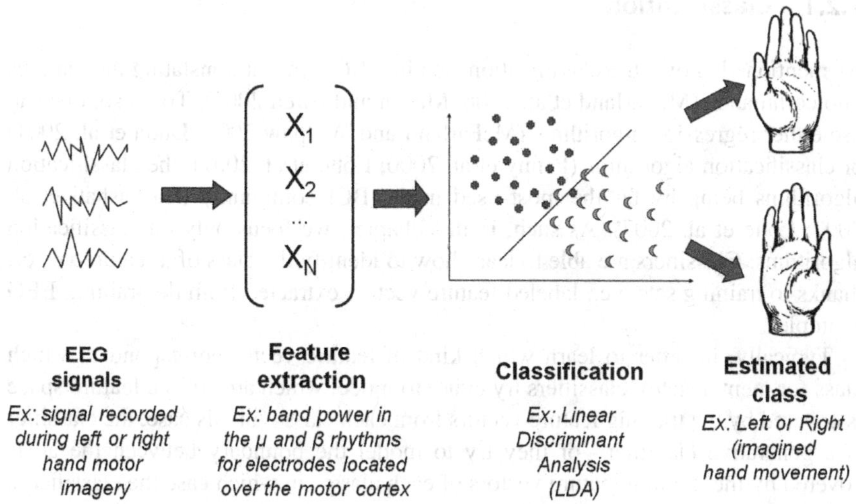

Fig. 7.1 A classical EEG signal-processing pipeline for BCI, here in the context of a motor imagery-based BCI, i.e., a BCI that can recognized imagined movements from EEG signals

frequency bands, for electrode localized over the motor cortex areas of the brain (around locations C3 and C4 for right and left hand movements, respectively) (Pfurtscheller and Neuper 2001). Such features are then typically classified using a linear discriminant analysis (LDA) classifier.

It should be mentioned that EEG signal processing is often built using machine learning. This means the classifier and/or the features are automatically tuned, generally for each user, according to examples of EEG signals from this user. These examples of EEG signals are called a training set and are labeled with their class of belonging (i.e., the corresponding mental state). Based on these training examples, the classifier will be tuned in order to recognize as appropriately as possible the class of the training EEG signals. Features can also be tuned in such a way, e.g., by automatically selecting the most relevant channels or frequency bands to recognized the different mental states. Designing BCI based on machine learning (most current BCI are based on machine learning) therefore consists of two phases:

- **Calibration** (a.k.a., training) phase: This consists in (1) acquiring training EEG signals (i.e., training examples) and (2) optimizing the EEG signal-processing pipeline by tuning the feature parameters and/or training the classifier.
- **Use** (a.k.a., test) phase: This consists in using the model (features and classifier) obtained during the calibration phase in order to recognize the mental state of the user from previously unseen EEG signals, in order to operate the BCI.

Feature extraction and classification are discussed in more details hereafter.

7.2.1 Classification

As mentioned above, the classification step in a BCI aims at translating the features into commands (McFarland et al. 2006; Mason and Birch 2003). To do so, one can use either regression algorithms (McFarland and Wolpaw 2005; Duda et al. 2001) or classification algorithms (Penny et al. 2000; Lotte et al. 2007), the classification algorithms being by far the most used in the BCI community (Bashashati et al. 2007; Lotte et al. 2007). As such, in this chapter, we focus only on classification algorithms. Classifiers are able to learn how to identify the class of a feature vector, thanks to training sets, i.e., labeled feature vectors extracted from the training EEG examples.

Typically, in order to learn which kind of feature vector correspond to which class (or mental state), classifiers try either to model which area of the feature space is covered by the training feature vectors from each class—in this case, the classifier is a generative classifier—or they try to model the boundary between the areas covered by the training feature vectors of each class—in which case the classifier is a discriminant classifier. For BCI, the most used classifiers so far are discriminant classifiers, and notably linear discriminant analysis (LDA) classifiers.

The aim of LDA (also known as Fisher's LDA) was to use hyperplanes to separate the training feature vectors representing the different classes (Duda et al. 2001; Fukunaga 1990). The location and orientation of this hyperplane are determined from training data. Then, for a two-class problem, the class of an unseen (a.k.a., test) feature vector depends on which side of the hyperplane the feature vector is (see Fig. 7.2). LDA has very low computational requirements which makes it suitable for online BCI system. Moreover, this classifier is simple which makes it naturally good at generalizing to unseen data, hence generally providing good results in practice (Lotte et al. 2007). LDA is probably the most used classifier for BCI design.

Another very popular classifier for BCI is the support vector machine (SVM) (Bennett and Campbell 2000). An SVM also uses a discriminant hyperplane to identify classes (Burges 1998). However, with SVM, the selected hyperplane is the

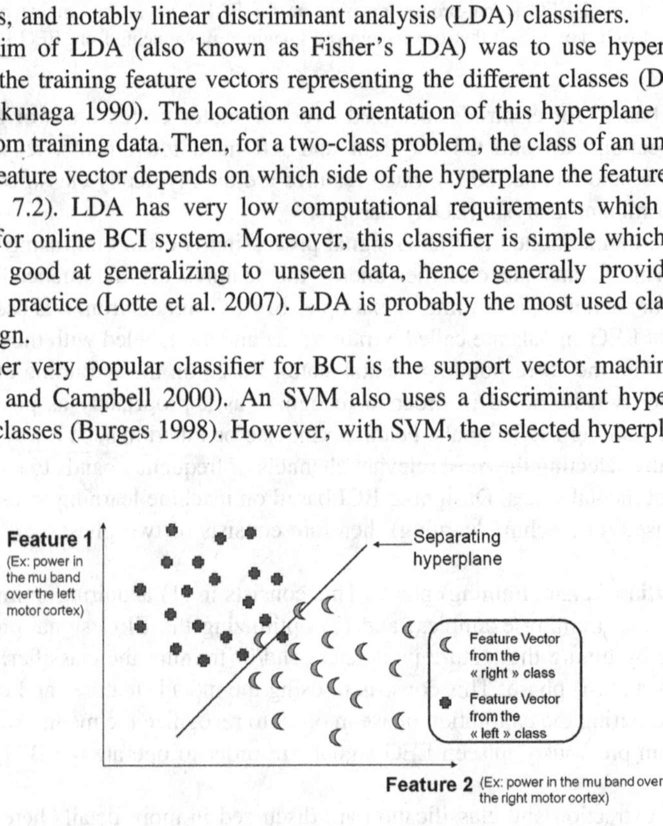

Fig. 7.2 Discriminating two types of motor imagery with a linear hyperplane using a linear discriminant analysis (LDA) classifier

one that maximizes the margins, i.e., the distance from the nearest training points, which has been found to increase the generalization capabilites (Burges 1998; Bennett and Campbell 2000).

Generally, regarding classification algorithms, it seems that very good recognition performances can be obtained using appropriate off-the-shelf classifiers such as LDA or SVM (Lotte et al. 2007). What seems to be really important is the design and selection of appropriate features to describe EEG signals. With this purpose, specific EEG signal-processing tools have been proposed to design BCI. In the rest of this chapter, we will therefore focus on EEG feature extraction tools for BCI. For readers interested to learn more about classification algorithms, we refer them to (Lotte et al. 2007), a review paper on this topic.

7.2.2 Feature Extraction

As mentioned before, feature extraction aims at representing raw EEG signals by an ideally small number of relevant values, which describe the task-relevant information contained in the signals. However, classifiers are able to learn from data which class corresponds to which input features. As such, why not using directly the EEG signals as input to the classifier? This is due to the so-called curse-of-dimensionality, which states that the amount of data needed to properly describe the different classes increases exponentially with the dimensionality of the feature vectors (Jain et al. 2000; Friedman 1997). It has been recommended to use from 5 to 10 times as many training examples per class as the input feature vector dimensionality[1] (Raudys and Jain 1991). What would it mean to use directly the EEG signals as input to the classifier? Let us consider a common setup with 32 EEG sensors sampled at 250 Hz, with one trial of EEG signal being 1 s long. This would mean a dimensionality of 32 * 250 = 8,000, which would require at least 40,000 training examples. Obviously, we cannot ask the BCI user to perform each mental task 40,000 times to calibrate the BCI before he/she could use it. A much more compact representation is therefore needed, hence the necessity to perform some form of feature extraction.

With BCI, there are three main sources of information that can be used to extract features from EEG signals:

- **Spatial information**: Such features would describe where (spatially) the relevant signal comes from. In practice, this would mean selecting specific EEG channels, or focusing more on specific channels than on some other. This amounts to focusing on the signal originating from specific areas of the brain.
- **Spectral (frequential) information**: Such features would describe how the power in some relevant frequency bands varies. In practice, this means that the features will use the power in some specific frequency bands.

[1] Note that this was estimated before SVM were invented and that SVM are generally less sensitive—although not completely immune—to this curse-of-dimensionality.

- **Temporal information**: Such features would describe how the relevant signal varies with time. In practice, this means using the EEG signals values at different time points or in different time windows.

Note that these three sources of information are not the only ones, and alternatives can be used (see Sect. 7.5). However, they are by far the most used one, and, at least so far, the most efficient ones in terms of classification performances. It should be mentioned that so far, nobody managed to discover nor to design a set of features that would work for all types of BCI. As a consequence, different kinds of BCI currently use different sources of information. Notably, BCI based on oscillatory activity (e.g., BCI based on motor imagery) mostly need and use the spectral and spatial information whereas BCI based on ERP (e.g., BCI based on the P300) mostly need and use the temporal and spatial information. The next sections detail the corresponding tools for these two categories of BCI.

7.3 EEG Signal-processing Tools for BCI Based on Oscillatory Activity

BCI based on oscillatory activity are BCI that use mental states which lead to changes in the oscillatory components of EEG signals, i.e., that lead to change in the power of EEG signals in some frequency bands. Increase of EEG signal power in a given frequency band is called an event-related synchronization (ERS), whereas a decrease of EEG signal power is called an event-related desynchronization (ERD) (Pfurtscheller and da Silva 1999). BCI based on oscillatory activity notably includes motor imagery-based BCI (Pfurtscheller and Neuper 2001), steady-state visual evoked potentials (SSVEP)-based BCI (Vialatte et al. 2010) as well as BCI based on various cognitive imagery tasks such as mental calculation, mental geometric figure rotation, mental word generation, etc. (Friedrich et al. 2012; Millán et al. 2002). As an example, imagination of a left hand movement leads to a contralateral ERD in the motor cortex (i.e., in the right motor cortex for left hand movement) in the μ and β bands during movement imagination, and to an ERS in the β band (a.k.a., beta rebound) just after the movement imagination ending (Pfurtscheller and da Silva 1999). This section first describes a basic design for oscillatory activity-based BCI. Then, due to the limitations exhibited by this design, it exposes more advanced designs based on multiple EEG channels. Finally, it presents a key tool to design such BCIs: the common spatial pattern (CSP) algorithm, as well as some of its variants.

7.3.1 Basic Design for an Oscillatory Activity-based BCI

Oscillatory activity-based BCI are based on change in power in some frequency bands, in some specific brain areas. As such, they naturally need to exploit both the spatial and spectral information. As an example, a basic design for a motor-imagery

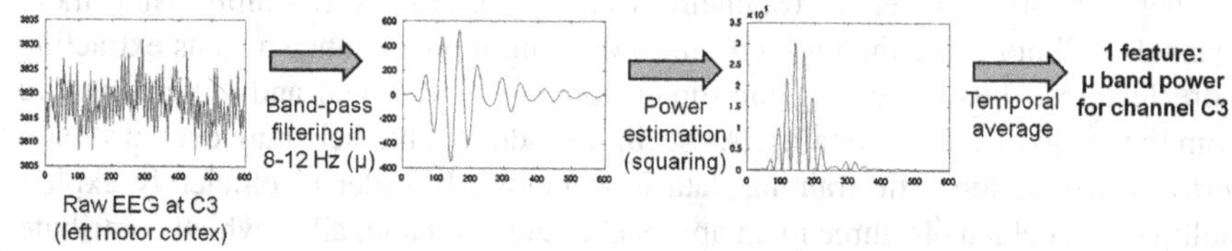

Fig. 7.3 Signal-processing steps to extract band-power features from raw EEG signals. The EEG signal displayed here was recorded during *right hand* motor imagery (the instruction to perform the imagination was provided at $t = 0$ s on the plots). The contralateral ERD during imagination is here clearly visible. Indeed, the signal power in channel C3 (*left motor cortex*) in 8–12 Hz clearly decreases during this imagination of a right hand movement

BCI would exploit the spatial information by extracting features only from EEG channels localized over the motor areas of the brain, typically channels C3 for right hand movements, Cz for foot movements and C4 for left hand movements. It would exploit the spectral information by focusing on frequency bands μ (8–12 Hz) and β (16–24 Hz). More precisely, for a BCI that can recognize left hand MI versus right hand MI, the basic features extracted would be the average band power in 8–12 and 16–24 Hz from both channels C3 and C4. Therefore, the EEG signals would be described by only four features.

There are many ways to compute band-power features from EEG signals (Herman et al. 2008; Brodu et al. 2011). However, a simple, popular, and efficient one is to first band-pass filter the EEG signal from a given channel into the frequency band of interest, then to square the resulting signal to compute the signal power, and finally to average it over time (e.g., over a time window of 1 s). This is illustrated in Fig. 7.3.

Unfortunately, this basic design is far from being optimal. Indeed, it uses only two fixed channels. As such, relevant information, measured by other channels might be missing, and C3 and C4 may not be the best channels for the subject at hand. Similarly, using the fixed frequency bands 8–12 Hz and 16–24 Hz may not be the optimal frequency bands for the current subject. In general, much better performances are obtained when using subject-specific designs, with the best channels and frequency bands optimized for this subject. Using more than two channels is also known to lead to improved performances, since it enables to collect the relevant information spread over the various EEG sensors.

7.3.2 Toward Advanced BCI Using Multiple EEG Channels

Both the need to use subject-specific channels and the need to use more than two channels lead to the necessity to design BCI based on multiple channels. This is confirmed by various studies which suggested that, for motor imagery, eight channels is a minimum to obtain reasonable performances (Sannelli et al. 2010; Arvaneh et al. 2011), with optimal performances achieved with a much larger number, e.g., 48 channels in (Sannelli et al. 2010). However, simply using more channels will not solve the problem. Indeed, using more channels means extracting more features, thus increasing the dimensionality of the data and suffering more from the curse-of-dimensionality. As such, just adding channels may even decrease performances if too little training data is available. In order to efficiently exploit multiple EEG channels, three main approaches are available, all of which contribute to reducing the dimensionality:

- **Feature selection** algorithm: These are methods to select automatically a subset of relevant features, among all the features extracted.
- **Channel selection** algorithms: These are similar methods that select automatically a subset of relevant channels, among all channels available.
- **Spatial Filtering** algorithms: These are methods that combine several channels into a single one, generally using weighted linear combinations, from which features will be extracted.

They are described below.

7.3.2.1 Feature Selection

Feature selection are classical algorithms widely used in machine learning (Guyon and Elisseeff 2003; Jain and Zongker 1997) and as such also very popular in BCI design (Garrett et al. 2003). There are too main families of feature selection algorithms:

- *Univariate* algorithms: They evaluate the discriminative (or descriptive) power of each feature individually. Then, they select the N best individual features (*N* needs to be defined by the BCI designer). The usefulness of each feature is typically assessed using measures such as Student t-statistics, which measures the feature value difference between two classes, correlation-based measures such as R^2, mutual information, which measures the dependence between the feature value and the class label, etc. (Guyon and Elisseeff 2003). Univariate methods are usually very fast and computationally efficient but they are also suboptimal. Indeed, since they only consider the individual feature usefulness, they ignore possible redundancies or complementarities between features. As such, the best subset of *N* features is usually not the *N* best individual features. As an example, the *N* best individual features might be highly redundant and measure almost the same information. As such using them together would add very little discriminant power. On the other hand, adding a feature that is individually not very good but which measures a different information from that of the best individual ones is likely to improve the discriminative power much more.
- *Multivariate* algorithms: They evaluate subsets of features together and keep the best subset with *N* features. These algorithms typically use measures of global performance for the subsets of features, such as measures of classification

performances on the training set (typically using cross-validation (Browne 2000)) or multivariate mutual information measures, see, e.g., (Hall 2000; Pudil et al. 1994; Peng et al. 2005). This global measure of performance enables to actually consider the impact of redundancies or complementarities between features. Some measures also remove the need to manually select the value of N (the number of features to keep), the best value of N being the number of features in the best subset identified. However, evaluating the usefulness of subsets of features leads to very high computational requirements. Indeed, there are many more possible subsets of any size than individual features. As such there are many more evaluations to perform. In fact, the number of possible subsets to evaluate is very often far too high to actually perform all the evaluations in practice. Consequently, multivariate methods usually rely on heuristics or greedy solutions in order to reduce the number of subsets to evaluate. They are therefore also suboptimal but usually give much better performances than univariate methods in practice. On the other hand, if the initial number of features is very high, multivariate methods may be too slow to use in practice.

7.3.2.2 Channel Selection

Rather than selecting features, one can also select channels and only use features extracted from the selected channels. While both channel and feature selection reduce the dimensionality, selecting channels instead of features has some additional advantages. In particular, using less channels means a faster setup time for the EEG cap and also a lighter and more comfortable setup for the BCI user. It should be noted, however, that with the development of dry EEG channels, selecting channels may become less crucial. Indeed the setup time will not depend on the number of channel used, and the BCI user will not have more gel in his/her hair if more channels are used. With dry electrodes, using less channels will still be lighter and more comfortable for the user though.

Algorithms for EEG channel selection are usually based or inspired from generic feature selection algorithm. Several of them are actually analogous algorithms that assess individual channel usefulness or subsets of channels discriminative power instead of individual features or subset of features. As such, they also use similar performance measures and have similar properties. Some other channel selection algorithms are based on spatial filter optimization (see below). Readers interested to know more about EEG channel selection may refer to the following papers and associated references (Schröder et al. 2005; Arvaneh et al. 2011; Lal et al. 2004; Lan et al. 2007), among many other.

7.3.2.3 Spatial Filtering

Spatial filtering consists in using a small number of new channels that are defined as a linear combination of the original ones:

$$\tilde{x} = \sum_i w_i x_i = wX \tag{7.1}$$

with $\tilde{x}$ the spatially filtered signal, x_i the EEG signal from channel i, w_i the weight given to that channel in the spatial filter, and X a matrix whose ith row is x_i, i.e., X is the matrix of EEG signals from all channels.

It should be noted that spatial filtering is useful not only because it reduces the dimension from many EEG channels to a few spatially filtered signals (we typically use much less spatial filters than original channels), but also because it has a neurophysiological meaning. Indeed, with EEG, the signals measured on the surface of the scalp are a blurred image of the signals originating from within the brain. In other words, due to the smearing effect of the skull and brain (a.k.a., volume conduction effect), the underlying brain signal is spread over several EEG channels. Therefore, spatial filtering can help recovering this original signal by gathering the relevant information that is spread over different channels.

There are different ways to define spatial filters. In particular, the weights w_i can be fixed in advance, generally according to neurophysiological knowledge, or they can be data driven, that is, optimized on training data. Among the fixed spatial filters, we can notably mention the bipolar and Laplacian which are local spatial filters that try to locally reduce the smearing effect and some of the background noise (McFarland et al. 1997). A bipolar filter is defined as the difference between two neighboring channels, while a Laplacian filter is defined as 4 times the value of a central channel minus the values of the four channels around. For instance, a bipolar filter over channel C3 would be defined as $C3_{\text{bipolar}} = FC3 - CP3$, while a Laplacian filter over C3 would be defined as $C3_{\text{Laplacian}} = 4C3 - FC3 - C5 - C1 - CP3$, see also Fig. 7.4. Extracting features from bipolar or Laplacian spatial filters rather than from the single corresponding electrodes has been shown to significantly increase classification performances (McFarland et al. 1997). An inverse solution is another kind of fixed spatial filter (Michel et al. 2004; Baillet et al. 2001). Inverse solutions are algorithms that enable to estimate the signals originating from sources within the brain based on the measurements taken from the scalp. In other words, inverse solutions enable us to look into the activity of specific brain regions. A word of caution though: Inverse solutions do not provide more information than what is already available in scalp EEG signals. As such, using inverse solutions will NOT make a noninvasive BCI as accurate and efficient as an invasive one. However, by focusing on some specific brain areas, inverse solutions can contribute to reducing background noise, the smearing effect and irrelevant information originating from other areas. As such, it has been shown than extracting features from the signals spatially filtered using inverse solutions (i.e., from the sources within the brain) leads to higher classification performances than extracting features directly from scalp EEG signals (Besserve et al. 2011; Noirhomme et al. 2008). In general, using inverse solutions has been shown to lead to high classification performances (Congedo et al. 2006; Lotte et al. 2009b; Qin et al. 2004; Kamousi et al. 2005; Grosse-Wentrup et al. 2005). It should be noted that since the

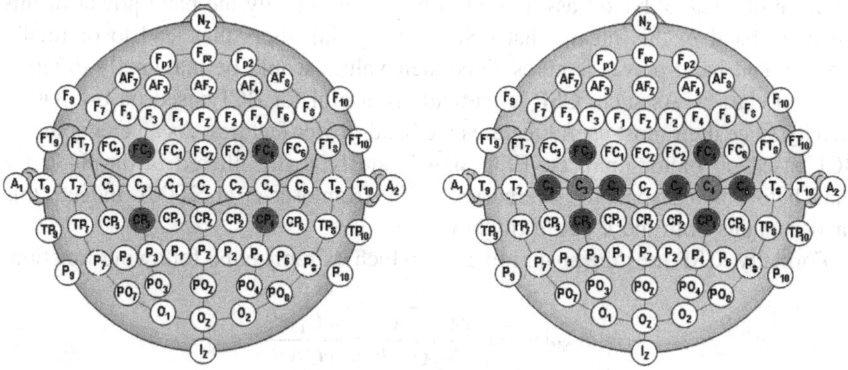

Fig. 7.4 *Left* channels used in bipolar spatial filtering over channels C3 and C4. *Right* channels used in Laplacian spatial filtering over channels C3 and C4

number of source signals obtained with inverse solutions is often larger than the initial number of channels, it is necessary to use feature selection or dimensionality reduction algorithms.

The second category of spatial filters, i.e., data-driven spatial filters, is optimized for each subject according to training data. As any data-driven algorithm, the spatial filter weights w_i can be estimated in an unsupervised way, that is without the knowledge of which training data belong to which class, or in a supervised way, with each training data being labeled with its class. Among the unsupervised spatial filters, we can mention principal component analysis (PCA), which finds the spatial filters that explain most of the variance of the data, or independent component analysis (ICA), which find spatial filters whose resulting signals are independent from each other (Kachenoura et al. 2008). The later has been shown rather useful to design spatial filters able to remove or attenuate the effect of artifacts (EOG, EMG, etc. (Fatourechi et al. 2007)) on EEG signals (Tangermann et al. 2009; Xu et al. 2004; Kachenoura et al. 2008; Brunner et al. 2007). Alternatively, spatial filters can be optimized in a supervised way, i.e., the weights will be defined in order to optimize some measure of classification performance. For BCI based on oscillatory EEG activity, such a spatial filter has been designed: the common spatial patterns (CSP) algorithm (Ramoser et al. 2000; Blankertz et al. 2008b). This algorithm has greatly contributed to the increase of performances of this kind of BCI and thus has become a standard tool in the repertoire of oscillatory activity-based BCI designers. It is described in more details in the following section, together with some of its variants.

7.3.3 Common Spatial Patterns and Variants

Informally, the CSP algorithm finds spatial filters w such that the variance of the filtered signal is maximal for one class and minimal for the other class. Since the variance of a signal band-pass filtered in band b is actually the band power of this signal in band b, this means that CSP finds spatial filters that lead to optimally discriminant band-power features since their values would be maximally different between classes. As such, CSP is particularly useful for BCI based on oscillatory activity since their most useful features are band-power features. As an example, for BCI based on motor imagery, EEG signals are typically filtered in the 8–30 Hz band before being spatially filtered with CSP (Ramoser et al. 2000). Indeed, this band contains both the μ and β rhythms.

Formally, CSP uses the spatial filters w which extremize the following function:

$$J_{CSP}(w) = \frac{wX_1X_1^Tw^T}{wX_2X_2^Tw^T} = \frac{wC_1w^T}{wC_2w^T} \tag{7.2}$$

where T denotes transpose, X_i is the training band-pass filtered signal matrix for class i (with the samples as columns and the channels as rows), and C_i the spatial covariance matrix from class i. In practice, the covariance matrix C_i is defined as the average covariance matrix of each trial from class i (Blankertz et al. 2008b). In this equation, wX_i is the spatially filtered EEG signal from class i, and $wX_iX_i^Tw^T$ is thus the variance of the spatially filtered signal, i.e., the band power of the spatially filtered signal. Therefore, extremizing $J_{CSP}(w)$, i.e., maximizing and minimizing it, indeed leads to spatially filtered signals whose band power is maximally different between classes. $J_{CSP}(w)$ happens to be a Rayleigh quotient. Therefore, extremizing it can be solved by generalized eigenvalue decomposition (GEVD). The spatial filters w that maximize or minimize $J_{CSP}(w)$ are thus the eigenvectors corresponding to the largest and lowest eigenvalues, respectively, of the GEVD of matrices C_1 and C_2. Typically, six filters (i.e., three pairs), corresponding to the three largest and three lowest eigenvalues are used. Once these filters obtained, a CSP feature f is defined as follows:

$$f = \log(wXX^Tw^T) = \log(wCw^T) = \log(\mathrm{var}(wX)) \tag{7.3}$$

i.e., the features used are simply the band power of the spatially filtered signals. CSP requires more channels than fixed spatial filters such as Bipolar or Laplacian, however in practice, it usually leads to significantly higher classification performances (Ramoser et al. 2000). The use of CSP is illustrated in Fig. 7.5. In this figure, the signals spatially filtered with CSP clearly show difference in variance (i.e., in band power) between the two classes, hence ensuring high classification performances.

The CSP algorithm has numerous advantages: First, it leads to high classification performances. CSP is also versatile, since it works for any ERD/ERS BCI. Finally, it is computationally efficient and simple to implement. Altogether this makes CSP one of the most popular and efficient approach for BCI based on oscillatory activity (Blankertz et al. 2008b).

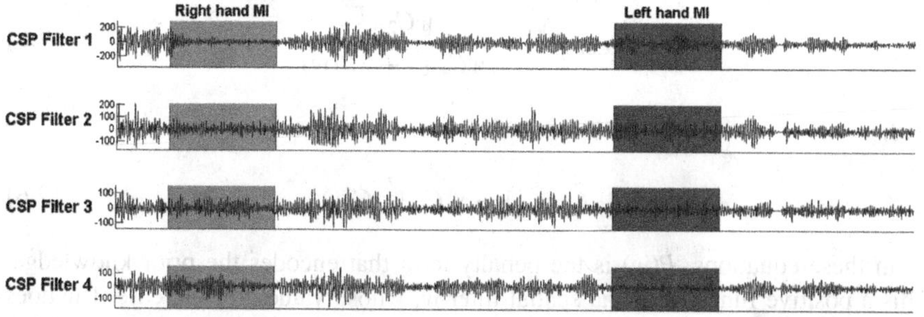

Fig. 7.5 EEG signals spatially filtered using the CSP algorithm. The first two spatial filters (*top filters*) are those maximizing the variance of signals from class "left hand motor imagery" while minimizing that of class "right hand motor imagery." They correspond to the largest eigenvalues of the GEVD. The last two filters (*bottom filters*) are the opposite, they maximize the variance of class "right hand motor imagery" while minimizing that of class "left hand motor imagery" (they correspond to the lowest eigenvalues of the GEVD). This can be clearly seen during the periods of *right* or *left hand* motor imagery, in *light* and *dark gray*, respectively

Nevertheless, despite all these advantages, CSP is not exempt from limitations and is still not the ultimate signal-processing tool for EEG-based BCI. In particular, CSP has been shown to be non-robust to noise, to non-stationarities and prone to overfitting (i.e., it may not generalize well to new data) when little training data is available (Grosse-Wentrup and Buss 2008; Grosse-Wentrup et al. 2009; Reuderink and Poel 2008). Finally, despite its versatility, CSP only identifies the relevant spatial information but not the spectral one. Fortunately, there are ways to make CSP robust and stable with limited training data and with noisy training data. An idea is to integrate prior knowledge into the CSP optimization algorithm. Such knowledge could represent any information we have about what should be a good spatial filter for instance. This can be neurophysiological prior, data (EEG signals) or meta-data (e.g., good channels) from other subjects, etc. This knowledge is used to guide and constraint the CSP optimization algorithm toward good solutions even with noise, limited data, and non-stationarities (Lotte and Guan 2011). Formally, this knowledge is represented in a regularization framework that penalizes unlikely solutions (i.e., spatial filters) that do not satisfy this knowledge therefore enforcing it. Similarly, prior knowledge can be used to stabilize statistical estimates (here, covariance matrices) used to optimize the CSP algorithm. Indeed, estimating covariance matrices from few training data usually leads to poor estimates (Ledoit and Wolf 2004).

Formally, a regularized CSP (RCSP) can be obtained by maximizing both Eqs. 7.4 and 7.5:

$$J_{\mathrm{RCSP1}}(w) = \frac{w\tilde{C}_1 w^T}{w\tilde{C}_2 w^T + \lambda P(w)} \tag{7.4}$$

$$J_{\mathrm{RCSP2}}(w) = \frac{w\tilde{C}_2 w^T}{w\tilde{C}_1 w^T + \lambda P(w)} \tag{7.5}$$

with

$$\tilde{C}_i = (1-\gamma)C_i + \gamma G_i \tag{7.6}$$

In these equations, $P(w)$ is the penalty term that encodes the prior knowledge. This a positive function of the spatial filter w, whose value will increase if w does not satisfy the knowledge encoded. Since the filters are obtained by maximizing $J_{\mathrm{RCSP}i}$, this means that the numerator (which is positive) must be maximized and the denominator (which is also positive) must be minimized. Since $P(w)$ is positive and part of the denominator, this means that $P(w)$ will be minimized as well, hence enforcing that the spatial filters w satisfy the prior knowledge. Matrix G_i is another way of using prior knowledge, in order to stabilize the estimates of the covariance matrices C_i. If we have any idea about how these covariance matrices should be, this can be encoded in G_i in order to define a new covariance matrix $\tilde{C}_i$ which is a mix of the matrix C_i estimated on the data and of the prior knowledge G_i. We will present below what kind of knowledge can be encoded in $P(w)$ and G_i.

For the penalty term $P(w)$, a kind of knowledge that can be used is spatial knowledge. For instance, from a neurophysiological point of view, we know that neighboring neurons tend to have similar functions, which supports the idea that neighboring electrodes should measure similar brain signals (if the electrodes are close enough to each other), notably because of the smearing effect. Thus, neighboring electrodes should have similar contributions in the spatial filters. In other words, spatial filters should be spatially smooth. This can be enforced by using the following penalty term:

$$P(w) = \sum_{i,j} \mathrm{Prox}(i,j)(w_i - w_j)^2 \tag{7.7}$$

where $\mathrm{Prox}(i,j)$ measures the proximity of electrodes i and j, and $(w_i - w_j)^2$ is the weight difference between electrodes i and j, in the spatial filter. Thus, if two electrodes are close to each other and have very different weights, the penalty term $P(w)$ will be high, which would prevent such solutions to be selected during the optimization of the CSP (Lotte and Guan 2010b). Another knowledge that can be used is that for a given mental task, not all the brain regions are involved and useful. As such, some electrodes are unlikely to be useful to classify some specific mental tasks. This can be encoded in $P(w)$ as well:

$$P(w) = wDw^T \quad \text{with} \quad D(i,j) = \begin{cases} \text{channel } i \text{ ``uselessness''} & \text{if } i = j \\ 0 & \text{otherwise} \end{cases} \tag{7.8}$$

Basically, the value of $D(i,i)$ is the penalty for the ith channel. The higher this penalty, the less likely this channel will have a high contribution in the CSP filters. The value of this penalty can be defined according to neurophysiological prior knowledge for instance, large penalties being given to channels unlikely to be useful and small or no penalty being given to channels that are likely to genuinely contribute to the filter. However, it may be difficult to precisely define the extent of the penalty from the literature. Another alternative is the use data previously recorded from other subjects. Indeed, the optimized CSP filters already obtained from previous subject give information about which channels have large contributions on average. The inverse of the average contribution of each channel can be used as the penalty, hence penalizing channels with small average contribution (Lotte and Guan 2011). Penalty terms are therefore also a nice way to perform subject-to-subject transfer and re-use information from other subjects. These two penalties are examples that have proven useful in practice. This usefulness is notably illustrated in Fig. 7.6, in which spatial filters obtained with the basic CSP are rather noisy, with strong contributions from channels not expected from a neurophysiological point of view. On the contrary, the spatial filters obtained using the two RCSP penalties described previously are much cleaner, spatially smoother and with strong contributions localized in neurophysiologically relevant areas. This in turns led to higher classification performances, with CSP obtaining 73.1 % classification accuracy versus 78.7 % and 77.6 % for the regularized versions (Lotte and Guan 2011). It should be mentioned, however, that strong contributions from non-neurophysiologically relevant brain areas in a CSP spatial filter may be present to perform noise cancelation, and as such does not mean the spatial filter is bad per se (Haufe et al. 2014). It should also be mentioned that other interesting penalty terms have been proposed, in order to deal with known noise sources (Blankertz et al. 2008a), non-stationarities (Samek et al. 2012) or to perform simultaneous channel selection (Farquhar et al. 2006; Arvaneh et al. 2011).

Matrix G_i in Eq. 7.6 is another way to add prior knowledge. This matrix can notably be defined as the average covariance matrix obtained from other subjects who performed the same task. As such it enables to define a good and stable estimate of the covariance matrices, even if few training EEG data are available for the target subject. This has been shown to enable us to calibrate BCI system with 2–3 times less training data than with the basic CSP, while maintaining classification performances (Lotte and Guan 2010a).

Regularizing CSP using a priori knowledge is thus a nice way to deal with some limitations of CSP such as its sensitivity to overfitting and its non-robustness to noise. However, these regularized algorithms cannot address the limitation that CSP only optimizes the use of the spatial information, but not that of the spectral one. In general, independently of the use of CSP, there are several ways to optimize the use of the spectral information. Typically, this consists in identifying, in one way or another, the relevant frequency bands for the current subject and mental tasks

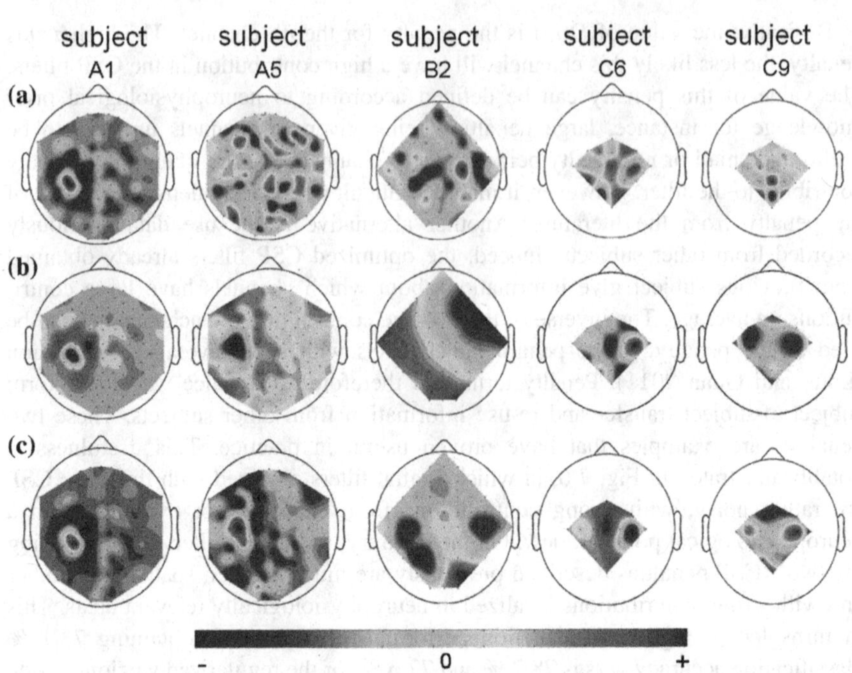

Fig. 7.6 Spatial filters (i.e., weight attributed to each channel) obtained to classify *left hand* versus *right hand* motor imagery. The electrodes, represented by *black dots*, are here seen from *above*, with the subject nose on *top*. **a** basic CSP algorithm, **b** RCSP with a penalty term imposing spatial smoothness, **c** RCSP with a penalty term penalizing unlikely channels according to EEG data from other subjects

performed. For instance, this can be done manually (by trial and errors), or by looking at the average EEG frequency spectrum in each class. In a more automatic way, possible methods include extracting band-power features in multiple frequency bands and then selecting the relevant ones using feature selection (Lotte et al. 2010), by computing statistics on the spectrum to identify the relevant frequencies (Zhong et al. 2008), or even by computing optimal band-pass filters for classification (Devlaminck 2011). These ideas can be used within the CSP framework in order to optimize the use of both the spatial and spectral information. Several variants of CSP have been proposed in order to optimize spatial and spectral filters at the same time (Lemm et al. 2005; Dornhege et al. 2006; Tomioka et al. 2006; Thomas et al. 2009). A simple and computationally efficient method is worth describing: The filter bank CSP (FBCSP) (Ang et al. 2012). This method, illustrated in Fig. 7.7, consists in first filtering EEG signals in multiple frequency bands using a filter bank. Then, for each frequency band, spatial filters are optimized using the classical CSP algorithm. Finally, among the multiple spatial filters obtained, the best resulting features are selected using feature selection algorithms (typically mutual information-based feature selection). As such, this selects both the

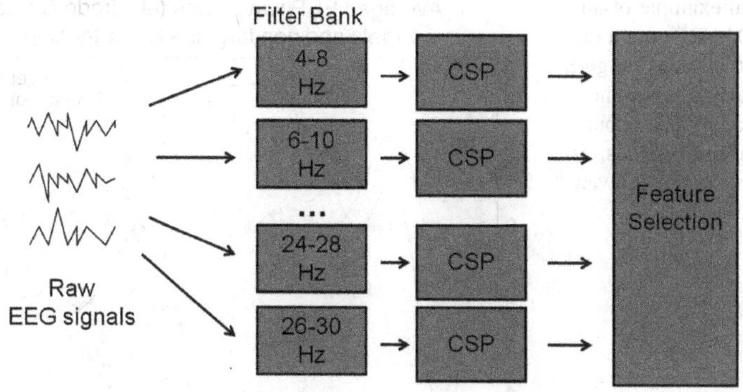

Fig. 7.7 Principle of filter bank common spatial patterns (FBCSP): (1) band-pass filtering the EEG signals in multiple frequency bands using a filter bank; (2) optimizing CSP spatial filter for each band; (3) selecting the most relevant filters (both spatial and spectral) using feature selection on the resulting features

best spectral and spatial filters since each feature corresponds to a single frequency band and CSP spatial filter. This algorithm, although simple, has proven to be very efficient in practice. It was indeed the algorithm used in the winning entries of all EEG data sets from the last BCI competition[2] (Ang et al. 2012).

7.3.4 Summary for Oscillatory Activity-based BCI

In summary, when designing BCI aiming at recognizing mental states that involve oscillatory activity, it is important to consider both the spectral and the spatial information. In order to exploit the spectral information, using band-power features in relevant frequency bands is an efficient approach. Feature selection is also a nice tool to find the relevant frequencies. Concerning the spatial information, using or selecting relevant channels is useful. Spatial filtering is a very efficient solution for EEG-based BCI in general, and the CSP algorithm is a must-try for BCI based on oscillatory activity in particular. Moreover, there are several variants of CSP that are available in order to make it robust to noise, non-stationarity, limited training data sets, or to jointly optimize spectral and spatial filters. The next section will address the EEG signal-processing tools for BCI based on evoked potentials, which are different from the ones described so far, but share some general concepts.

[2] BCI competitions are contests to evaluate the best signal processing and classification algorithms on given brain signals data sets. See http://www.bbci.de/competition/ for more info.

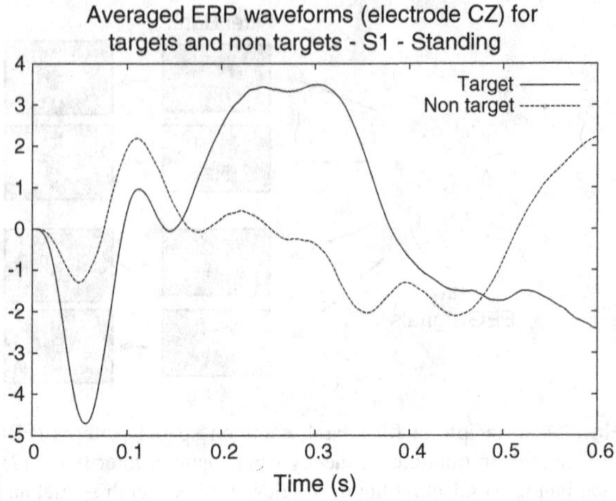

Fig. 7.8 An example of an average P300 ERP after a rare and relevant stimulus (target). We can clearly observe the increase in amplitude about 300 ms after the stimulus, as compared to the non-relevant stimulus (nontarget)

7.4 EEG Signal-processing Tools for BCI Based on Event-related Potentials

An event-related potential (ERP) is a brain responses due to some specific stimulus perceived by the BCI user. A typical ERP used for BCI design is the P300, which is a positive deflection of the EEG signal occurring about 300 ms after the user perceived a rare and relevant stimulus (Fazel-Rezai et al. 2012) (see also Fig. 7.8).

ERP are characterized by specific temporal variations with respect to the stimulus onset. As such, contrary to BCI based on oscillatory activity, ERP-based BCI exploit mostly a temporal information, but rarely a spectral one. However, as for BCI based on oscillatory activity, ERP-based can also benefit a lot from using the spatial information. Next section illustrates how the spatial and temporal information is used in basic P300-based BCI designs.

7.4.1 Basic Signal-processing Tools for P300-based BCI

In P300-based BCI, the spatial information is typically exploited by focusing mostly on electrodes located over the parietal lobe (i.e., by extracting features only for these electrodes), where the P300 is know to originate. As an example, Krusienski et al. recommend to use a set of eight channels, in positions Fz, Cz, P3, Pz, P4, PO7, Oz, PO8 (see Fig. 7.9) (Krusienski et al. 2006).

Once the relevant spatial information identified, here using, for instance, only the electrodes mentioned above, features can be extracted for the signal of each of them. For ERP in general, including the P300, the features generally exploit the temporal information of the signals, i.e., how the amplitude of the EEG signal varies with time. This is typically achieved by using the values of preprocessed

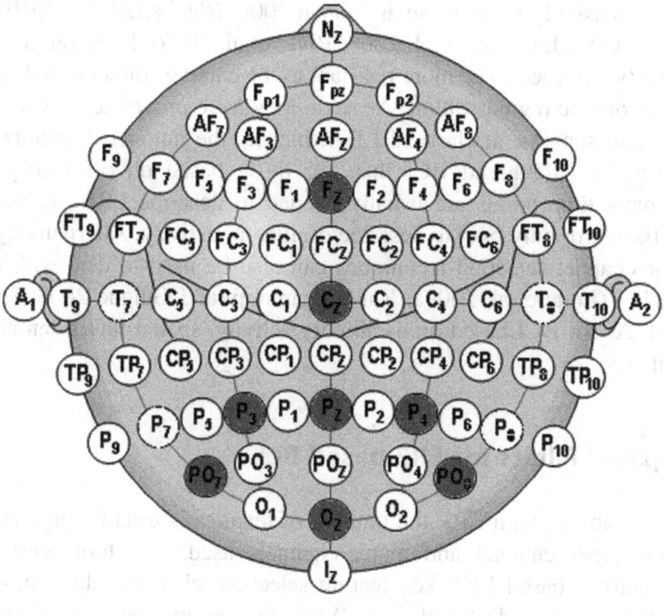

Fig. 7.9 Recommended electrodes for P300-based BCI design, according to (Krusienski et al. 2006)

EEG time points as features. More precisely, features for ERP are generally extracted by (1) low-pass or band-pass filtering the signals (e.g., in 1–12 Hz for the P300), ERP being generally slow waves, (2) downsampling the filtered signals, in order to reduce the number of EEG time points and thus the dimensionality of the problem, and (3) gathering the values of the remaining EEG time points from all considered channels into a feature vector that will be used as input to a classifier. This process is illustrated in Fig. 7.10 to extract features from channel Pz for a P300-based BCI experiment.

Once the features extracted, they can be provided to a classifier which will be trained to assigned them to the target class (presence of an ERP) or to the nontarget class (absence of an ERP). This is often achieved using classical classifiers such as LDA or SVM (Lotte et al. 2007). More recently, automatically regularized LDA

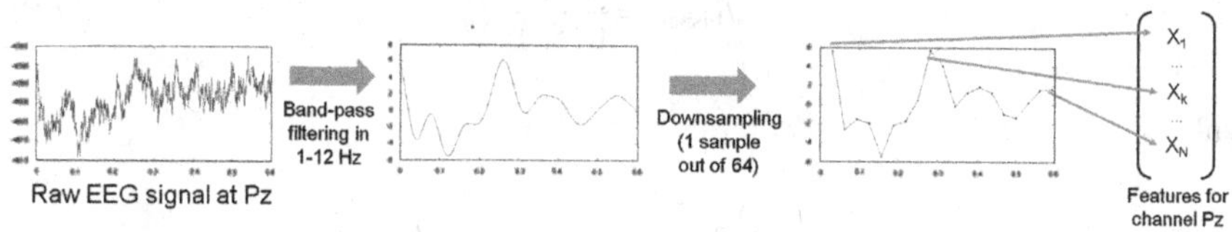

Fig. 7.10 Typical process to extract features from a channel of EEG data for a P300-based BCI design. On this picture, we can see the P300 becoming more visible with the different processing steps

have been increasingly used (Lotte and Guan 2009; Blankertz et al. 2010), as well as Bayesian LDA (Hoffmann et al. 2008; Rivet et al. 2009). Both variants of LDA are specifically designed to be more resistant to the curse-of-dimensionality through the use of automatic regularization. As such, they have proven to be very effective in practice, and superior to classical LDA. Indeed, the number of features is generally higher for ERP-based BCI than for those based on oscillatory activity. Actually, many time points are usually needed to describe ERP but only a few frequency bands (or only one) to describe oscillatory activity. Alternatively, feature selection or channel selection techniques can also be used to deal with this high dimensionality (Lotte et al. 2009a; Rakotomamonjy and Guigue 2008; Krusienski et al. 2006). As for BCI based on oscillatory activity, spatial filters can also prove very useful.

7.4.2 Spatial Filters for ERP-based BCI

As mentioned above, with ERP the number of features is usually quite large, with many features per channel and many channels used. The tools described for oscillatory activity-based BCI, i.e., feature selection, channel selection, or spatial filtering can be used to deal with that. While feature and channel selection algorithms are the same (these are generic algorithms), spatial filtering algorithms for ERP are different. One may wonder why CSP could not be used for ERP classification. This is due to the fact that a crucial information for classifying ERP is the EEG time course. However, CSP completely ignores this time course as it only considers the average power. Therefore, CSP is not suitable for ERP classification. Fortunately, other spatial filters have been specifically designed for this task.

One useful spatial filter available is the Fisher spatial filter (Hoffmann et al. 2006). This filter uses the Fisher criterion for optimal class separability. Informally, this criterion aims at maximizing the between-class variance, i.e., the distance between the different classes (we want the feature vectors from the different classes to be as far apart from each other as possible, i.e., as different as possible) while minimizing the within-class variance, i.e., the distance between the feature vectors from the same class (we want the feature vectors from the same class to be as similar as possible). Formally, this means maximizing the following objective function:

$$J_{\text{Fisher}} = \frac{\text{tr}(S_b)}{\text{tr}(S_w)} \tag{7.9}$$

with

$$S_b = \sum_{k=1}^{N_c} p_k(\bar{x}_k - \bar{x})(\bar{x}_k - \bar{x})^T \tag{7.10}$$

and

$$S_w = \sum_{k=1}^{N_c} p_k \sum_{i \in C_k} (x_i - \bar{x}_k)(x_i - \bar{x}_k)^T \quad (7.11)$$

In these equations, S_b is the between-class variance, S_w the within-class variance, N_c is the number of classes, x_i is the ith feature vector, $\bar{v}$ is the average of all vectors v, C_k is the kth class, and p_k the probability of class k.

This criterion is widely used in machine learning in general (Duda et al. 2001) and can be used to find spatial filters such that the resulting features maximize this criterion and thus the discriminability between the classes. This is what the Fisher spatial filter does. It finds the spatial filters such that the spatially filtered EEG time course (i.e., the feature vector) is maximally different between classes, according to the Fisher criterion. This is achieved by replacing x_i (the feature vector) by wX_i (i.e., the spatially filtered signal) in Eqs. 7.10 and 7.11. This gives an objective function of the form $J(w) = \frac{w\hat{S}_b w^T}{w\hat{S}_w w^T}$, which, like the CSP algorithm, can be solved by GEVD. This has been showed to be very efficient in practice (Hoffmann et al. 2006).

Another option, that has also proved very efficient in practice, is the xDAWN spatial filter (Rivet et al. 2009). This spatial filter, also dedicated to ERP classification, uses a different criterion from that of the Fisher spatial filter. xDAWN aims at maximizing the signal-to-signal plus noise ratio. Informally, this means that xDAWN aims at enhancing the ERP response, at making the ERP more visible in the middle of the noise. Formally, xDAWN finds spatial filters that maximize the following objective function:

$$J_{\mathrm{xDAWN}} = \frac{wADD^T A^T w^T}{wXX^T w^T} \quad (7.12)$$

where A is the time course of the ERP response to detect for each channel (estimated from data, usually using a least square estimate) and D is a matrix containing the positions of target stimuli that should evoke the ERP. In this equation, the numerator represents the signal, i.e., the relevant information we want to enhance. Indeed, $wADD^T A^T w^T$ is the power of the time course of the ERP responses after spatial filtering. On the contrary, in the denominator, $wXX^T w^T$ is the variance of all EEG signals after spatial filtering. Thus, it contains both the signal (the ERP) plus the noise. Therefore, maximizing J_{xDAWN} actually maximizes the signal, i.e., it enhances the ERP response, and simultaneously minimizes the signal plus the noise, i.e., it makes the noise as small as possible (Rivet et al. 2009). This has indeed been shown to lead to much better ERP classification performance.

In practice, spatial filters have proven to be useful for ERP-based BCI (in particular for P300-based BCI), especially when little training data are available. From a theoretical point of view, this was to be expected. Actually, contrary to CSP and band power which extract nonlinear features (the power of the signal is a quadratic operation), features for ERP are all linear and linear operations are commutative. Since BCI classifiers, e.g., LDA, are generally also linear, this means that the classifier could theoretically learn the spatial filter as well. Indeed, both linearly combining the original features X for spatial filtering ($F = WX$), then linearly combining the spatially filtered signals for classification ($y = wF = w(WX) = \hat{W}X$) or directly linearly combining the original features for classification ($y = WX$) are overall a simple linear operation. If enough training data are available, the classifier, e.g., LDA, would not need spatial filtering. However, in practice, there is often little training data available, and first performing a spatial filtering eases the subsequent task of the classifier by reducing the dimensionality of the problem. Altogether, this means that with enough training data, spatial filtering for ERP may not be necessary, and leaving the classifier learn everything would be more optimal. Otherwise, if few training data are available, which is often the case in practice, then spatial filtering can benefit a lot to ERP classification (see also Rivet et al. (2009) for more discussion of this topic).

7.4.3 Summary of Signal-processing Tools for ERP-based BCI

In summary, when designing ERP-based BCI, it is important to use the temporal information. This is mostly achieved by using the amplitude of preprocessed EEG time points as features, with low-pass or band-pass filtering and downsampling as preprocessing. Feature selection algorithms can also prove useful. It is also important to consider the spatial information. To do so, either using or selecting relevant channels is useful. Using spatial filtering algorithms such as xDAWN or Fisher spatial filters can also prove a very efficient solution, particularly when little training data are available. In the following, we will briefly describe some alternative signal-processing tools that are less used but can also prove useful in practice.

7.5 Alternative Methods

So far, this chapter has described the main tools used to recognize mental states in EEG-based BCI. They are efficient and usually simple tools that have become part of the standard toolbox of BCI designers. However, there are other signal-processing tools, and in particular other kinds of features or information sources that can be exploited to process EEG signals. Without being exhaustive, this section briefly presents some of these tools for interested readers, together with corresponding references. The alternative EEG feature representations that can be used include the following four categories:

- **Temporal** representations: Temporal representations measure how the signal varies with time. Contrary to basic features used for ERP, which simply consist in the EEG time points over time, some measures have been developed in order to characterize and quantify those variations. The corresponding features include Hjorth parameters (Obermeier et al. 2001) or time domain parameters (TDP) (Vidaurre et al. 2009). Recent research results have even suggested that TDP could be more efficient that the gold-standard band-power features (Vidaurre et al. 2009; Ofner et al. 2011).

- **Connectivity** measures: They measure how much the signal from two channels are correlated, synchronized or even if one signal may be the cause of the other one. In other words, connectivity features measure how the signal of two channels are related. This is particularly useful for BCI since it is known that, in the brain, there are many long distance communications between separated areas (Varela et al. 2001). As such, connectivity features are increasingly used for BCI and seem to be a very valuable complement to traditional features. Connectivity features include coherence, phase locking values or directed transfer function (DFT) (Krusienski et al. 2012; Grosse-Wentrup 2009; Gouy-Pailler et al. 2007; Caramia et al. 2014).
- **Complexity** measures: They naturally measure how complex the EEG signal may be, i.e., they measure its regularity or how predictable it can be. This has also been shown to provide information about the mental state of the user and also proved to provide complementary information to classical features such as band-power features. The features from this category used in BCI include approximate entropy (Balli and Palaniappan 2010), predictive complexity (Brodu et al. 2012) or waveform length (Lotte 2012).
- **Chaos theory**-inspired measures: Another category of features that has been explored is chaos-related measures, which assess how chaotic the EEG signal can be, or which chaotic properties it can have. This has also been shown to extract relevant information. Examples of corresponding features include fractal dimension (Boostani and Moradi 2004) or multi-fractal cumulants (Brodu et al. 2012).

While these various alternative features may not be as efficient as the standards tools such as band-power features, they usually extract a complementary information. Consequently, using band-power features together with some of these alternative features has led to increase classification performances, higher that the performances obtained with any of these features used alone (Dornhege et al. 2004; Brodu et al. 2012; Lotte 2012).

It is also important to realize that while several spatial filters have been designed for BCI, they are optimized for a specific type of feature. For instance, CSP is the optimal spatial filter for band-power features and xDAWN or Fisher spatial filters are optimal spatial filters for EEG time points features. However, using such spatial filters with other features, e.g., with the alternative features described above, would be clearly suboptimal. Designing and using spatial filters dedicated to these alternative features are therefore necessary. Results with waveform length features indeed suggested that dedicated spatial filters for each feature significantly improve classification performances (Lotte 2012).

7.6 Discussion

Many EEG signal-processing tools are available in order to classify EEG signals into the corresponding user's mental state. However, EEG signal processing is a very difficult task, due to the noise, non-stationarity, complexity of the signals as well as due to the limited amount of training data available. As such, the existing tools are still not perfect, and many research challenges are still open. In particular, it is necessary to explore and design EEG features that are (1) more *informative*, in order to reach better performances, (2) *robust*, to noise and artifacts, in order to use the BCI outside laboratories, potentially with moving users, (3) *invariant*, to deal with non-stationarity and session-to-session transfer and (4) *universal*, in order to design subject-independent BCI, i.e., BCI that can work for any user, without the need for individual calibration. As we have seen, some existing tools can partially address, or at least, mitigate such problems. Nevertheless, there is so far no EEG signal-processing tool that has simultaneously all these properties and that is perfectly robust, invariant, and universal. Therefore, there are still exciting research works ahead.

7.7 Conclusion

In this chapter, we have provided a tutorial and overview of EEG signal-processing tools for users' mental-state recognition. We have presented the importance of the feature extraction and classification components. As we have seen, there are three main sources of information that can be used to design EEG-based BCI: (1) the spectral information, which is mostly used with band-power features; (2) the temporal information, represented as the amplitude of preprocessed EEG time points, and (3) the spatial information, which can be exploited by using channel selection and spatial filtering (e.g., CSP or xDAWN). For BCI based on oscillatory activity, the spectral and spatial information are the most useful, while for ERP-based BCI, the temporal and spatial information are the most relevant. We have also briefly explored some alternative sources of information that can also complement the 3 main sources mentioned above.

This chapter aimed at being didactic and easily accessible, in order to help people not already familiar with EEG signal processing to start working in this area or to start designing and using BCI in their own work or activities. Indeed, BCI being such a multidisciplinary topic, it is usually difficult to understand enough of the different scientific domains involved to appropriately use BCI systems. It should also be mentioned that several software tools are now freely available to help users design BCI systems, e.g., Biosig (Schlögl et al. 2007), BCI2000 (Mellinger and Schalk 2007) or OpenViBE (Renard et al. 2010). For instance, with OpenViBE, it is possible to design a new and complete BCI system without writing a single line of code. With such tools and this tutorial, we hope to make BCI design and use more accessible, e.g., to design brain-computer music interfaces (BCMI).

7.8 Questions

Please find below 10 questions to reflect on this chapter and try to grasp the essential messages:

1. Do we need feature extraction? In particular why not using the raw EEG signals as input to the classifier?
2. What part of the EEG signal-processing pipeline can be trained/optimized based on the training data?
3. Can we design a BCI system that would work for all users (a so-called subject-independent BCI)? If so, are BCI designed specifically for one subject still relevant?
4. Are univariate and multivariate feature selection methods both suboptimal in general? If so, why using one type or the other?
5. By using an inverse solution with scalp EEG signals, can I always reach a similar information about brain activity as I would get with invasive recordings?
6. What would be a good reason to avoid using spatial filters for BCI?
7. Which spatial filter to you have to try when designing an oscillatory activity-based BCI?
8. Let us assume that you want to design an EEG-based BCI, whatever its type: Can CSP be always useful to design such a BCI?
9. Among typical features for oscillatory activity-based BCI (i.e., band-power features) and ERP-based BCI (i.e., amplitude of the preprocessed EEG time points), which ones are linear and which ones are not (if applicable)?
10. Let us assume you want to explore a new type of features to classify EEG data: Could they benefit from spatial filtering and if so, which one?

References

Ang K, Chin Z, Wang C, Guan C, Zhang H (2012) Filter bank common spatial pattern algorithm on BCI competition IV datasets 2a and 2b. Front Neurosci 6. doi:10.3389/fnins.2012.00039

Arvaneh M, Guan C, Ang K, Quek H (2011) Optimizing the channel selection and classification accuracy in eeg-based BCI. IEEE Trans Biomed Eng 58:1865–1873

Baillet S, Mosher J, Leahy R (2001) Electromagnetic brain mapping. IEEE Signal Process Mag 18 (6):14–30

Balli T, Palaniappan R (2010) Classification of biological signals using linear and nonlinear features. Physiol Meas 31(7):903

Bashashati A, Fatourechi M, Ward RK, Birch GE (2007) A survey of signal processing algorithms in brain-computer interfaces based on electrical brain signals. J Neural Eng 4(2):R35–R57

Bennett KP, Campbell C (2000) Support vector machines: hype or hallelujah? ACM SIGKDD Explor Newslett 2(2):1–13

Besserve M, Martinerie J, Garnero L (2011) Improving quantification of functional networks with eeg inverse problem: evidence from a decoding point of view. Neuroimage 55(4):1536–1547

Blankertz B, Kawanabe M, Tomioka R, Hohlefeld F, Nikulin V, Müller KR (2008a) Invariant common spatial patterns: alleviating nonstationarities in brain-computer interfacing. In: Advances in neural information processing systems, vol 20. MIT Press, Cambridge

Blankertz B, Tomioka R, Lemm S, Kawanabe M, Müller KR (2008b) Optimizing spatial filters for robust EEG single-trial analysis. IEEE Signal Proc Mag 25(1):41–56

Blankertz B, Lemm S, Treder M, Haufe S, Müller KR (2010) Single-trial analysis and classification of ERP components—a tutorial. Neuroimage 51(4):1303–1309

Boostani R, Moradi MH (2004) A new approach in the BCI research based on fractal dimension as feature and adaboost as classifier. J Neural Eng 1(4):212–217

Brodu N, Lotte F, Lécuyer A (2011) Comparative study of band-power extraction techniques for motor imagery classification. In: IEEE symposium on computational intelligence, cognitive algorithms, mind, and brain (CCMB) 2011, IEEE, pp 1–6

Brodu N, Lotte F, Lécuyer A (2012) Exploring two novel features for EEG-based brain-computer interfaces: Multifractal cumulants and predictive complexity. Neurocomputing 79(1):87–94

Browne MW (2000) Cross-validation methods. J Math Psychol 44(1):108–132

Brunner C, Naeem M, Leeb R, Graimann B, Pfurtscheller G (2007) Spatial filtering and selection of optimized components in four class motor imagery EEG data using independent components analysis. Pattern Recogn Lett 28(8):957–964. doi:10.1016/j.patrec.2007.01.002. http://www.sciencedirect.com/science/article/B6V15-4MV74WJ-1/2/525b7adff6f9a8a71984d1a2e083e365

Burges CJC (1998) A tutorial on support vector machines for pattern recognition. Knowledge Discovery and Data Mining 2:121–167

Caramia N, Ramat S, Lotte F (2014) Optimizing spatial filter pairs for EEG classification based on phase synchronization. In: International conference on audio, speech and signal processing (ICASSP'2014)

Congedo M, Lotte F, Lécuyer A (2006) Classification of movement intention by spatially filtered electromagnetic inverse solutions. Phys Med Biol 51(8):1971–1989

Devlaminck D (2011) Optimization of brain-computer interfaces. PhD thesis, University of Ghent

Dornhege G, Blankertz B, Curio G, Müller K (2004) Boosting bit rates in non-invasive EEG single-trial classifications by feature combination and multi-class paradigms. IEEE Trans Biomed Eng 51(6):993–1002

Dornhege G, Blankertz B, Krauledat M, Losch F, Curio G, Müller KR (2006) Combined optimization of spatial and temporal filters for improving brain-computer interfacing. IEEE Trans Biomed Eng 53(11):2274–2281

Duda RO, Hart PE, Stork DG (2001) Pattern recognition, 2nd edn. Wiley, New York

Farquhar J, Hill N, Lal T, Schölkopf B (2006) Regularised CSP for sensor selection in BCI. In: Proceedings of the 3rd international BCI workshop

Fatourechi M, Bashashati A, Ward R, Birch G (2007) EMG and EOG artifacts in brain computer interface systems: a survey. Clin Neurophysiol 118(3):480–494

Fazel-Rezai R, Allison B, Guger C, Sellers E, Kleih S, Kübler A (2012) P300 brain computer interface: current challenges and emerging trends. Frontiers Neuroeng 5(14). doi:10.3389/fneng.2012.00014

Friedman JHK (1997) On bias, variance, 0/1-loss, and the curse-of-dimensionality. Data Min Knowl Disc 1(1):55–77

Friedrich E, Scherer R, Neuper C (2012) The effect of distinct mental strategies on classification performance for brain-computer interfaces. Int J Psychophysiol. doi:10.1016/j.ijpsycho.2012.01.014. http://www.sciencedirect.com/science/article/pii/S0167876012000165

Fukunaga K (1990) Statistical pattern recognition, 2nd edn. Academic Press Inc., San Diego

Garrett D, Peterson DA, Anderson CW, Thaut MH (2003) Comparison of linear, nonlinear, and feature selection methods for EEG signal classification. IEEE Trans Neural Syst Rehabil Eng 11:141–144

Gouy-Pailler C, Achard S, Rivet B, Jutten C, Maby E, Souloumiac A, Congedo M (2007) Topographical dynamics of brain connections for the design of asynchronous brain-computer

interfaces. In: Proceedings of the international conference on IEEE engineering in medicine and biology society (IEEE EMBC), pp 2520–2523
Grosse-Wentrup M (2009) Understanding brain connectivity patterns during motor imagery for brain-computer interfacing. In: Advances in neural information processing systems (NIPS), vol 21. MIT Press, Cambridge
Grosse-Wentrup M, Buss M (2008) Multi-class common spatial pattern and information theoretic feature extraction. IEEE Trans Biomed Eng 55(8):1991–2000
Grosse-Wentrup M, Gramann K, Wascher E, Buss M (2005) EEG source localization for brain-computer-interfaces. In: 2nd International IEEE EMBS conference on neural engineering, pp 128–131
Grosse-Wentrup M, Liefhold C, Gramann K, Buss M (2009) Beamforming in non invasive brain computer interfaces. IEEE Trans Biomed Eng 56(4):1209–1219
Guyon I, Elisseeff A (2003) An introduction to variable and feature selection. J Mach Learn Res 3:1157–1182
Hall M (2000) Correlation-based feature selection for discrete and numeric class machine learning. In: Proceedings of the 17th international conference on machine learning, pp 359–366
Haufe S, Meinecke F, Görgen K, Dähne S, Haynes JD, Blankertz B, Bießmann F (2014) On the interpretation of weight vectors of linear models in multivariate neuroimaging. NeuroImage 87:96–110
Herman P, Prasad G, McGinnity T, Coyle D (2008) Comparative analysis of spectral approaches to feature extraction for EEG-based motor imagery classification. IEEE Trans Neural Syst Rehabil Eng 16(4):317–326
Hoffmann U, Vesin J, Ebrahimi T (2006) Spatial filters for the classification of event-related potentials. In: European symposium on artificial neural networks (ESANN 2006)
Hoffmann U, Vesin JM, Ebrahimi T, Diserens K (2008) An efficient P300-based brain-computer interface for disabled subjects. J Neurosci Methods 167:115–125
Jain A, Zongker D (1997) Feature selection: evaluation, application, and small sample performance. IEEE Trans Pattern Anal Mach Intell 19(2):153–158
Jain A, Duin R, Mao J (2000) Statistical pattern recognition : a review. IEEE Trans Pattern Anal Mach Intell 22(1):4–37
Kachenoura A, Albera L, Senhadji L, Comon P (2008) ICA: a potential tool for BCI systems. IEEE Signal Process Mag 25(1):57–68
Kamousi B, Liu Z, He B (2005) Classification of motor imagery tasks for brain-computer interface applications by means of two equivalent dipoles analysis. IEEE Trans Neural Syst Rehabil Eng 13(2):166–171
Krusienski D, Sellers E, Cabestaing F, Bayoudh S, McFarland D, Vaughan T, Wolpaw J (2006) A comparison of classification techniques for the P300 speller. J Neural Eng 3:299–305
Krusienski D, McFarland D, Wolpaw J (2012) Value of amplitude, phase, and coherence features for a sensorimotor rhythm-based brain–computer interface. Brain Res Bull 87(1):130–134
Lal T, Schröder M, Hinterberger T, Weston J, Bogdan M, Birbaumer N, Schölkopf B (2004) Support vector channel selection in BCI. IEEE TBME 51(6):1003–1010
Lan T, Erdogmus D, Adami A, Mathan S, Pavel M (2007) Channel selection and feature projection for cognitive load estimation using ambulatory EEG. Comput Intell Neurosci 2007 (74895):12. doi:10.1155/2007/74895
Ledoit O, Wolf M (2004) A well-conditioned estimator for large-dimensional covariance matrices. J Multivar Anal 88(2):365–411
Lemm S, Blankertz B, Curio G, Müller KR (2005) Spatio-spectral filters for improving classification of single trial EEG. IEEE Trans Biomed Eng 52(9):1541–1548
Lotte F (2012) A new feature and associated optimal spatial filter for EEG signal classification: waveform length. In: International conference on pattern recognition (ICPR), pp 1302–1305
Lotte F, Guan C (2009) An efficient P300-based brain-computer interface with minimal calibration time. In: Assistive machine learning for people with disabilities symposium (NIPS'09 symposium)

Lotte F, Guan C (2010a) Learning from other subjects helps reducing brain-computer interface calibration time. In: International conference on audio, speech and signal processing (ICASSP'2010), pp 614–617

Lotte F, Guan C (2010b) Spatially regularized common spatial patterns for EEG classification. In: International conference on pattern recognition (ICPR)

Lotte F, Guan C (2011) Regularizing common spatial patterns to improve BCI designs: unified theory and new algorithms. IEEE Trans Biomed Eng 58(2):355–362

Lotte F, Congedo M, Lécuyer A, Lamarche F, Arnaldi B (2007) A review of classification algorithms for EEG-based brain-computer interfaces. J Neural Eng 4:R1–R13

Lotte F, Fujisawa J, Touyama H, Ito R, Hirose M, Lécuyer A (2009 a) Towards ambulatory brain-computer interfaces: A pilot study with P300 signals. In: 5th advances in computer entertainment technology conference (ACE), pp 336–339

Lotte F, Lécuyer A, Arnaldi B (2009b) FuRIA: an inverse solution based feature extraction algorithm using fuzzy set theory for brain-computer interfaces. IEEE Trans Signal Process 57 (8):3253–3263

Lotte F, Langhenhove AV, Lamarche F, Ernest T, Renard Y, Arnaldi B, Lécuyer A (2010) Exploring large virtual environments by thoughts using a brain-computer interface based on motor imagery and high-level commands. Presence: teleoperators and virtual environments 19 (1):54–70

Mason S, Birch G (2003) A general framework for brain-computer interface design. IEEE Trans Neural Syst Rehabil Eng 11(1):70–85

McFarland DJ, Wolpaw JR (2005) Sensorimotor rhythm-based brain-computer interface (BCI): feature selection by regression improves performance. IEEE Trans Neural Syst Rehabil Eng 13 (3):372–379

McFarland DJ, McCane LM, David SV, Wolpaw JR (1997) Spatial filter selection for EEG-based communication. Electroencephalogr Clin Neurophysiol 103(3):386–394

McFarland DJ, Anderson CW, Müller KR, Schlögl A, Krusienski DJ (2006) BCI meeting 2005-workshop on BCI signal processing: feature extraction and translation. IEEE Trans Neural Syst Rehabil Eng 14(2):135–138

Mellinger J, Schalk G (2007) BCI2000: a general-purpose software platform for BCI research. In: Dornhege G, Millán JR et al (eds) Toward brain-computer interfacing. MIT Press, Cambridge, pp 372–381, 21

Michel C, Murray M, Lantz G, Gonzalez S, Spinelli L, de Peralta RG (2004) EEG source imaging. Clin Neurophysiol 115(10):2195–2222

Millán J, Mourino J, Franzé M, Cincotti F, Varsta M, Heikkonen J, Babiloni F (2002) A local neural classifier for the recognition of EEG patterns associated to mental tasks. IEEE Trans Neural Networks 13(3):678–686

Miranda E, Magee W, Wilson J, Eaton J, Palaniappan R (2011) Brain-computer music interfacing (BCMI) from basic research to the real world of special needs. Music Med 3(3):134–140

Noirhomme Q, Kitney R, Macq B (2008) Single trial EEG source reconstruction for brain-computer interface. IEEE Trans Biomed Eng 55(5):1592–1601

Obermeier B, Guger C, Neuper C, Pfurtscheller G (2001) Hidden markov models for online classification of single trial EEG. Pattern Recogn Lett 22:1299–1309

Ofner P, Muller-Putz G, Neuper C, Brunner C (2011) Comparison of feature extraction methods for brain-computer interfaces. In: International BCI conference 2011

Peng H, Long F, Ding C (2005) Feature selection based on mutual information: criteria of max-dependency, max-relevance, and min-redundancy. IEEE Trans Pattern Anal Mach Intell 27 (8):1226–1238

Penny W, Roberts S, Curran E, Stokes M (2000) EEG-based communication: a pattern recognition approach. IEEE Trans Rehabilitation Eng 8(2):214–215

Pfurtscheller G, Neuper C (2001) Motor imagery and direct brain-computer communication. Proc IEEE 89(7):1123–1134

Pfurtscheller G, da Silva FHL (1999) Event-related EEG/MEG synchronization and desynchronization: basic principles. Clin Neurophysiol 110(11):1842–1857

Pudil P, Ferri FJ, Kittler J (1994) Floating search methods for feature selection with nonmonotonic criterion functions. Pattern Recogn 2:279–283

Qin L, Ding L, He B (2004) Motor imagery classification by means of source analysis for brain computer interface applications. J Neural Eng 1(3):135–141

Rakotomamonjy A, Guigue V (2008) BCI competition III: dataset II—ensemble of SVMs for BCI P300 speller. IEEE Trans Biomed Eng 55(3):1147–1154

Ramoser H, Muller-Gerking J, Pfurtscheller G (2000) Optimal spatial filtering of single trial EEG during imagined hand movement. IEEE Trans Rehabilitation Eng 8(4):441–446

Raudys SJ, Jain AK (1991) Small sample size effects in statistical pattern recognition: recommendations for practitioners. IEEE Trans Pattern Anal Mach Intell 13(3):252–264

Renard Y, Lotte F, Gibert G, Congedo M, Maby E, Delannoy V, Bertrand O, Lécuyer A (2010) OpenViBE: an open-source software platform to design, test and use brain-computer interfaces in real and virtual environments. Presence: teleoperators and virtual environments 19(1):35–53

Reuderink B, Poel M (2008) Robustness of the common spatial patterns algorithm in the BCI-pipeline. Tech. rep., HMI, University of Twente

Rivet B, Souloumiac A, Attina V, Gibert G (2009) xDAWN algorithm to enhance evoked potentials: application to brain computer interface. IEEE Trans Biomed Eng 56(8):2035–2043

Samek W, Vidaurre C, Müller KR, Kawanabe M (2012) Stationary common spatial patterns for brain–computer interfacing. J Neural Eng 9(2):026013

Sannelli C, Dickhaus T, Halder S, Hammer E, Müller KR, Blankertz B (2010) On optimal channel configurations for SMR-based brain-computer interfaces. Brain Topogr 23(2):186–193, 32

Schlögl A, Brunner C, Scherer R, Glatz A (2007) BioSig—an open source software library for BCI research. In: Dornhege G, Millán JR, Hinterberger T, McFarland DJ, Müller K-R (eds) Towards brain-computer interfacing, MIT press, Cambridge, pp 347–358, 20

Schröder M, Lal T, Hinterberger T, Bogdan M, Hill N, Birbaumer N, Rosenstiel W, Schölkopf B (2005) Robust EEG channel selection across subjects for brain-computer interfaces. EURASIP J Appl Signal Process 19:3103–3112

Tangermann M, Winkler I, Haufe S, Blankertz B (2009) Classification of artifactual ICA components. Int J Bioelectromagnetism 11(2):110–114

Thomas K, Guan C, Chiew T, Prasad V, Ang K (2009) New discriminative common spatial pattern method for motor imagery brain computer interfaces. IEEE Trans Biomed Eng 56 (11):2730–2733

Tomioka R, Dornhege G, Aihara K, Müller KR (2006) An iterative algorithm for spatio-temporal filter optimization. In: Proceedings of the 3rd international brain-computer interface workshop and training course 2006, pp 22–23

Varela F, Lachaux J, Rodriguez E, Martinerie J (2001) The brainweb: phase synchronization and large-scale integration. Nat Rev Neurosci 2(4):229–239

Vialatte F, Maurice M, Dauwels J, Cichocki A (2010) Steady-state visually evoked potentials: Focus on essential paradigms and future perspectives. Prog Neurobiol 90:418–438

Vidaurre C, Krämer N, Blankertz B, Schlögl A (2009) Time domain parameters as a feature for EEG-based brain computer interfaces. Neural Networks 22:1313–1319

Xu N, Gao X, Hong B, Miao X, Gao S, Yang F (2004) BCI competition 2003–data set IIb: enhancing P300 wave detection using ICA-based subspace projections for BCI applications. IEEE Trans Biomed Eng 51(6):1067–1072

Zhong M, Lotte F, Girolami M, Lécuyer A (2008) Classifying EEG for brain computer interfaces using Gaussian processes. Pattern Recogn Lett 29:354–359

8 An Introduction to EEG Source Analysis with an Illustration of a Study on Error-Related Potentials

Marco Congedo, Sandra Rousseau and Christian Jutten

Abstract

Over the last twenty years, blind source separation (BSS) has become a fundamental signal processing tool in the study of human electroencephalography (EEG), other biological data, as well as in many other signal processing domains such as speech, images, geophysics, and wireless. This chapter introduces a short review of brain volume conduction theory, demonstrating that BSS modeling is grounded on current physiological knowledge. Then, it illustrates a general BSS scheme requiring the estimation of second-order statistics (SOS) only. A simple and efficient implementation based on the approximate joint diagonalization of covariance matrices (AJDC) is described. The method operates in the same way in the time or frequency domain (or both at the same time) and is capable of modeling explicitly physiological and experimental source of variations with remarkable flexibility. Finally, this chapter provides a specific example illustrating the analysis of a new experimental study on error-related potentials.

8.1 Introduction

Over the last twenty years, blind source separation (BSS) has become a fundamental signal processing tool in the study of human electroencephalography (EEG), other biological data, as well as in many other signal processing domains such as speech, images, geophysics, and wireless communication (Comon and Jutten 2010). Without relying on head modeling, BSS aims at estimating both the

M. Congedo (✉) · S. Rousseau · C. Jutten
GIPSA-lab, CNRS and Grenoble University, 11 rue des Mathématiques,
Domaine universitaire—BP 46, Grenoble 38402, France
e-mail: marco.congedo@gmail.com

E.R. Miranda and J. Castet (eds.), *Guide to Brain-Computer Music Interfacing*,
DOI 10.1007/978-1-4471-6584-2_8

waveform and the scalp spatial pattern of the intracranial dipolar current responsible for the observed EEG, increasing the sensitivity and specificity of the signal received from the electrodes on the scalp. This chapter begins with a short review of brain volume conduction theory, demonstrating that BSS modeling is grounded on current physiological knowledge. We then illustrate a general BSS scheme requiring the estimation of second-order statistics (SOS) only. A simple and efficient implementation based on the approximate joint diagonalization of covariance matrices (AJDC) is described. The method operates in the same way in the time or frequency domain (or both at the same time) and is capable of modeling explicitly physiological and experimental source of variations with remarkable flexibility. Finally, we provide a specific example illustrating the analysis of a new experimental study on error-related potentials.

The AJDC method for EEG data has been reviewed and described in details in Congedo et al. (2008), based upon theoretical bases to be found in Pham (2002) and Pham and Cardoso (2001). Typically, it has been used on continuously recorded EEG (*spontaneous activity*, e.g., Van der Loo et al. 2007). An extension of the method to treat group EEG data and normative EEG data has been proposed in Congedo et al. (2010). Such group BSS approach has been used in a clinical study on obsessive-compulsive disorder in Kopřivová et al. (2011) and in a cognitive study on spatial navigation in White et al. (2012). The AJDC method has also been employed for motor imagery-based brain–computer interfaces in Gouy-Pailler et al. (2010), showing that it can be applied purposefully to event-related (de)synchronization data (*induced activity*). Extension of the method to the analysis of simultaneous multiple-subject EEG data is a current line of research in our laboratory (Chatel-Goldman et al. 2013; Congedo et al. 2011, 2012). This chapter contributes demonstrating that the AJDC method can be used purposefully on event-related potential (ERP) data as well (*evoked activity*).

8.2 Physiological Ground of BSS Modeling

It is well established that the generators of brain electric fields recordable from the scalp are macroscopic postsynaptic potentials created by assemblies of pyramidal cells of the neocortex (Speckmann and Elger 2005). Pyramidal cells are aligned and oriented perpendicularly to the cortical surface. Their synchrony is possible thanks to a dense net of local horizontal connections (mostly <1 mm). At recording distances larger than about three/four times the diameter of the synchronized assemblies, the resulting potential behaves as if it were produced by electric dipoles; all higher terms of the multipole expansion vanish, and we obtain the often invoked dipole approximation (Lopes Da Silva and Van Rotterdam 2005; Nunez and Srinivasan 2006, Chap. 3). Three physical phenomena are important for the arguments we advocate in this study. First, unless dipoles are moving, there is no appreciable delay in the scalp sensor measurement (Lopes da Silva and Van Rotterdam 2005). Second, in brain electric fields, there is no appreciable electromagnetic coupling (magnetic induction) in the frequencies up to about 1 MHz; thus, the quasi-static approximation of Maxwell equations holds throughout the spectrum of interest (Nunez and Srinivasan 2006, p. 535–540). Finally, for source oscillations below 40 Hz, it has been verified experimentally that capacitive effects are also negligible, implying that potential difference is in phase with the corresponding generator (Nunez and Srinivasan 2006, p. 61). These phenomena strongly support the *superposition principle*, according to which the relation between neocortical dipolar fields and scalp potentials may be approximated by a system of linear equations (Sarvas 1987). We can therefore employ a *linear BSS model*. Because of these properties of volume conduction, scalp EEG potentials describe an *instantaneous mixture* of the fields emitted by several dipoles extending over large cortical areas. Whether this is a great simplification, we need to keep in mind that it does not hold true for all cerebral phenomena. Rather, it does at the macroscopic spatial scale concerned by EEG.

The goal of EEG blind source separation (BSS) is to "isolate" in space and time the generators of the observed EEG as much as possible, counteracting the mixing caused by volume conduction and maximizing the signal-to-noise ratio (SNR). First explored in our laboratory during the first half of the 1980s (Ans et al. 1985; Hérault and Jutten 1986), BSS has enjoyed considerable interest worldwide only starting a decade later, inspired by the seminal papers of Jutten and Hérault (1991), Comon (1994), and Bell and Sejnowski (1995). Thanks to its flexibility and power, BSS has today greatly expanded encompassing a wide range of applications such as speech enhancement, image processing, geophysical data analysis, wireless communication, and biological signal analysis (Comon and Jutten, 2010).

8.3 The BSS Problem for EEG, ERS/ERD, and ERP

For N scalp sensors and $M \leq N$ EEG dipolar fields with fixed location and orientation in the analyzed time interval, the linear BSS model simply states the superposition principle discussed above, i.e.,

$$\boldsymbol{v}(t) = \boldsymbol{A}\boldsymbol{s}(t) + \boldsymbol{\eta}(t) \tag{8.1}$$

$\boldsymbol{v}(t) \in \Re^N$ is the *sensor measurement vector* at sample t, $\boldsymbol{A} \in \Re^{N \cdot M}$is a time-invariant full column rank *mixing matrix*, $\boldsymbol{s}(t) \in \Re^M$ holds the time course of the source components, and $\boldsymbol{\eta}(t) \in \Re^N$ is additive noise, temporally white, possibly uncorrelated with $\boldsymbol{s}(t)$ and with spatially uncorrelated components. Equation (8.1) states that each observation $\boldsymbol{v}(t)$ (EEG) is a linear combination (mixing) of sources $\boldsymbol{s}(t)$, given by the coefficients in the corresponding column of matrix$\boldsymbol{A}$. Neither $\boldsymbol{s}(t)$ nor $\boldsymbol{A}$ is known, that is why the problem is said to be *blind*. Our source estimation is given by

$$\hat{\boldsymbol{s}}(t) = \hat{\boldsymbol{B}}\boldsymbol{v}(t) \tag{8.2}$$

where $\boldsymbol{B} \in \Re^{M \cdot N}$ is called the *demixing* or *separating matrix*. This is what we want to estimate in order to recover the sources from EEG. Hereafter, the hat indicates a statistical estimation. Although this is the classical BSS model, we need a few clarifications for the EEG case: By $\boldsymbol{\eta}(t)$, we model *instrumental* noise only. In the following, we drop the $\boldsymbol{\eta}(t)$ term because the instrumental (and quantization) noise of modern EEG equipment is typically low (<1 μV). On the other hand, *biological* noise (extra-cerebral artifacts such as eye movements and facial muscle contractions) and *environmental* noise (external electromagnetic interference) may obey a mixing process as well; thus, they are generally modeled as components of $\boldsymbol{s}(t)$, along with cerebral ones. Notice that while biological and environmental noise can be identified as separated components of $\boldsymbol{s}(t)$, hence removed, source estimation will be affected by the underlying cerebral *background noise* propagating with the same coefficients as the signal (Belouchrani and Amin 1998).

8.4 A Suitable Class of Solutions to the Brain BSS Problem

To tackle problem (8.2) assuming knowledge of sensor measurement only, we need to reduce the number of admissible solutions. In this paper, we are interested in weak restrictions converging toward condition

$$\hat{\boldsymbol{s}}(t) = \boldsymbol{G}\boldsymbol{s}(t), \tag{8.3}$$

where $\boldsymbol{s}(t)$ holds the time course of the true (unknown) source processes and $\hat{\boldsymbol{s}}(t)$ our estimation, and the *system matrix*

$$\boldsymbol{G} = \hat{\boldsymbol{B}}\boldsymbol{A} \approx \boldsymbol{\Lambda}\boldsymbol{P} \tag{8.4}$$

approximates a signed scaling (a diagonal matrix $\boldsymbol{\Lambda}$) and permutation ($\boldsymbol{P}$) of the rows of $\boldsymbol{s}(t)$. Equation (8.3) is obtained by substituting (8.1) in (8.2) ignoring the noise term in the former. Whether condition (8.3) may be satisfied is a problem of *identifiability*, which establishes the theoretical ground of BSS theory (Tong, Inouye and Liu 1993; Cardoso 1998; Pham and Cardoso 2001; Pham, 2002). We will come back on how identifiability is sought in practice with the proposed BSS approach. Matching condition (8.3) implies that we can recover faithfully the source *waveform*, but only out of a *scale* (including sign) and *permutation* (order) indeterminacy. This limitation is not constraining for EEG, since it is indeed the waveform that bears meaningful physiological and clinical information. Notice the correspondence between the *m*th source, its *separating vector* (*m*th *row* of $\hat{\mathbf{B}}$), and its *scalp spatial pattern* (mixing vector), given by the *m*th *column* of $\hat{\boldsymbol{A}} = \hat{\boldsymbol{B}}^{+}$. Hereafter, superscript + indicates the Moore–Penrose pseudo-inverse. The monodimensionality of those vectors and their sign/energy indeterminacy implies the explicit modeling of the orientation and localization parameters of the *mth* source, but not its moment. This is also the case of inverse solutions with good source localization performance (Greenblatt et al. 2005). On the other hand, when we estimate current density by EEG inverse solutions, we estimate current flowing in the three orthogonal directions (hence, the filter is given by three vectors, not one as here), resulting in a considerable loss of spatial resolution. Linearity allows switching back from the source space into the sensor space. Substituting (8.2) into (8.1) and dropping the noise term in the latter yield *BSS filtering*

$$\boldsymbol{v}'(t) = \hat{\boldsymbol{A}}\boldsymbol{R}\hat{\boldsymbol{s}}(t) = \hat{\boldsymbol{A}}\boldsymbol{R}\hat{\boldsymbol{B}}\boldsymbol{v}(t),$$

where $\boldsymbol{R}$ is a diagonal matrix with *mth* diagonal element equal to 1 if the *mth* component is to be retained and equal to 0 if it is to be removed. BSS filtering is common practice to remove artifacts from the EEG data.

8.5 An Approach for Solving the BSS Problem Based on Second-Order Statistics Only

It has been known for a long time that in general, the BSS problem cannot be solved for sources that are Gaussian, independent, and identically distributed (iid) (Darmois 1953). EEG data are clearly non-iid; thus, we may proceed assuming that source components are all pair-wise uncorrelated and that either (a) within each source component, the successive samples are temporally correlated[1] (Molgedey and Schuster 1994; Belouchrani et al. 1997) or (b) samples in successive time intervals do not have the same statistical distribution, i.e., they are non-stationary (Matsuoka et al. 1995; Souloumiac 1995; Pham and Cardoso 2001). Provided that *source components have non-proportional spectra or the time courses of their variance (energy) vary differently*, one can show that SOS are sufficient for solving the source separation problem (Yeredor 2010). Since SOS are sufficient, the method is able to separate also Gaussian sources, contrary to another well-known BSS approach named *independent component analysis* (ICA: Comon and Jutten 2010). If these assumptions are fulfilled, the separating matrix can be identified uniquely; thus, source can be recovered *regardless of the true mixing process* (uniform performance property: see, e.g., Cardoso 1998) and regardless of the distribution of sources, which is a remarkable theoretical advantage. The fundamental question is therefore whether or not the above assumptions fit EEG, ERS/ERD, and ERP data.

- **Sources are uncorrelated**: This assumption may be conceived as a working assumption. In practice, the BSS output is never exactly uncorrelated, but just as uncorrelated as possible. What we try to estimate is the coherent signal of large cortical patches, enough separated in space one from the other. BSS may be conceived as a spatial filter minimizing the correlation of the observed mixtures and recovering the signal emitted from the most energetic and uncorrelated

[1] Such processes are called *colored*, in opposition to iid processes, which are called *white*.

cortical patches. For EEG data, this is an effective way to counteract the effect of volume conduction. In fact, we have seen that the brain tissue behaves approximately as a linear conductor; thus, observed potentials (mixtures) must be more correlated than the generating dipolar fields.

- **Sources are colored and/or their energy varies over time**: Observed potentials are the summation of postsynaptic potentials over large cortical areas caused by trains of action potentials carried by afferent fibers. The action potentials come in trains/rest periods, resulting in sinusoidal oscillations of the scalp potentials, with negative shifts during the train discharges and positive shifts during rest. The periodicity of trains/rest periods is deemed responsible for high-amplitude EEG rhythms (oscillations) up to about 12 Hz, whereas higher-frequency (>12 Hz) low-amplitude rhythms may result from sustained (tonic) afferent discharges (Speckmann and Elegr, 2005). There is no doubt that an important portion of spontaneous EEG activity is rhythmic, whence strongly colored (Niedermeyer 2005a; Steriade 2005; Buzsáki 2006, Chap. 6, 7). Some rhythmic waves come in more or less short bursts. Typical examples are sleep spindles (7–14 Hz) (Niedermeyer 2005b; Steriade 2005) and frontal theta (4– Hz) and beta (13–35 Hz) waves (Niedermeyer 2005a). Others are more sustained, as it is the case for slow delta (1– Hz) waves during deep sleep stages III and IV (Niedermeyer 2005b), the Rolandic mu rhythms (around 10 Hz and 20 Hz), and posterior alpha rhythms (8–12 Hz) (Niedermeyer 2005a). In all cases, brain electric oscillations are not everlasting and one can always define time intervals when rhythmic activity is present and others when it is absent or substantially reduced. Such intervals may be precisely defined based on known reactivity properties of the rhythms. For example, in event-related synchronization/desynchronization (ERD/ERS: Pfurtscheller and Lopes da Silva 1999), which are time-locked, but not phase-locked, increases and decreases of the oscillating energy (Steriade 2005) intervals may be defined before and after event onset. On the other hand, event-related potentials (ERP: Lopes Da Silva 2005), which are both time-locked and phase-locked, can be further partitioned in several successive intervals comprising the different peaks. Such source energy variation signatures can be modeled precisely by SOS, as we will show with the ensuing ErrP study.

8.6 Approximate Joint Diagonalization of Covariance Matrices (AJDC)

The SOS BSS method we are considering is consistently solved by *approximate joint diagonalization* algorithms (Cardoso and Souloumiac 1993[2]; Tichavsky and Yeredor 2009). Given a set of covariance matrices $\{\boldsymbol{C}_1, \boldsymbol{C}_2, \ldots\}$, the AJD seeks a

[2] This paper does not consider SOS but fourth-order statistics; however, the algorithms are based on approximate joint diagonalization of matrices which are the slices of the tensor of fourth-order cumulants and thus can be used for SOS matrices as well.

matrix $\hat{\boldsymbol{B}}$ such that the products $\hat{\boldsymbol{B}}\boldsymbol{C}_1\hat{\boldsymbol{B}}^T$, $\hat{\boldsymbol{B}}\boldsymbol{C}_2\hat{\boldsymbol{B}}^T, \ldots$ are as diagonal as possible (subscript "*T*" indicates matrix transposition). Given an appropriate choice of the *diagonalization set* $\{\boldsymbol{C}_1, \boldsymbol{C}_2, \ldots\}$, such matrix $\hat{\boldsymbol{B}}$ is indeed an estimation of the separating matrix in (8.2) and one obtains an estimate of the mixing matrix as $\hat{\boldsymbol{A}} = \hat{\boldsymbol{B}}^+$. Matrices in $\{\boldsymbol{C}_1, \boldsymbol{C}_2, \ldots\}$ are chosen so as to hold in the off-diagonal entries statistics describing some form of *correlation* among the sensor measurement channels; then, the AJD will vanish those terms resulting in linear combination vectors (the rows of $\hat{\boldsymbol{B}}$) extracting uncorrelated components from the observed mixture via (8.2). More particularly, the joint diagonalization is applied on matrices that *change* according to the assumptions about the source. They are those changes, when available, that provide enough information to solve the BSS problem. Formally, for AJDC, the identifiability of sources discussed above, that is, matching condition (8.3), is described by the fundamental AJD-based BSS theorem (Afsari 2008; see also Aïssa-El-Bey et al. 2008): Let matrices $\boldsymbol{S}_1$, $\boldsymbol{S}_2$, … be the K (unknown) covariance matrices of sources corresponding to the covariance matrices included in the diagonalization set and $s_{k(ij)}$ their elements. The diagonal elements of these matrices $s_{k(ii)}$ hold the source variance. The off-diagonal elements $s_{k(ij)}, i \neq j$, are null as sources are assumed to be uncorrelated. Let

$$\boldsymbol{Y} = (\boldsymbol{y}_1 \cdots \boldsymbol{y}_M)^T = \begin{pmatrix} s_{1(11)} & \cdots & s_{k(11)} \\ \vdots & \ddots & \vdots \\ s_{1(MM)} & \cdots & s_{k(MM)} \end{pmatrix} \tag{8.5}$$

be the matrix formed by stacking one below the other row vectors $\boldsymbol{y}_1, \boldsymbol{y}_2, \ldots \boldsymbol{y}_M$ constructed as shown in Fig. 8.1. Each vector $\boldsymbol{y}_m = (s_{1(mm)}, \ldots, s_{K(mm)})$ holds the *energy profile along the diagonalization set* for each source, with m:1…M and M the number of estimated sources. *The fundamental theorem says that the* m*th source can be separated as long as its energy profile vector* $\boldsymbol{y}_m$ *is not collinear*[3] *with any other vector in* $\boldsymbol{Y}$. Said differently, the wider the angle between $\boldsymbol{y}_m$ and any other vector in $\boldsymbol{Y}$, the greater the chance to separate the mth source. Even if two vectors are collinear, the other sources can still be identified.

Table 8.1 reports useful information to define an appropriate diagonalization set so as to ensure identifiability of sources.

Importantly, the two basic theoretical frameworks for working in a SOS framework reported in Table 8.1, the *coloration* and the *non-stationary*, can be combined in any reasonable way: One may estimate covariance matrices in different blocks (and/or conditions) for different frequency band-pass regions, effectively increasing the uniqueness of the source energy profile. This is for instance the path we have followed for solving the problem of separating sources generating error potentials, as we will demonstrate here below. In fact, AJDC method can be applied

[3] Two vectors are collinear if they are equal out of a scaling factor, that is, the energy profile is proportional.

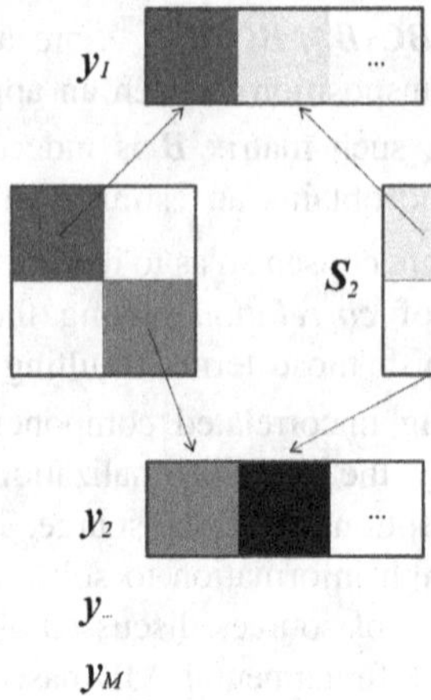

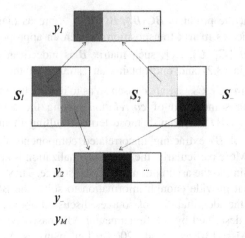

Fig. 8.1 Graphical illustration of the construction of the source energy profile vectors $\mathbf{y}_m$

in different representation spaces; applying to (1) any invertible and linearity-preserving transform T leads to

$$T[v(t)] = AT[s(t)],$$

which preserves the mixing model. Then, solving source separation in the transformed space still provides estimation of the matrix A or of its inverse B, which can be used directly in Eq. (8.2) for recovering the source $s(t)$ in the initial space. For example, the transform T may be a discrete Fourier transform, a time–frequency transform such as the Wigner–Ville transform or a wavelet transform. AJDC can be easily and conveniently transposed in the frequency domain, thence in the time-–frequency domain, whether we perform the frequency expansion for several time segments.

It is important to consider that the number of matrices should be high enough to help non-collinearity of source energy profiles. One may want to have at least as many matrices in the diagonalization set as sources to be estimated. On the other hand, one should not try to increase the number of matrices indefinitely to the detriment of the goodness of their estimation, i.e., selecting too many discrete frequencies or blocks of data that are too shorts. In summary, the key for succeeding with BSS by AJDC is the definition of an adequate size and content of the diagonalization set; it should include matrices estimated on data as homogeneous as possible for each matrix, with enough samples to allow a proper estimation, in frequency region and time blocks when the signal-to-noise ratio is high and with a high probability to uncover unique source energy profiles.

Table 8.1 Criteria to achieve identifiability of sources in BSS methods based on AJD of SOS

Assumption on the sources	Covariance matrices (CM) estimation	What is the energy profile	Sufficient condition for identifiability	Examples of data
Coloration	i. Lagged covariance matrices, ii. Fourier cospectral matrices, iii. CM estimated with a filter bank	i. The source *autocorrelation*, ii. The source *power spectrum*, iii. As in *ii.*	The power spectrum of the source is non-proportional to the power spectrum of any other sources	Spontaneous oscillation with characteristic power spectrum such as posterior dominant rhythms (alpha), somatosensory mu rhythms, frontal midline theta, beta bursts, etc.
Non-Stationary	CM estimated on i. Different *time blocks* of data ii. Different *experimental conditions*	The variation of the energy of the source along the (i) *blocks or* (ii) *experimental conditions*	The variation of the source energy along (i) *blocks or* (ii) *experimental conditions* do not correlate with the same variation of any other sources	• Blocks of data according to physiological reactivity of EEG oscillations (e.g., eyes close vs. eyes open) • CM estimated before and after the event in ERD/ERS • CM estimated on different peaks in ERP (after averaging the ERP) • Active vs. control condition –

8.7 A Study on Error-Related Potentials

We now turn to the illustration of the AJDC method by means of a new study on error-related potentials (ErrP). We show that BSS analysis increases the specificity and sensitivity that can be obtained working at the sensor level, increasing as a consequence the single-trial classification rate. ErrPs are a family of event-related potential (ERP) that can be elicited after the commission of an error, firstly reported in Miltner et al. (1997) as associated to receiving external negative feedback after error commission. This feedback error-related potential (ErrPf) is characterized by a negative deflection peaking between 250 and 400 ms with a fronto-central scalp distribution. The authors named it the feedback-related negativity (FRN) and put it in relation to the response error-related negativity (ERN) that had been previously reported (Felkenstein et al. 1991; Gehring et al. 1993), also characterized by a negative deflection. Initially, the ErrPf has been studied prevalently in the case of gambling tasks with monetary gain and loss. More recently, it has attracted much attention in the brain–computer interface (BCI) community because its online detection provides a unique opportunity to automatically correct erroneous BCI operations, effectively increasing the consistency and transfer rate of a BCI system (Farquhar and Hill 2013). In order to do so, accurate online single-trial ErrP detection is necessary. Here, we contribute along this direction in two ways: (1) We design a new experimental protocol in order to study single-trial ErrPf detection in a controlled situation that mimics actual BCI operation and (2) we apply the AJDC source analysis in order to better characterize this potential, hence increasing the accuracy of its online single-trial detection.

(1) *New Experimental Protocol*

In all previous studies on single-trial detection of ErrP for integration of a control loop in a BCI system, the involvement of the participants is very far from the involvement of participants during BCI operation, that is, as such, they lack ecological validity. In particular, in previous studies, the feedback is the main focus of the subject, while in actual BCI operations, receiving such a feedback is only a small part of a complex cognitive task. Furthermore, previous studies have mainly returned shame feedback, that is, feedback completely unrelated to the performance of the subject. Finally, the subject-specific control capability of a BCI system has not been taken into consideration. Here, we study the feedback-related potential in the case of a memory task, with no monetary gain or loss. The feedback is returned when the subject gives the answer, and no reward is given to the subject except a score; thus, our participants have no other interest besides their own performance. Such an experimental protocol allows us to study the ErrPf in a real "error versus correct" condition. The protocol we use is a memory task inducing a high cognitive load. The subject is continuously engaged in a demanding task (and not only on the feedback presentation), mimicking the actual conditions of a BCI use, where focus, concentration, and attention are essential requisite for successful BCI operation.

Then, in this study, the feedback corresponds to the actual performance achieved in the task, again approximating the actual operation of a BCI. Finally, the memory task continuously adapts to the ability of the participants during the whole experiment. This ensures that the cognitive load is approximately constant across the duration of the experiment, that it is comparable across individuals regardless of their memory span, and that the error rate across subjects is approximately equal. This latter point is particularly important in ErrP studies since it is known that the error rate affects the ErrP ([8]). In this study, the adaptive algorithm is tuned to engender an error rate of about 20 %, which amount approximately to the reasonable accuracy of a reactive BCI operation in real-world situations.

(2) *New Multivariate Signal processing Analysis*

Some of the previous studies on single-trial ErrP classification (correct vs. error) have reached encouraging results (around 70 % of overall accuracy) using only little a priori knowledge on this potential. As usual, a more profound knowledge of the electrophysiological characteristics of the ErrPf can be used to select more relevant and robust features for the purpose of single-trial online detection. Previous studies showed that the ErrP can be characterized in the temporal domain both as an ERP (time- and phase-locked event) and as an event-related synchronization, or ERS (time- but non-phase-locked event). The ERP is characterized by a negative deflection, named Ne, sometimes followed by a positive one named Pe (Gentsch et al. 2009; Steinhauser and Kiesel 2011). The ERS is characterized by an increased oscillatory activity in the theta frequency band-pass region (4–7.5 Hz) occurring approximately in the same time window and spatial location as the Ne (Trujillo and Allen 2007). Source localization of the FRN using dipole analysis has suggested generators in the anterior cingulate cortex (ACC) and the supplementary motor area (Gehring and Willoughby 2002; Miltner et al. 1997). Similar results have been obtained for the ErrPr. *Hereby, we propose a sharp spatial filtering approach based on the blind source separation approach described above with the aim to disentangling the sources responsible for the ERP and the ERS;* if this proves feasible, then the ERP and ERS components will yield independent features to feed the classifier, hence potentially increasing the online accuracy.

As a first objective, we identify the different components of the ErrP along dimensions time, space, and frequency by means of a multivariate analysis both in the sensor space and in the source space. We jointly estimate the brain sources at the origin of the ERP and ERS components and assess their different roles in error reaction. Then, we study the role of these components on the ErrP with respect to the expectation of participants. Finally, we look at how these results impact on ErrP single-trial classification, which is the essential step in integrating ErrPs in BCI systems.

8.8 Method

8.8.1 Participants

Twenty-two healthy volunteers participated in this experiment. All subjects were BCI naive at the time of the experiment, and none of them reported neurological or psychiatric disorders in their lifetime. Due to the presence of excessive artifacts in the EEG data, three subjects were subsequently excluded from all analyses, leaving 19 participants, of which 9 females and 10 males, with age ranging from 20 to 30 with a mean and a standard deviation of 24 and 2.52, respectively.

8.8.2 Experimental Design

The experiment involved two sessions lasting altogether approximately half an hour. Each session consisted of six blocks of six trials, for a total of $6 \times 6 \times 2 = 72$ trials. Participants seated comfortably 80 cm in front of a 21-inch computer screen. Nine square boxes were arranged in circle on the screen. Each trial consisted of the same memory retrieval task: The trial started with the display of the current score for 3,000 ms (initialized at zero), followed by a fixation cross, also displayed for 3,000 ms (Fig. 8.2a). Then, the memorization sequence started; each memorization comprised a random sequence of two to nine digits appearing sequentially in random positions, with each digit of the sequence randomly assigned to a different box for each sequence (Fig. 8.2b). Subjects were instructed to retain positions of all digits. At the end of the sequence, the target digit (always contained in the previous sequence) was displayed (Fig. 8.2c) and subjects had to click with the aid of a mouse on the box where it had appeared. Once the subject had answered, the interface waited for 1,500 ms in order to avoid any contamination of ErrP by beta rebound motor phenomena linked to mouse clicking (Pfurtscheller 1981). Then, if the answer was correct, the chosen box background color turned into green ("correct" feedback); otherwise, it turned into red ("error" feedback). Subjects were then asked to report if the feedback (error/correct) matched their expectation by a mouse click ("yes"/"no") (Fig. 8.2d). Following this answer, a random break of 1,000–1,500 ms preceded the beginning of the new trial.

In order to keep the subjects motivated throughout the experiment, the accumulated score was computed at the beginning of each trial. When subjects localized correctly the target digits, their score increased; otherwise, it remained unchanged. The number of digits in the sequence was always between two and nine, fixed within blocks and updated, at the beginning of each block, according to the change in performance from the block just finished and the previous one, as assessed online by means of statistical t tests. The first block started always with four digits for all subjects. The parameters of the adaptation were set, thanks to a pilot study and a computer simulation, and were chosen to yield about 20 % of errors, regardless of the working memory ability. Moreover, our learning approach is capable of adapting to

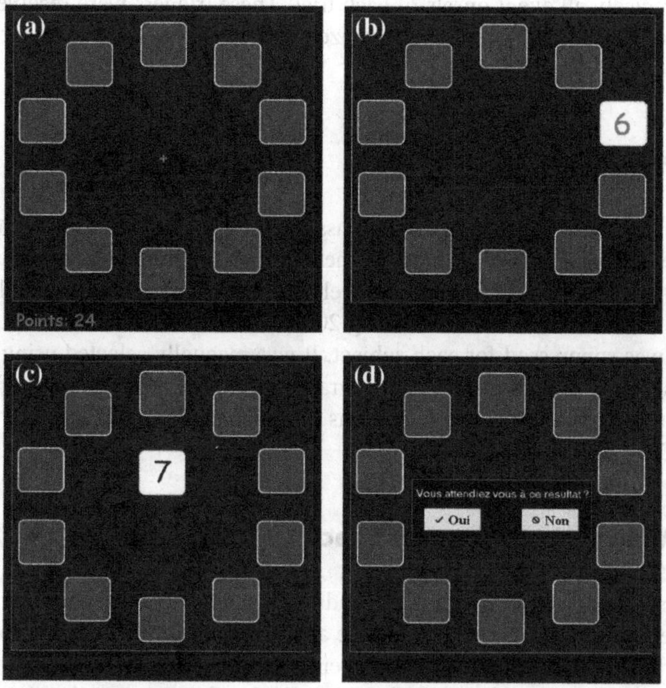

Fig. 8.2 Screenshots from the experiment representing different steps of the experiment. **a** Fixation cross. **b** One digit appearing in the memorization sequence. **c** Target digit appearing. **d** Feedback report question: 'Vous attendiez-vous à ce resultat' = 'Did you expect this result?', 'Oui' = 'Yes' and 'Non' = 'No'

fatigue as well as other possible nuisance intervening during the experiment. A random rest break was allowed between blocks, during which the boxes performed a colorful animation chosen each time at random among four preset animations. Between the two sessions, the screen was shut down to allow a rest break of 2–3 min.

8.8.3 Data Acquisition

EEG recordings were acquired from 31 silver/chloride electrodes positioned according to the extended 10/20 system (FP1, FPz, FP2, F7, F3, Fz, F4, F8, FT7, FC3, FCz, FC4, FT8, T7, C3, Cz, C4, T8, TP7, CP3, CPz, CP4, TP8, P7, P3, Pz, P4, P8, O1, Oz, O2) with the aid of a standard elastic cap. Both earlobes, digitally linked, were used as electrical reference. The ground sensor was positioned on the forehead. The impedance of each sensor was kept below 5 k. The EEG was band-pass-filtered in the range 0.1–70 Hz and digitized at 500 Hz using the Mitsar 202 DC EEG acquisition system (Mitsar Co. Ltd., Saint Petersburg, Russia). During recording, the stimulation program continuously sent to the Mitsar system triggers to track precisely all event onsets of each trial. These triggers were received by the Mitsar system as a logic signal, synchronized with the EEG stream, and recorded as a supplementary data channel.

8.8.4 Preprocessing

Data were filtered in the 1–40 Hz band-pass region using an order four Butterworth FIR filter with linear phase response in the band-pass region. Ocular artifacts were extracted using the SOBI algorithm (Belouchrani et al. 1997) available in the EEGLAB toolbox (Delorme and Makeig 2004). One EOG source corresponding to eyeblinks was suppressed for each subject. It was manually selected using both the temporal shape of the source and its topography. All other artifacts were left into the signal, so as to approximate the conditions of online analysis of EEG data acquired during BCI operation.

8.8.5 Analysis in the Sensor Space

The analysis in the sensor space is the traditional analysis of the signal as recorded at each electrode. We are interested in the analysis of the error versus correct trials. We performed both the analysis of the event-related potential (ERP: both time- and phase-locked: Lopes Da Silva 2005) and analysis of the event-related synchronization (ERS: time-locked, but not necessarily phase-locked: Pfurtscheller and Lopes da Silva 1999). ERPs were analyzed contrasting the average potential obtained from each subject at each electrode and time sample. ERSs were analyzed contrasting the average time–frequency map obtained on each trial from each subject at each electrode. In order to compute ERS, we employed a multitapering Hanning sliding window (frequency dependent, with the taper equal to four cycles for each frequency) over the 2–32 Hz band using a 1 Hz step, as implemented in the Fieldtrip software (Oostenveld et al. 2011). ERSs were computed on time window [−0.5 s 1.2 s] using a time step of 0.03 s and a baseline defined as [−1 s 0 s] prestimulus.

The statistical analysis in the sensor space for contrasting "error" versus "correct" trials needs to be performed for each electrode, discrete frequency, and time segment in the case of ERS and for each electrode and time segment for ERP data. In order to account for the extreme multiple-comparison nature of the test, we employed a permutation strategy. The test chosen is a slight modification of the supra-threshold cluster size permutation test originally proposed for neuroimaging data by Holmes et al. (1996). Here, the statistic is not the supra-threshold cluster size, but the supra-threshold cluster intensity, defined as the sum of the t values within the supra-threshold clusters. As compared to the test described by Holmes et al. (1996), such a statistic is influenced not only by the spatial extent of the clusters, but also by the strength of the effect. The test is sensitive to effects that are contiguous in space (adjacent electrodes), frequency, and time, in line with physiological considerations. The family-wise error rate for multiple comparisons was set to 0.05, meaning that the probability of falsely rejecting even only one hypothesis is less than 0.05. All permutation tests were approximated by the use of 5,000 random permutations.

8.8.6 Analysis in the Source Space

As we have seen that a spatial filter computes a weighted sum (linear combination) of the signal obtained at each electrode, potentially isolating delimited dipolar sources from each other. We apply here the method introduced above adapting it to ERP data. Our goal is to separate the source of the Ne (ERP) and the source for the theta ERS. We need to separate them one from the other, but also from background EEG activity. For our purpose, we need to include in the diagonalization set matrices holding (a) the spatial structure of the ERP component, (b) the spatial structure of the ERS component, and (c) the spatial structure of the spontaneous EEG oscillations and persistent artifacts such as lateral and horizontal eye movements, jaw muscle contractions, etc. For (a) and (b), we compute the relevant covariance matrices both on error trials and on correct trials so to exploit variations of source energy between the two conditions (Table 8.1). We define an exactly determined BSS model, that is to say, we estimate as many sources (M in the formula above) as electrodes ($N = M = 31$). For the ERP components (a), we estimate the covariance matrix of the average ERP in the three time windows where the ERP analysis in the sensor space revealed significant results (see next section). Covariance matrices were separately computed for error and correct conditions, providing $3 \times 2 = 6$ matrices. *These six matrices provide unique source energy profile about ERP that have different potential in error versus correct trials.* For the ERS component (b), we estimate the averaged covariance matrix in the time–frequency region where the sensor space analysis revealed significant results (see next section). These matrices were computed as the covariance matrices of the EEG filtered in the frequency band of interest. Again, matrices were computed separately for error and correct conditions, providing two additional matrices. *These two matrices provide unique source energy profile about ERS that display different power in the theta band in error versus correct trials.* Notice that matrices for the ERP and the ERS components are substantially different: For the ERP components, EEG trials are averaged before computing the covariance matrix (thus only both time-locked and phase-locked signals are preserved), while for the ERS components, trials are averaged only after computing covariance matrices on single-trial data (thus, non-phase-locked signals are preserved as long as they are time-locked). To separate possible sources of ERP and ERS from spontaneous EEG oscillations and artifacts (c), we include in the set all cospectral matrices (Bloomfield 2000) of the signal during the fixation cross sequence in the frequency range 2–20 Hz using a frequency step of 2 Hz, providing 10 additional matrices. *These latter 10 matrices provide unique source energy profile to separate all spontaneous sources having non-proportional power spectrum* (Table 8.1). In summary, our BSS algorithm jointly diagonalizes a total of 18 matrices. For solving the approximate joint diagonalization, we employ the iterative algorithm proposed by Tichavsky and Yeredor (2009), which is fast and in our long-lasting practice has proven robust.

Once estimated the 31 sources, they were inspected analyzing their ERP, ERS, topographies, and the mutual information criterion between the source and the error class (Grosse-Wentrup and Buss 2008). Meaningful sources were localized in a standard brain using the sLORETA inverse solution (Pascual-Marqui 2002) as implemented in the LORETA-Key software. This software makes use of revisited realistic electrode coordinates (Jurcak et al. 2007) and the head model (and corresponding lead-field matrix) produced by Fuchs et al. (2002), applying the boundary element method on the MNI-152 (Montreal neurological institute, Canada) template of Mazziotta et al. (2001). The sLORETA-key anatomical template divides and labels the neocortical (including hippocampus and anterior cingulate cortex) MNI-152 volume in 6,239 voxels of dimension 5 mm^3, based on probabilities returned by the Demon Atlas (Lancaster et al. 2000). The coregistration makes use of the correct translation from the MNI-152 space into the Talairach and Tournoux (1988) space (Brett et al. 2002). Source localization was conducted on each participant separately, normalized to unit global current density (the input of the inverse solution is a vector estimated by BSS up to a scale indeterminacy) and summed up over participants in the brain space.

8.8.7 Classification of Single Trials

For classifying single trials, data were band-pass-filtered using an order four Butterworth FIR filter with linear phase response between 1 and 10 Hz for the ERP component and 4–8 Hz for the ERS component. Data were then spatially filtered using the results of the BSS analysis. Only samples corresponding to 250–750 ms were kept. For the ERP component, we used the temporal signal down-sampled at 32 Hz, providing 16 samples (features) for the classification. For the ERS component, we used the square of the temporal signal (power) down-sampled at 32 Hz, providing 16 samples (features) for the classification as well. This procedure assigns to each component equal chance for classification. As a classifier, we employed a linear discriminant analysis (LDA). One hundred random cross-validations were performed with the classifier trained on a randomly selected set containing 80 % of the data (both errors and corrects) and then tested on the remaining data.

8.9 Results

8.9.1 Behavioral Results

All subjects performed the task with a convenient error rate, with mean (sd) = 22.2 (4) % and a quasi-equal repartition of expected and unexpected errors, with mean (sd) = 10.4 (4.3) % and 11.8(3) %, respectively. Reaction time was higher for error trials as compared to correct trials in 80 % of the subjects (all t tests with $p < 0.05$). The maximum number of digits to memorize for each subject was highly variable, ranging from 4 to 10, with mean (sd) = 6.5 (1.37). These results demonstrate that our presentation software succeeded in equalizing the cognitive load across subjects, despite the great intersubject variability of digit memory span.

8.9.2 Sensor Space Analysis

The ERP in the error trials differed from the correct trials in three time windows with different timing and/or electrode location (Fig. 8.3). A significant positivity for errors was found at time window [320 ms 400 ms] at electrode Cz ($p < 0.01$), a significant negativity for errors at time window [450 ms 550 ms] at clustered electrodes Fz, FCz, Cz ($p < 0.01$), and a significant positivity for errors at time [650 ms 775 ms] at clustered electrodes Fz, FCz ($p = 0.025$).

An ERS (power increase as compared to baseline) could be seen in the theta band in both correct and error feedback at fronto-midline locations. This synchronization unfolds from around 250–600 ms poststimulus. In some subject, it goes up to more than 200 % of power increase for error trials. Albeit present in both conditions, this ERS is significantly more intense for error trials as compared to correct ones (Fig. 8.4) in the frequency band-pass region 5–8 Hz and time window [350 ms 600 ms] poststimulus over the clustered electrodes Fz and FCz ($p = 0.015$).

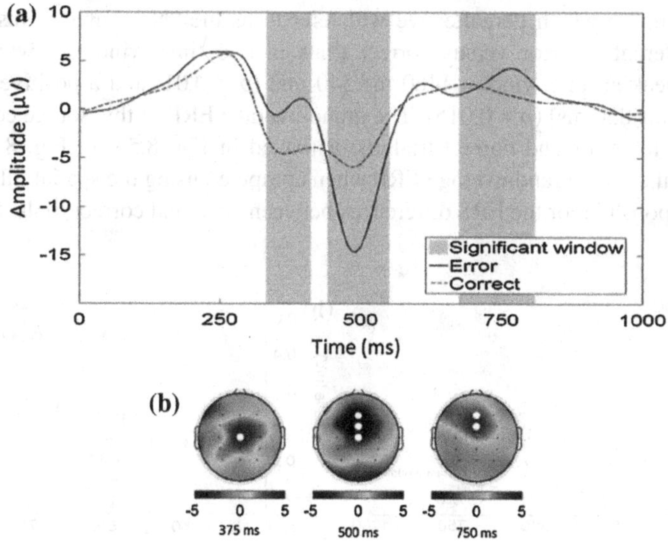

Fig. 8.3 **a** Grand-average (N = 19) ERP for correct (*pointed line*) and error (*solid line*) trials. Time windows where the difference in amplitude between the two conditions is significant (*gray panels*) and **b** scalp topographies of t values computed within the three significant windows. *White disks* show the significant clustered electrodes

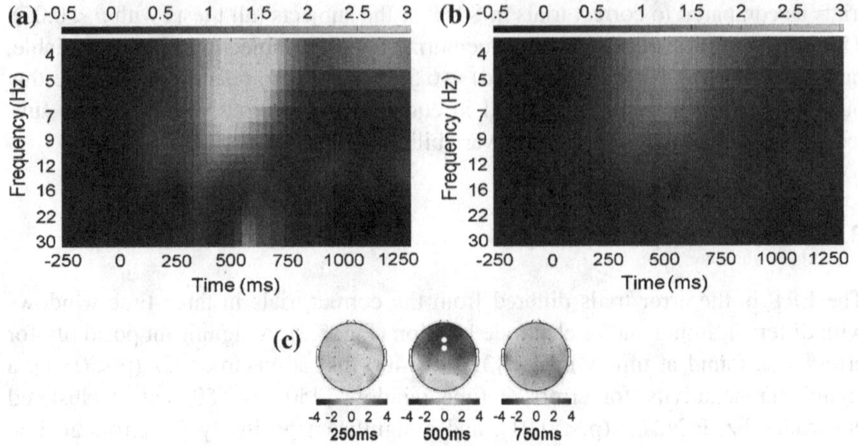

Fig. 8.4 Grand-average (N = 19) ERS averaged at electrodes (Fz, FCz, Cz, CPz) for error **a** and correct **b** trials. **c** Topographic maps of t values averaged over the theta band and time window [350 ms 600 ms]. White disks show the significant clustered electrodes

8.9.3 Source Analysis

BSS analysis revealed two uncorrelated sources with variable sensitivity and specificity, however, clearly responsible one for the ERP findings and one for the ERS findings. The source responsible for the ERP differences between error and correct trials, to which hereafter we will refer to as the "Ne source," was significantly different in error versus correct trials in two time windows, with a first negative peak at time window [460 ms 540 ms] ($p < 0.01$) and a positive peak at time [750 ms 830 ms] ($p = 0.015$). The grand-average ERP of this source computed separately for error and correct trials is displayed in Fig. 8.5a. In Fig. 8.5b, it is displayed the same grand-average ERP when computed using the spatial filter of the source responsible for the ERS differences between error and correct trials, to which

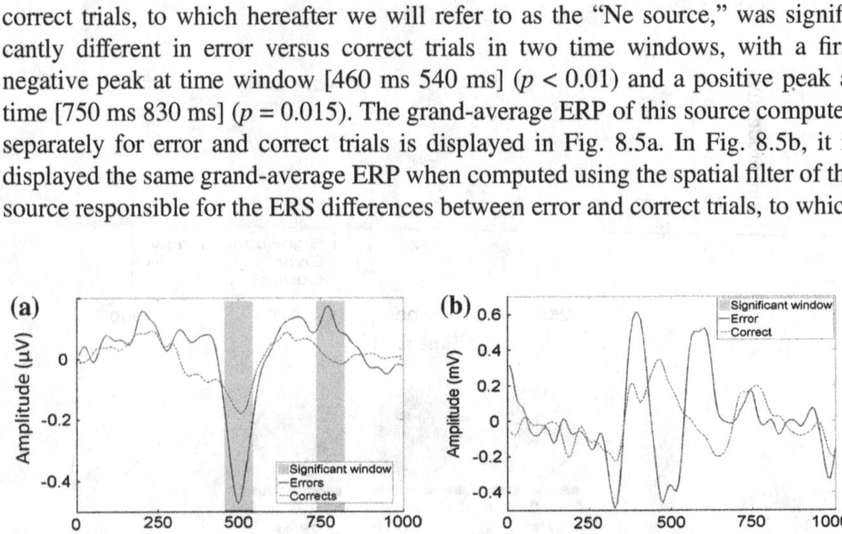

Fig. 8.5 Grand average (N = 19) of the ERP generated by the Ne source **a** and by the theta source **b** for error (*solid line*) and correct (*pointed line*) trials. Time windows where the difference in amplitude between the two conditions is significant are highlighted by *gray panels*

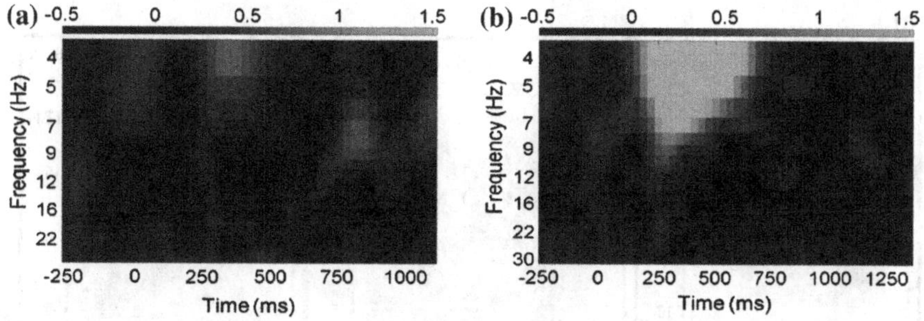

Fig. 8.6 Grand average ($N = 19$) of the ERS generated by the Ne source **a** and by the theta source **b** for error trials

hereafter we will refer to as the "theta source"; although differences in amplitude exist also for this latter source, they are not significant.

On the other hand, the theta source power increase was significant in frequency band-pass region [5 Hz 8 Hz] for time window [300 ms 600 ms] ($p < 0.01$). The ERS generated by this source is shown in Fig. 8.6b. In Fig. 8.6a, it is displayed the same ERS when computed using the spatial filter of the Ne source instead; the ERS in this case disappears. These results suggest that the Ne source and the theta source correspond to separate phenomena generated by different brain structures with different dynamics. The source responsible for the ERS (theta source) appears more specific.

We can now illustrate the advantage brought upon from the BSS analysis with these data. Compare Fig. 8.5a to 8.3 and Fig. 8.6 to 8.4. *Although in both cases, results in the sensor space are computed for the optimal cluster of electrodes, in both cases, it is clear that working in the source space allows a better sensitivity and specificity: In both cases, the difference between the error and correct trials is highlighted.*

8.9.4 Source Localization

The BSS source responsible for the ERP (Ne source) difference between correct and error trials was localized by sLORETA in the anterior cingulate gyrus (BA 24). The BSS source responsible for the ERS (theta source) was localized close to the supplementary motor area (BA 6) (Fig. 8.7). Keeping in mind the approximation of a source localization method applied on a standard head model, these anatomical results are in line with results reported by previous studies (Gehring and Willoughby 2002; Herrmann et al. 2004; Nieuwenhuis et al. 2003).

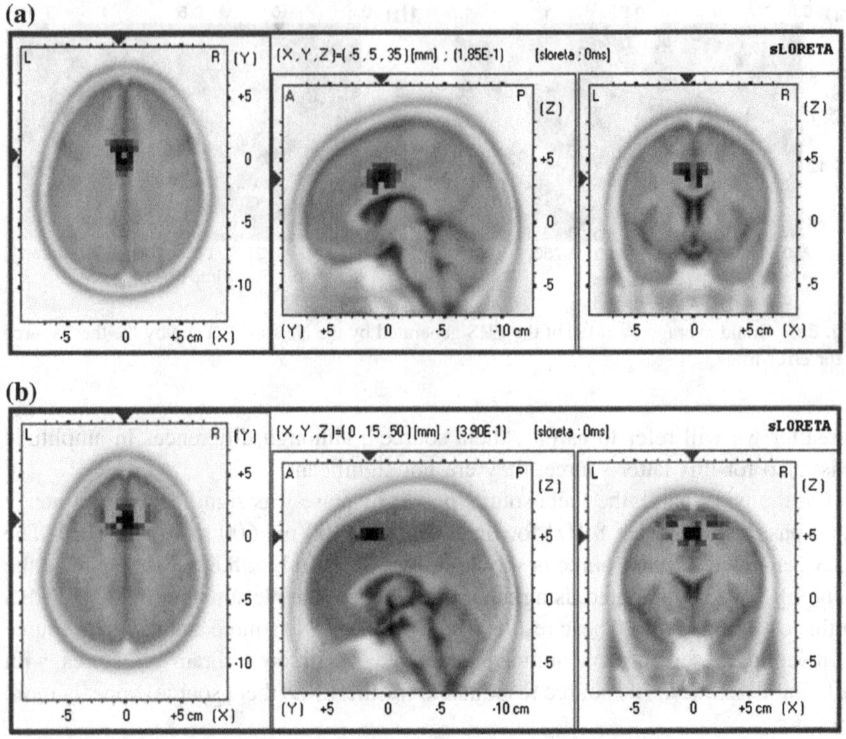

Fig. 8.7 **a** Ne source sLORETA localization. The source is localized in BA 32. **b** Theta source sLORETA localization. The source is localized in BA 6. For each image, from left to right are the axial, sagittal, and coronal views across the maximum. The images **a** and **b** are scaled to their own maximum. The activity is color-coded with *black* representing the maximum and transparent representing zero. *Legend A* = anterior; *P* = posterior; *S* = superior; *I* = inferior; *L* = left; *R* = right

8.9.5 Error Expectation

We then studied the impact of the error expectation on the two identified sources identified (Ne and ERS). Each trial could outcome one out of four results: unexpected errors (UE), expected errors (EE), expected corrects (EC), and unexpected corrects (UC). Since most subjects reported no trials from the UC condition, we only studied the first three outcomes. Only subjects providing at least four trials for each condition were kept. Three further subjects were therefore excluded from this analysis. For each component, a one-way repeated-measure ANOVA with factor "outcome feedback" at three levels was applied i) to the temporal signal averaged over time window [450 ms 520 ms] for the Ne source and ii) to the power signal filtered between 5 and 8 Hz and averaged over significant time window [300 ms 600 ms] for the theta source. For the Ne source, no significant result was found. For the theta source, the means of the three outcomes were not all equal ($F = 4.75$; $p = 0.0138$). All pair-wise post hoc tests corrected by Bonferroni method showed that the ERS engendered by this source is in this relationship: ERS(UE) > ERS (EE) > ERS(EC), with both inequality signs indicating a significant difference ($p < 0.05$).

8.9.6 Classification of Single Trials

The Ne source alone leads to better accuracy in classifying error trials as compared to the theta source alone ($p < 0.01$). The theta source leads to better accuracy for classifying correct trials ($p = 0.028$). These corroborate the conclusion that the ERP and ERS represent different phenomena of the ErrP. When looking at the average classification rate (Te + Tc)/2, with Te the classification rate of error trials and Tc the classification rate of correct trials, one see that the use of both components leads to better results for 14 subjects out of 19. The use of both components increases the mean classification rate on the 19 subjects from 67 % up to 71 %. We performed a repeated-measure two-way ANOVA with factor "type" (error vs. correct) and "feature" (Ne source ERP, theta source ERS, both). It revealed a main effect on the "type" factor ($p < 0.01$) with correct trials being better classified than error trials and a "type" x "feature" interaction (p = 0.013), demonstrating that the use of both the ERP feature and the ERS feature in the source space improves the performance of single-trial classification. It should be noticed that with a total of 72 trials per subject, training set included only a mean of 17 single trials for the error condition; thus, the classification task for this data set is hard since the training sets include very few examples of error trials.

Knowing that the error expectation has an influence on the theta ERS, we have looked at classification results for expected and unexpected errors, for the theta ERS components, and the Ne ERP components. Classification performance is higher for unexpected errors (mean Te = 62 %) than for expected errors (mean Te = 47 %) ($p = 0.011$) when using the theta ERS component. On the other hand, results are equivalent (mean Te = 63 % for unexpected errors and mean Te = 64 % for expected errors) using the Ne ERP component for classification. Thus, classification using the theta ERS component performs poorly on error trials only for expected errors. As a consequence, in the case of a system where errors are unexpected, the classification using the theta ERS component as compared to using the Ne ERP component would allow similar results for error trials and better results for corrects trials, leading to a better average classification accuracy.

8.10 Conclusions and Discussion

We have described a blind source separation approach requiring only estimation of second-order statistics of data, that is, covariance matrices. The method, which is well grounded on current theory of volume conduction, is consistently solved by means of approximate joint diagonalization of a set of covariance matrices, whence the name (AJDC). AJDC is simple, fast, and flexible, allowing explicit modeling of physiological and experimental a priori knowledge. We have argued that the success of the source separation depends solely on

- An appropriate choice of covariance matrices to form the diagonalization set;
- Their appropriate estimation.

To fulfill the first requirement, we have provided guidance for the analysis of continuously recorded EEG, event-related (de)synchronizations (ERS/D), and ERP. While studies for the first two cases are already available and well established, we have here presented for the first time the use of AJDC for ERD. We have conducted a source analysis by means of BSS of the feedback ErrP in high cognitive load conditions. In this experiment, we have used conditions that resemble those one can find on real BCI experiments. Our results showed that the feedback-related potential observed here shares the same characteristics as the FRN observed in gambling tasks and the ERN observed in reaction time tasks. Indeed, all three error potentials are notably characterized by a negative deflection generated by the dorsal ACC, but with different time of activation. A sharp analysis in the source space by means of approximate diagonalization of covariance matrices has allowed the identification of three main components accounting for the differentiation between error and correct trials. Two temporal (ERP) characteristics were identified: a first sharp negativity (Ne) and a broad positivity (Pe). One frequential (ERS) characteristic was identified as theta ERS at the same time that the Ne. This observation is in accordance with previous findings (Luu et al. 2004; Trujillo and Allen 2007) which also pointed to the implication to oscillations in the theta band as an indicator of response error-related potentials. Luu et al. (Luu et al. 2004) reported that the theta band (4–7 Hz) is responsible for most variability of the ERN (57 %); meanwhile, Trujillo and Allen (2007) reported a power increase in the theta band at a time course similar to the Ne for erroneous responses. In this paper, we have observed that the ErrPf is characterized by an important ERS in the theta band. This ERS seems to occur at the same time as the negative evoked potential. This observation leads to the question of the independence of these two components. Indeed, even if they occur simultaneously, they may represent different manifestation of the same neuronal process. Blind source separation coupled with source localization (sLORETA) has allowed the identification of two spatially distinct sources, one accounting for the temporal component (BA24) and the other for the frequency component (BA6). Statistical analysis at source level validated this separation with a significant temporal activity only for the first source exhibiting a significant ERP at the time of Ne and Pe and a significant ERS only for the second source in the theta band. The fact that these two sources are uncorrelated and spatially segregated suggests that these two phenomena do not reflect the same neuronal process. This point is of great interest for BCI applications and for the online detection of the ErrPf since they may therefore provide independent information for classification. In fact, up to now in BCI, only the negative wave (Ne) has been used as a feature for classifying the ErrPf. Our results suggest that one could use both the ERP component and the ERS component. Indeed, our classification results showed that the theta ERS brings independent information and allows better classification results (as compared with using the ERP alone). It has to be noted that while the Ne was clearly identifiable in all subjects, the Pe was not strong enough to be clearly identified in some subjects. This might explain why our BSS approach has not been successful in finding separated sources for the Pe peaks (poor signal-to-noise ratio and/or high interindividual variability).

Interestingly, we have found that the expectation of the outcome has a direct impact on the theta ERS, but not on the Ne; the more the error is expected, the weaker is the theta ERS. To our knowledge, no such effect has been reported so far. We conclude that the error-related potential may depend on two factors: the value of the observation (erroneous or correct) and the expectation of the outcome. Thus, the error-related potential may be the combination of two reactions, one to the error and the other to the surprising character of the observation. Further studies may now try to investigate this new aspect of the error-related potential and try to determine whether these two components are physiologically separated or interlaced. Within the frame of a BCI application, the more accurate the BCI is, the more unexpected the error will be. Classification results showed that when using theta component, performance is higher for unexpected errors as compared to expected errors. If the subject is concentrated and performs well the task, the occurrence of an error will be less expected, since it would result mainly from a nuisance such as an artifact decreasing the signal-to-noise ratio. Under these circumstances, the theta ERS component will be more efficient in detecting errors coming directly from the interface. In order to improve ErrP recognition in a real BCI system, the performance of the system should be maximized, so that the ErrP can be more easily detected. We conclude that the theta ERS will be stronger for high-performance BCIs and therefore that the error can be more easily detected for high-performance BCI. This fact should be taken into consideration in ensuing attempts to integrate a control loop based on ErrP detection in a BCI. More in general, the error potential should not be seen as a panacea for correcting BCI operation errors, since a high number of errors will lead to a poor detection of ErrP.

In conclusion, the AJDC method proves at the same time flexible and powerful. We hope that it turns out useful for extracting meaningful information to be used in the studies at the crossroad of music and brain electrophysiology.

8.11 Questions

1. What are the physical generators of brain electric fields recordable from the scalp?
2. Why a linear mixing model is a good approximation for the genesis of observable scalp potentials?
3. What is the relation between the mixing matrix and the demixing matrix?

4. List the main sources of instrumental, biological, and environmental noise affecting EEG recordings
5. What is an error-related potential (ErrP)?
6. The ERS associated with the ErrP is temporally related to a positive or to a negative evoked potential?
7. What are the advantages of a source-level analysis via blind source separation as compared to a sensor-level analysis?
8. Why the blind source separation method is said to be "blind"?
9. Create an experimental design where a blind source separation method exploiting source non-stationary would be appropriate for data analysis.
10. Why error-related potentials are of interest in the field of brain–computer interface?

References

Afsari B (2008) Sensitivity analysis for the problem of matrix joint diagonalization. SIAM J Matrix Ana Appl 30(3):1148–1171

Aïssa-El-Bey A, Abed-Meraim K, Grenier Y, Hua Y (2008) A general framework for second order blind separation of stationary colored sources. Sig Process 88(9):2123–2137

Ans B, Hérault J, Jutten C (1985) Adaptive neural architectures: detection of primitives. In: Proceedings of COGNITIVA, pp 593–597

Bell AJ, Sejnowski TJ (1995) An information-maximization approach to blind separation and blind deconvolution. Neural Comput 7:1129–1159

Belouchrani A, Amin MG (1998) Blind source separation based on time-frequency signal representations. IEEE Trans Signal Process 46(11):2888–2897

Belouchrani A, Abed-Meraim K, Cardoso JF, Moulines E (1997) A blind source separation technique using second order statistics'. IEEE Trans Signal Process 45(2):434–444

Bloomfield P (2000) Fourier analysis of time series. Wiley, New York

Brett M, Anton J-L, Valabregue R, Poline J-B (2002) Region of interest analysis using an SPM toolbox [abstract]. In: Presented at the 8th international conference on functional mapping of the human brain, 2–6 June 2002, Sendai, Japan. Available on CD-ROM in NeuroImage 16(2)

Buzsaki G (2006) Rhythms of the Brain. Oxford University Press, London

Cardoso J-F (1998) Blind signal separation: statistical principles. IEEE Proc 9(10):2009–2025

Cardoso JF, Souloumiac A (1993) Blind beamforming for non-Gaussian signals. IEE Proc-F (Radar Signal Process) 140(6):362–370

Chatel-Goldman J, Congedo M, Phlypo R (2013) Joint BSS as a natural analysis framework for EEG-hyperscanning (in press)

Comon P (1994) Independent component analysis, a new concept? Sig Process 36(3):287–314

Comon P, Jutten C (2010) Handbook of Blind Source Separation: Independent component analysis and applications. Academic Press, London

Congedo C, Gouy-Pailler C, Jutten C (2008) On the blind source separation of human electroencephalogram by approximate joint diagonalization of second order statistics. Clin Neurophysiol 119(12):2677–2686

Congedo M, John ER, De Ridder D, Prichep L (2010) Group independent component analysis of resting-state EEG in large normative samples. Int J Psychophysiol 78:89–99

Congedo M, Phlypo R, Chatel-Goldman J (2012) Orthogonal and non-orthogonal joint blind source separation in the least-squares sense. In: 20th European signal processing conference (EUSIPCO), 27–31 Aug, Bucharest, Romania, pp 1885–1889

Congedo M, Phlypo R, Pham D-T (2011) Approximate joint singular value decomposition of an asymmetric rectangular matrix set. IEEE Trans Signal Process 59(1):415–424
Darmois G (1953) Analyse générale des liaisons stochastiques. Rev Inst Inter Stat 21:2–8
Delorme A, Makeig S (2004) EEGLAB: an open source toolbox for analysis of single-trial EEG dynamics including independent component analysis. J Neurosci Methods 134(1):9–21
Falkenstein M, Hohnsbein J, Hoormann J (1991) Effects of crossmodal divided attention on late ERP components. II. Error processing in choice reaction tasks. Electroencephalogr Clin Neurophysiol 78:447–455
Farquhar J, Hill NJ (2013) Interactions between pre-processing and classification methods for event-related-potential classification : best-practice guidelines for brain–computer interfacing. Neuroinformatics 11(2):175–192
Fuchs M, Kastner J, Wagner M, Hawes S, Ebersole JS (2002) A standardized boundary element method volume conductor model. Clin Neurophysiol 113(5):702–712
Gehring WJ, Goss B, Coles MGH, Meyer DE, Donchin E (1993) A neural system for error detection and compensation. Psychol Sci 4(Suppl 6):385–390
Gehring WJ, Willoughby AR (2002) The medial frontal cortex and the rapid processing of monetary gains and losses. Science 295(5563):2279–2282
Gentsch A, Ullsperger P, Ullsperger M (2009) Dissociable medial frontal negativities from a common monitoring system for self- and externally caused failure of goal achievement. Neuroimage 47(4):2023–2030
Gouy-Pailler C, Congedo M, Brunner C, Jutten C, Pfurtscheller G (2010) Nonstationary brain source separation for multiclass motor imagery. IEEE Trans Biomed Eng 57(2):469–478
Greenblatt RE, Ossadtchi A, Pflieger ME (2005) Local Linear Estimators for Bioelectromagnetic Inverse Problem. IEEE Trans Signal Process 53(9):3403–3412
Grosse-Wentrup M, Buss M (2008) Multiclass common spatial patterns and information theoretic feature extraction. IEEE Trans Biomed Eng 55(8):1991–2000
Hérault J, Jutten C (1986) Space or time adaptive signal processing by neural network models. In: Proceedings of international conference neural netw computing, Snowbird (Utah), vol 151, pp 206–211
Herrmann MJ, Römmler J, Ehlis AC, Heidrich A, Fallgatter AJ (2004) Source localization (LORETA) of the error-related-negativity (ERN/Ne) and positivity (Pe). Cogn Brain Res 20 (2):294–299
Holmes AP, Blair RC, Watson JDG, Ford I (1996) Non-parametric analysis of statistic images from functional mapping experiments. J Cereb Blood Flow Metab 16:7–22
Jurcak V, Tsuzuki D, Dan I (2007) 10/20, 10/10, and 10/5 systems revisited: their validity as relative head-surface-based positioning systems. Neuroimage 34(4):1600–1611
Jutten C, Hérault J (1991) Blind separation of sources, Part 1: an adaptive algorithm based on neuromimetic architecture. Signal Process 24(1):1–10
Kopřivová J, Congedo M, Horáček J, Praško J, Raszka M, Brunovský M, Kohútová B, Höschl C (2011) EEG source analysis in obsessive–compulsive disorder. Clin Neurophysiol 122 (9):1735–1743
Lancaster JL, Woldor MG, Parsons LM, Liotti M, Freitas CS, Rainey L, Kochunov PV, Nickerson D, Mikiten SA, Fox PT (2000) Automated Talairach atlas labels for functional brain mapping. Hum Brain Mapp 10(3):120–131
Lopes da Silva F (2005) Event related potentials: methodology and quantification. In: Niedermeyer E, Lopes da Silva F (eds) Electroencephalography. Basic principles, clinical applications, and related fields, 5th edn. Lippincott Williams & Wilkins, New York, pp 991–1001
Lopes da Silva F, Van Rotterdam A (2005) Biophysical aspects of EEG and magnetoencephalogram generation. In: Niedermeyer E, Lopes da Silva F (eds) Electroencephalography. Basic principles, clinical applications, and related fields, 5th edn. Lippincott Williams & Wilkins, New York, pp 107–125
Luu P, Tucker DM, Makeig S (2004) Frontal midline theta and the error-related negativity: neurophysiological mechanisms of action regulation. Clin Neurophysiol 115(8):1821–1835

Matsuoka K, Ohya M, Kawamoto M (1995) A neural net for blind separation of nonstationary signals. Neural Netw 8(3):411–419

Mazziotta J, Toga A, Evans A, Fox P, Lancaster J, Zilles K, Woods R, Paus T, Simpson G, Pike B et al (2001) A probabilistic atlas and reference system for the human brain: International consortium for brain mapping (icbm). Philos Trans R Soc Lond Ser B: Biol Sci 356 (1412):1293–1322

Miltner WHR, Braun CH, Coles MGH (1997) Event-related brain potentials following incorrect feedback in a time-estimation task: evidence for a generic neural system for error detection. J Cogn Neurosci 9(6):788–798

Molgedey L, Schuster HG (1994) Separation of a mixture of independent signals using time delayed correlations. Phys Rev Lett 72:3634–3636

Niedermeyer E (2005a) The normal EEG of the waking Adult. In: Niedermeyer E, Lopes da Silva F (eds) Electroencephalography. Basic principles, clinical applications, and related fields, 5th edn. Lippincott Williams & Wilkins, New York, pp 167–191

Niedermeyer E (2005b) Sleep and EEG. In: Niedermeyer E, Lopes da Silva F (eds) Electroencephalography. Basic principles, clinical applications, and related fields, 5th edn. Lippincott Williams & Wilkins, New York, pp 193–207

Nieuwenhuis S, Yeung N, Van Den Wildenberg W, Ridderinkhof KR (2003) Electrophysiological correlates of anterior cingulate function in a go/no-go task: effects of response conflict and trial type frequency. Cogn Affect Behav Neurosci 3(1):17–26

Nunez PL, Srinivasan R (2006) Electric field of the brain, 2nd edn. Oxford University Press, New York

Oostenveld R, Fries P, Maris E, Schoffelen JM (2011) Fieldtrip: open source software for advanced analysis of MEG, EEG, and invasive electrophysiological data. Comput Intell Neurosci 1:ID156869

Pascual-Marqui RD (2002) Standardized low-resolution brain electromagnetic tomography (sLORETA): technical details. Meth Find Exp Clin Pharma 24D:5–12

Pham D-T (2002) Exploiting source non stationary and coloration in blind source separation. Digital Signal Process 1:151–154

Pham D-T, Cardoso J-F (2001) Blind separation of instantaneous mixtures of non stationary sources. IEEE Trans Signal Process 49(9):1837–1848

Pfurtscheller G (1981) Central beta rhythm during sensorimotor activities in man. Electroencephalogr Clin Neurophysiol 51(3):253–264

Pfurtscheller G, Lopes da Silva F (1999) Event-related eeg/meg synchronization and desynchronization: basic principles. Clin Neurophysiol 110(11):1842–1857

Sarvas J (1987) Basic mathematical and electromagnetic concepts of the biomagnetic inverse problem. Phys Med Biol 32(1):11–22

Souloumiac A (1995) Blind source detection and separation using second order nonstationarity. In: Proceedings of ICASSP, pp 1912–1915

Speckmann E-J, Elger CE (2005) Introduction to the neurophysiological basis of the EEG and DC potentials. In: Niedermeyer E, Lopes da Silva F (eds) Electroencephalography. Basic principles, clinical applications, and related fields. 5th edn. Lippincott Williams & Wilkins, New York, pp 17–29

Steinhauser M, Kiesel A (2011) Performance monitoring and the causal attribution of errors. Cogn Affect Behav Neurosci 1–12

Steriade M (2005) Cellular substrates of brain rhythms. In: Niedermeyer E, Lopes da Silva F (eds) Electroencephalography. Basic principles, clinical applications, and related fields. 5th edn. Lippincott Williams & Wilkins, New York, pp 31–83

Talairach J, Tournoux P (1988) Co-planar stereotaxic atlas of the human brain: 3-dimensional proportional system: an approach to cerebral imaging. Thieme

Tichavsky T, Yeredor A ,(2009) Fast approximate joint diagonalization incorporating weight matrices. IEEE Trans Signal Process 57(3):878–889

Tong L, Inouye Y, Liu RW (1993) Waveform-preserving blind estimation of multiple independent sources. IEEE Trans Signal Process 41(7):2461–2470

Trujillo LT, Allen JJB (2007) Theta EEG dynamics of the error-related negativity. Clin Neurophysiol 118(3):645–668

Van Der Loo E, Congedo M, Plazier M, Van De Heyning P, De Ridder D (2007) Correlation between independent components of scalp EEG and intra-cranial EEG (iEEG) time series. Int J Bioelectromagn 9(4):270–275

Yeredor A (2010) Second order methods based on color. In: Comon P, Jutten C (eds) Handbook of blind source separation. Independent component analysis and application. Academic Press, Paris

White D, Congedo M, Ciorciari J, Silberstein R (2012) Brain oscillatory activity during spatial navigation: theta and gamma activity link medial temporal and parietal regions. J Cogn Neurosci 24(3):686–697

9 Feature Extraction and Classification of EEG Signals. The Use of a Genetic Algorithm for an Application on Alertness Prediction

Pierrick Legrand, Laurent Vézard, Marie Chavent, Frédérique Faïta-Aïnseba and Leonardo Trujillo

Abstract

This chapter presents a method to automatically determine the alertness state of humans. Such a task is relevant in diverse domains, where a person is expected or required to be in a particular state of alertness. For instance, pilots, security personnel, or medical personnel are expected to be in a highly alert state, and this method could help to confirm this or detect possible problems. In this work, electroencephalographic (EEG) data from 58 subjects in two distinct vigilance states (state of high and low alertness) was collected via a cap with 58 electrodes. Thus, a binary classification problem is considered. To apply the proposed approach in a real-world scenario, it is necessary to build a prediction method that requires only a small number of sensors (electrodes), minimizing the total cost and maintenance of the system while also reducing the time required to

P. Legrand (✉) · L. Vézard · M. Chavent
IMB, UMR CNRS 5251, INRIA Bordeaux Sud-Ouest, University of Bordeaux, Bordeaux, France
e-mail: pierrick.legrand@u-bordeaux.fr

L. Vézard
e-mail: laurentvezard@gmail.com

M. Chavent
e-mail: marie.chavent@u-bordeaux.fr

F. Faïta-Aïnseba
University of Bordeaux, Bordeaux, France
e-mail: frederique.faita@u-bordeaux2.fr

L. Trujillo
Instituto Tecnológico de Tijuana, Tijuana, BC, México
e-mail: leonardo.trujillo@tectijuana.edu.mx
URL: http://www.tree-lab.org

E.R. Miranda and J. Castet (eds.), *Guide to Brain-Computer Music Interfacing*,
DOI 10.1007/978-1-4471-6584-2_9

properly setup the EEG cap. The approach presented in this chapter applies a preprocessing method for EEG signals based on the use of discrete wavelet decomposition (DWT) to extract the energy of each frequency in the signal. Then, a linear regression is performed on the energies of some of these frequencies and the slope of this regression is retained. A genetic algorithm (GA) is used to optimize the selection of frequencies on which the regression is performed and to select the best recording electrode. Results show that the proposed strategy derives accurate predictive models of alertness.

9.1 Introduction

Over the last decade, human–computer interaction (HCI) has grown and matured as a field. Gone are the days when only a mouse and keyboard could be used to interact with a computer. The most ambitious of such interfaces are brain–computer interaction (BCI) systems. The goal in BCI is to allow a person to interact with an artificial system using only brain activity. The most common approach toward BCI is to analyze, categorize, and interpret electroencephalographic (EEG) signals, in such a way that they alter the state of a computer.

In particular, the objective of the present work is to study the development of computer systems for the automatic analysis and classification of mental states of vigilance; i.e., a person's state of alertness. Such a task is relevant to diverse domains, where a person is expected or required to be in a particular state. For instance, pilots, security personnel, or medical staffs are expected to be in a highly alert state, and a BCI could help confirm this or detect possible problems.

It is possible to assume that the specific topic presented in this chapter lies outside the scope of this book, entitled "Guide to Brain-Computer Music Interfacing." Nevertheless, from our point of view, many tasks have to be accomplished before any interaction between a person's brain and music can be done by using EEG signals. Suppose that we wish to develop a musical instrument that can generate music that is specifically related to the alertness of a subject. For such a system, a first objective should be to classify the EEG signals of a subject based on different levels of alertness. In order to reach this objective, informative features have to be extracted, particularly since processing raw EEG data is highly impractical, and then proceed to a final classification step using relevant mathematical concepts. However, this problem is by no means a trivial one. In fact, EEG signals are known to be highly noisy, irregular, and tend to vary significantly from person to person, making the development of general techniques a very difficult scientific endeavor. Then, it is important to find a method that is adaptable to different persons and that it provides a rapid and accurate prediction of the alertness state. For instance, a similar problem is presented by Lin et al. (2010), the authors developed a feature extraction and classification approach to classify emotional states and build an immersive multimedia system, where a user's mental states influences the musical playback. Examples such as these illustrate the importance of developing efficient and accurate recognition systems that can automatically interpret the mental state of a person through EEG measurements.

9.1.1 Electroencephalographic Signals and Previous Works

The electrical activity of the brain is divided into different oscillatory rhythms characterized by their frequency bands. The main rhythms in ascending order of frequency are delta (1–3.5 Hz), theta (4–8 Hz), alpha (8–12 Hz), and beta (12–30 Hz). Alpha waves are characteristic of a diffuse awake state for healthy subjects and can be used to discern the normal awake and relaxed states, which is the topic of this experimental study. The oscillatory alpha rhythm appears as visually observable puffs on the electroencephalogram, especially over the occipital brain areas at the back of the skull, but also under certain conditions in more frontal recordings sites. The distribution of cortical electrical activity is taken into account in the characterization of an oscillatory rhythm. This distribution can be compared between studies reported in the literature through the use of a conventional electrode placement; the international system defined by Jasper (1958) and shown in Fig. 9.1.

Furthermore, the brain electrical activity is non-stationary, as specified in Subasi et al. (2005); i.e., the frequency content of EEG signals is time varying. EEG signals are almost always pre-treated before any analysis is performed. In most cases, the Fourier transform or discrete wavelet decomposition (DWT) are used (see Sect. 9.4.1). In Subasi et al. (2005), authors use a DWT to pick out the wavelet subband frequencies (alpha, delta, theta, and beta) and use it as an input to a neural networks classifier. In Hazarika et al. (1997), coefficients of a DWT are used as features to describe the EEG signal. These features are given as an input to an artificial neural network.

In Ben Khalifa et al. (2005), the EEG signal is decomposed in 23 bands of 1 Hz (from 1 to 23 Hz) and a short term fast Fourier transformation (STFFT) is used to calculate the percentage of the power spectrum of each band. In Cecotti and Graeser (2008), a Fourier transform is used between hidden layers of a convolutional neural network to switch from the time domain to the frequency domain analysis in the network.

To predict the state of alertness, the most common method is neural networks [see for example Subasi et al. (2005) or Vuckovic et al. (2002)]. However, the disadvantage of this approach is that it requires having a large set of test subjects relative to the number of predictive variables. To avoid this problem, the authors of Subasi et al. (2005) and Vuckovic et al. (2002) split their signal into several segments of a few seconds, called "epochs." Other approaches use different statistical methods. For example, Yeo et al. (2009) uses support vector machine, Anderson and Sijercic (1996) uses autoregressive models (AR), and Obermaier et al. (2001) use hidden Markov chains.

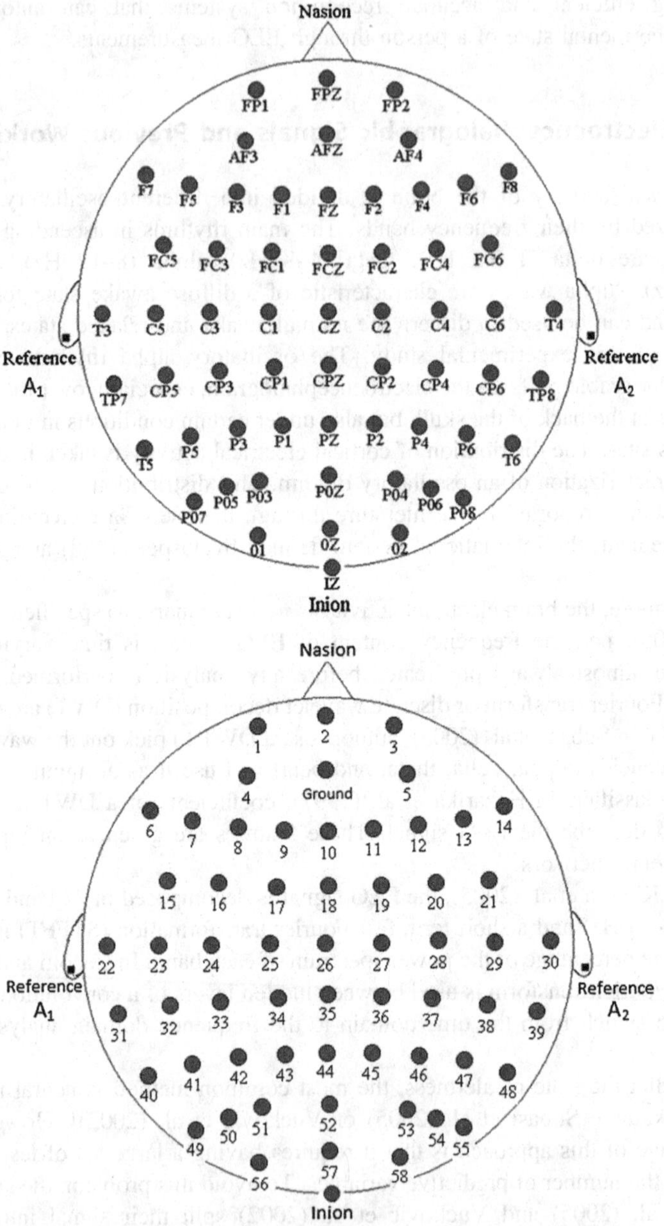

Fig. 9.1 Representation of the distribution of electrodes in the international system 10/10

9.1.2 Main Contributions

The aim of the work presented in this chapter was to construct a model that is able to predict the alertness state of a human using one electrode; and this model will be used in real time applications. That is why, the two main objectives are:

- Reduce the time needed to install the EEG cap on a participant using a variable selection method in order to choose the best electrode (based on classification rate). In fact, in real-world applications, it is necessary to reduce the number of electrodes needed because the cap installation process has to be short. A long installation of the EEG cap can cause a disturbance of the mental state of the person that we want to study (pilots or surgeons for example).
- To obtain a model (decision rule) that is able to give a reliable prediction of the alertness state of a new participant.

To achieve these objectives, we apply a wavelet decomposition as a preprocessing step and a new criterion for state discrimination is proposed. Then, several standard methods for supervised classification (binary decision tree, random forests, and others) are used to predict the state of alertness of the participants. The criterion is then refined using a genetic algorithm (GA) to improve the quality of the prediction. Finally, this work presents results that are part of a broader research program that is being investigated by the lead authors, focusing on the development of BCIs. In particular, this chapter contains a detailed description of the system originally presented in Vézard et al. (2014), where critical aspects were not discussed in detail.

The remainder of this chapter proceeds as follows. The data acquisition protocol is precisely detailed in the Sect. 9.2. The validation of the data is described in the Sect. 9.3. A data preprocessing is proposed in Sect. 9.4 and a feature extraction is performed Sect. 9.5 in order to compute a first attempt of classification of EEG signals. Section 9.6 contains the general principles of a GA and presents how this stochastic optimization method improves the results obtained in the previous section. Finally, Sect. 9.7 presents a summary of this work and discusses our main conclusions.

9.2 Data Acquisition

This work is based on real data that we have collected. This section will describe the data acquisition and data validation steps.

9.2.1 Participants

This work uses 44 participants, with ages between 18 and 35, all are right-handed, to avoid variations in the characteristics of the EEG due to age or handedness linked to a functional interhemispheric asymmetry.

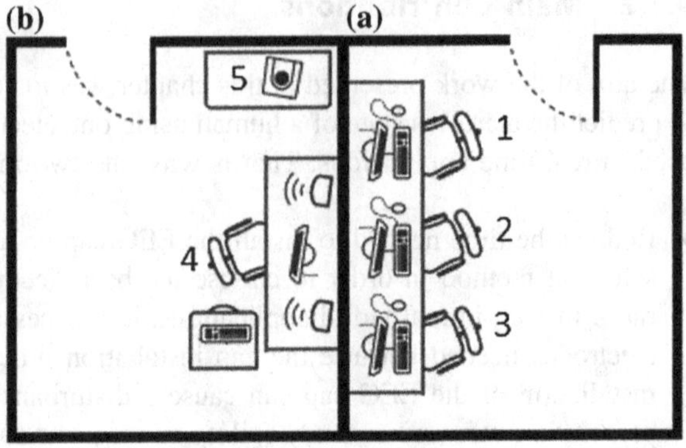

Fig. 9.2 Experimentation rooms. **a** Control room. **b** Room of the participant. *1* Recording computer. *2* Computer devoted to the relaxation process. *3* Control computer linked to the control camera. *4* Participant. *5* Control camera

9.2.2 Procedure

The experiment was conducted individually in a soundproof room, where the participant was comfortably seated in front of the computer screen (see Figs. 9.2 and 9.4).

It takes approximately 2 h and a half to place the EEG cap, to perform the experiment and to have a final explanatory interview with the participant. This interview occurred at the end of the whole data acquisition procedure to not affect EEG records. Data collection was controlled by the acquisition system Coherence 3NT (Deltamed, http://www.natus.com/). The data acquisition procedure is composed of five steps which are represented in Fig. 9.3:

1. First EEG: the participant has to look at a cross (fixation point) at the center of the screen to reduce eye movements. This first recording corresponds to the reference state, considered as the normal vigilance state of the participant. A photograph of a member of our team, took to represent the conditions of an EEG recording, is given in Fig. 9.4.

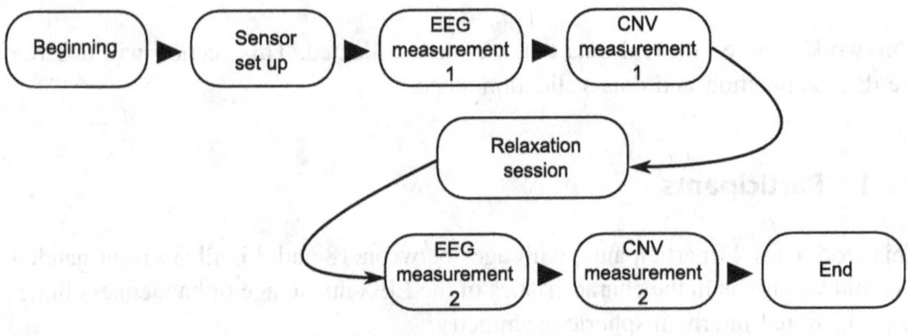

Fig. 9.3 Diagram of the data acquisition procedure

Fig. 9.4 Photograph that represents the conditions during an EEG recording

2. Attentional task devoted to collect contingent negative variation (CNV): the participant was instructed to press the space bar as quickly as possible after each time the cross was replaced by a square on the screen. For each appearance of this square, a warning sound (beep), presented 2.5 s before, allowed the participant to prepare his response. The experimental session included 50 pairs of stimuli (S1: beep, S2: square), with a random amount of time elapsing between each pair. The purpose of this task is specified in Sect. 9.3.1.
3. Relaxation session: the participant was fully guided by a soundtrack broadcast through loudspeakers placed in the room. The soundtrack suggested the participant to perform three successive exercises of self-relaxation, based on muscular relaxation and mental visualization. The first exercise is the autogenic training (Schultz 1958). In this exercise, the participant has to mentally repeating some sentences such as "I am calm" or "my arms and legs are heavy." The second exercise is the progressive relaxation (Jacobson 1974). It consists in tense and unflex some muscles of the body. The last exercise is the mental visualization. The participant imagines that he is moving in a familiar and lovely place. The purpose of this relaxation session is to try to bring the participant to a lower level of vigilance, qualified as the "relaxed" state.
4. Second EEG recording: 3 min of EEG were recorded with the same protocol as in the step 1. This second recording should reflect the relaxed state of the participant's brain if it was reached in the prior step.
5. Second CNV task: CNV is collected using exactly the same protocol as in step 2.

9.3 Data Validation

9.3.1 Contingent Negative Variation Extraction

For a given participant, the CNV analysis will allow us to determine whether the relaxation step was effective. CNV extraction has been performed by applying the event-related potentials (ERPs) method (Rosenblith 1959). It consists, in the present experimental design, on averaging the electrical activity recorded in synchrony with all warning signals (S1: beep) until the response stimulus (S2: square). Such average allows event-related brain activity components, reflecting stimulus processing, to emerge from the overall cortical electrical activity, unrelated to the task performed. Thus, in our paradigm, a negative deflection of the averaged waveform, called CNV, is obtained (Walter et al. 1964). This attentional component has the property of decreasing in amplitude when the participant is less alert, either because he is distracted (Tecce 1979), is deprived of sleep (Naitoh et al. 1971), or is falling asleep (Timsit-Berthier et al. 1981). This fundamental result is shown in Fig. 9.5. In this Figure, the CNV is plotted as a dotted line for an alert participant and as a solid line for a participant which is less alert. The amplitude of the CNV is proportional to the alertness of the subject.

That is why, although the instruction given to the participant during CNV acquisition was to press the space bar as quickly as possible after the square appearance, the reaction time is not investigated in this study. However, the way the participant prepares to perform the task is observed.

The comparison of the amplitude of the CNV between tasks performed in steps 2 and 5 is used to determine whether the alertness of a participant has changed. It allows us to know if he is actually relaxed. Only the positive cases, for which the amplitude of the CNV has significantly declined, were selected for comparative analysis of their raw EEG's (stages 1 and 4). Their EEGs were then tagged, respectively, as "normal" or "relaxed" state. An example of a participant kept after studying his CNV is shown in Fig. 9.6 and an example of a rejected participant is given in Fig. 9.7.

In these Figures, the solid curve represents the CNV recorded during step 2 and the dotted curve represents the CNV recorded in step 5. The solid vertical lines correspond to warning signals (S1: beep, S2: square). The area between the curve

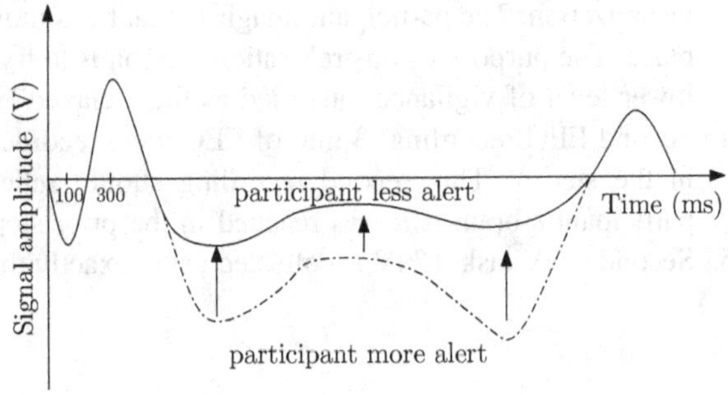

Fig. 9.5 Representation of the amplitude variation of the CNV with respect to the alertness of a participant

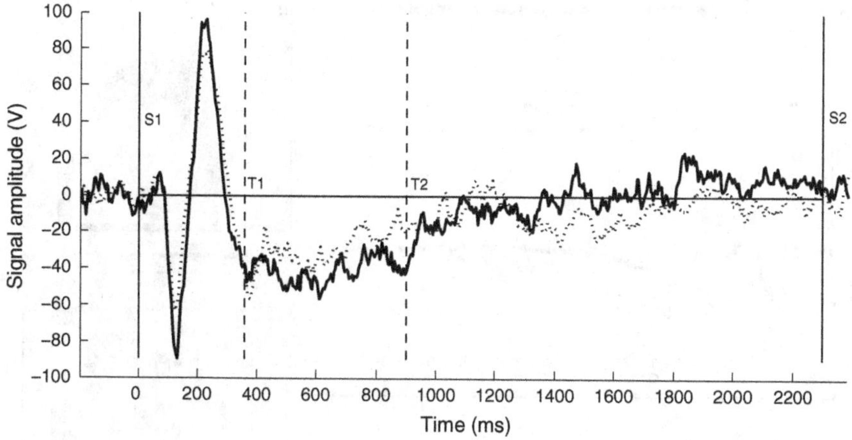

Fig. 9.6 Representation of CNV recorded on participant 4 during steps 2 (*solid curve*) and 5 (*dotted curve*). The *solid vertical lines* correspond to warning signals (S1: beep, S2: square). This participant is kept because the *solid curve* is mainly below the *dotted curve* between T1 and T2 (framed by the *dotted vertical lines*)

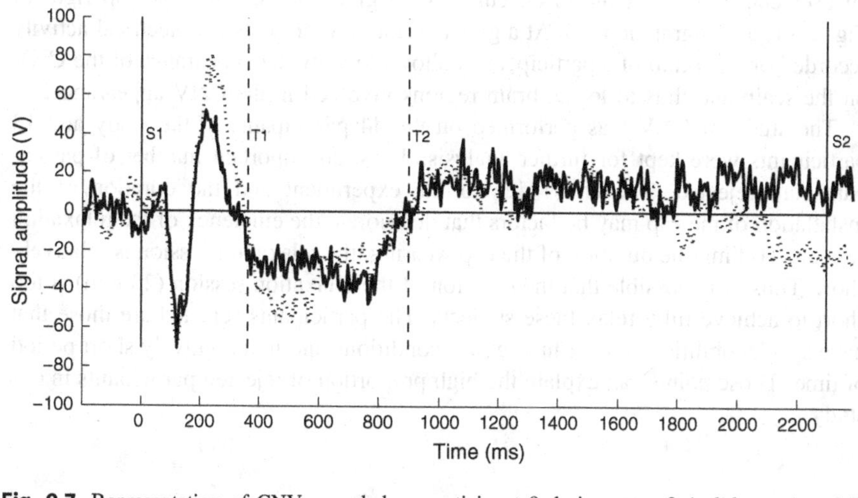

Fig. 9.7 Representation of CNV recorded on participant 9 during steps 2 (*solid curve*) and 5 (*dotted curve*). The *solid vertical lines* correspond to warning signals (S1: beep, S2: square). This participant is rejected because the *solid curve* is mainly above the *dotted curve* between T1 and T2 (framed by the *dotted vertical lines*)

and the x-axis is calculated between T1 and T2 (section framed by the dotted vertical lines). A participant is kept if the area calculated with the CNV recorded in step 5 is lower than the area calculated with the CNV recorded in step 2. To facilitate this validation step an allow a visual inspection of the curves, a graphical user interface was created. This interface is given in Fig. 9.8. Using this interface,

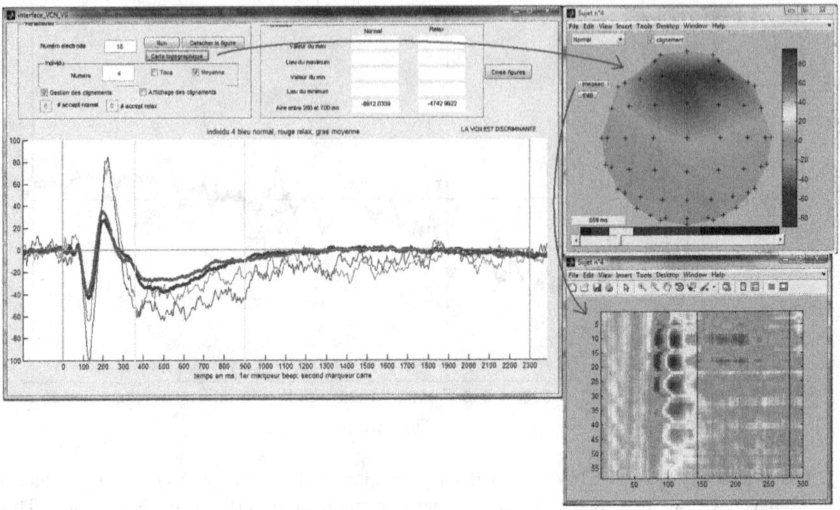

Fig. 9.8 Graphical user interface for the CNV display

an user can easily plot the CNV curve for a given participant. The top right of Fig. 9.8 is a topographic map. At a given period, it represents the electrical activity recorded on the scalp of a participant. It allows to view the appearance of the CNV on the scalp and thus to locate brain regions involved in the CNV appearance.

The study of CNV was performed on the 44 participants of the study and 13 participants were kept for further analysis. Thus, an important number of participants are rejected. The stress due to the experiment and the duration of the installation of the cap may be factors that deteriorate the efficiency of the relaxation session. To limit the duration of the cap wearing, the relaxation session is relatively short. Thus, it is possible that the duration of the relaxation session (20 min) is too short to achieve fully relax these subjects. The participants selected are those that have special abilities to relax in stressful conditions and in a relatively short period of time. Those points can explain the high proportion of rejected participants in our study.

9.3.2 Data

Finally, the data consist of 26 records of 3 min of raw EEG signals from the 13 selected participants (one "normal" EEG and one "relaxed" EEG for each participant). Each record contains variations of electric potential obtained with a sampling frequency of 256 Hz with 58 active electrodes placed on a cap (ElectroCap). Using this sampling frequency, each signal recorded by an electrode for a given subject in a given alertness state contains 46,000 data points. A representation of the data matrix is given in Fig. 9.9.

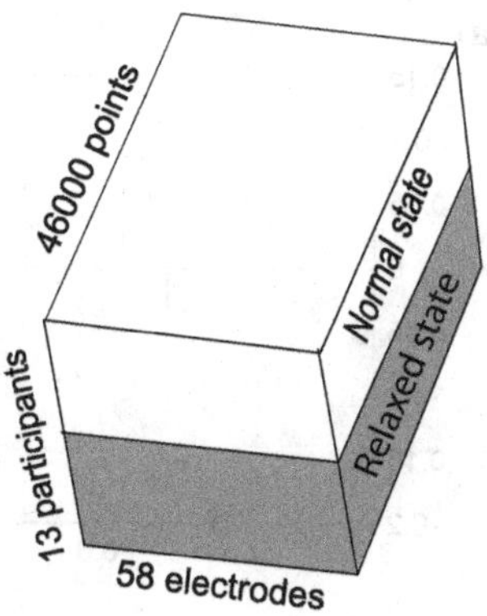

Fig. 9.9 Representation of the data matrix. There are three dimensions: one for the participants, one for the time (46,000 points corresponding to the number of points in each 3 min EEG signals recorded using a sampling frequency of 256 Hz), and one for the electrodes

9.4 Data Preprocessing

The data is specified in 3 dimensions (time, electrodes, and participants). The proposed approach is to extract a feature in 2 dimensions to implement common classification tools. To do this, the signal energy, obtained by the wavelet decomposition, is considered.

9.4.1 Wavelet Decomposition

Wavelet decomposition (Daubechies 1992; Mallat 2008) is a tool widely used in signal processing. Its main advantage is that it can be used to analyze the evolution of the frequency content of a signal in time. It is therefore more suitable than the Fourier transform for analyzing non-stationary signals.

A wavelet is a function $\psi \in L^2(\mathbb{R})$ such that $\int_{\mathbb{R}} \psi(t)\mathrm{d}t = 0$. The continuous wavelet transform of a signal X can be written as:

$$X(a,b) = \frac{1}{\sqrt{a}} \int_{-\infty}^{\infty} X(t)\psi\left(\frac{t-b}{a}\right)\mathrm{d}t \qquad (9.1)$$

where a is called the scale factor that represents the inverse of the signal frequency, b is a time-translation term and function ψ is called the mother wavelet. The mother wavelet is usually a continuous and differentiable function with compact support. Several families of wavelet mother exist such as Daubechies wavelets or Coiflets. Some wavelets are given in Fig. 9.10.

It is also possible to define the discrete wavelet transform, starting from the previous formula and discretizing parameters a and b. Then, let $a = a_0^j$, where a_0 is the resolution parameter such as $a_0 > 1$ and $j \in \mathbb{N}$ and let $b = kb_0a_0^j$, where $k \in \mathbb{N}$ and $b_0 > 0$. It is very common to consider the "dyadic" wavelet transform which

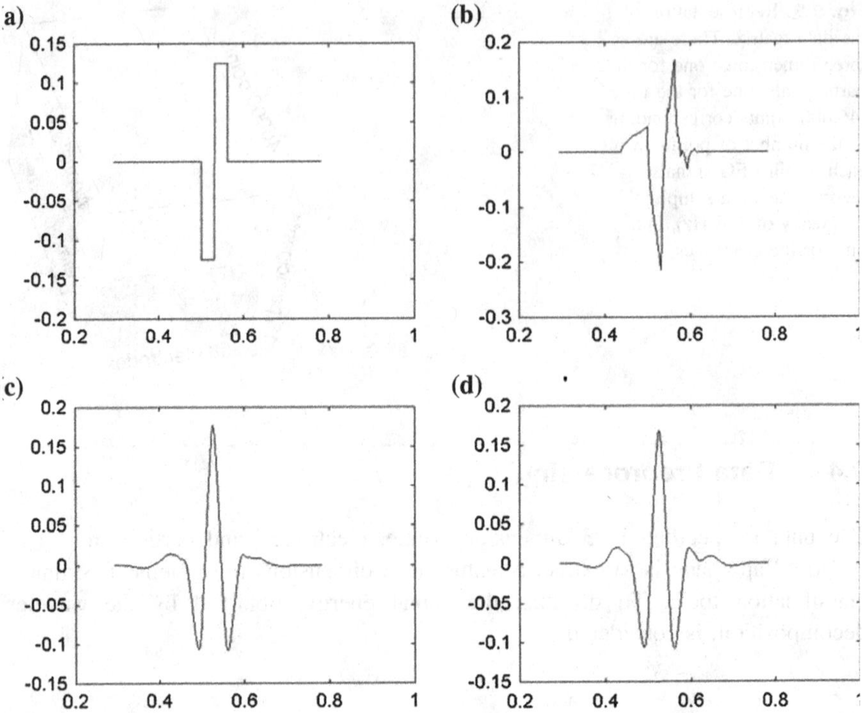

Fig. 9.10 Some wavelets. **a** Haar. **b** Daubechies4. **c** C3 Coiflet. **d** S8 Symmlet

corresponds to the case where $a_0 = 2$ and $b_0 = 1$. In this case, $j = 1, 2, \ldots, n$, where n is the base-2 logarithm of the number of points forming the signal and $k = 1, 2, \ldots, 2^{j-1}$. Then, the dyadic discrete wavelet transform is:

$$x_{j,k} = 2^{-\frac{j}{2}} \int_{-\infty}^{\infty} X(t)\psi(2^{-j}t - k)\mathrm{d}t \tag{9.2}$$

where j is the decomposition level (or scale) and k the time lag. The maximal number of decomposition levels, n, is the $\log_2$ of the number of points forming the signal. The discrete wavelet transform is faster than the continuous version and also allows for an exact reconstruction of the original signal by inverse transformation. The dyadic grid provides a spatial frequency representation of discrete dyadic wavelet transform (see Fig. 9.11). In this Figure, the x-axis corresponds to time, the y-axis represents the frequencies, and the circles correspond to the wavelet coefficients $x_{j,k}$. The signal points are represented below the last level of decomposition. At each additional level, the frequency is doubled.

The dyadic grid allows us to visualize the frequency content of the signal and to see when these frequencies appear. For example, Fig. 9.12 represents a signal and his DWT computed by the toolbox FracLab (Levy Vehel and Legrand 2004),

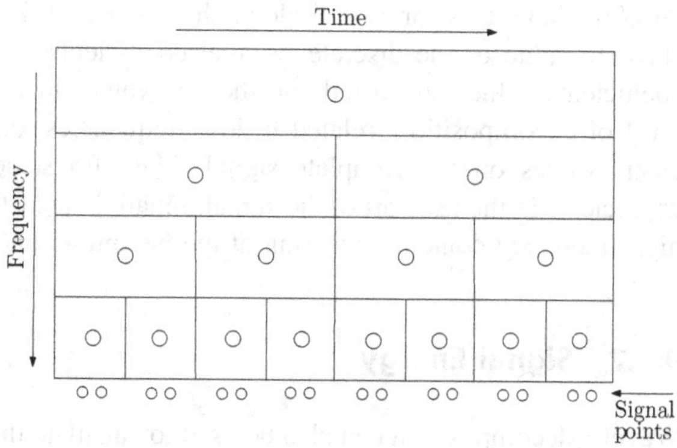

Fig. 9.11 Representation of the dyadic grid with 4 levels of decomposition (4 scales)

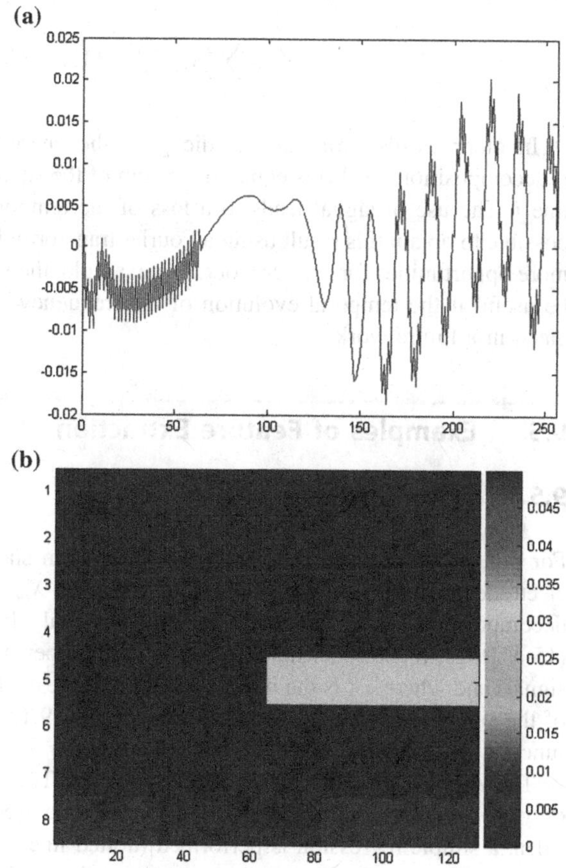

Fig. 9.12 A signal generated with the toolbox FracLab (*Top*). The dyadic grid, containing the absolute value of the discrete wavelet coefficients of the signal (*Bottom*). The large coefficients are in red and the smallest values in *blue*

http://fraclab.saclay.inria.fr/. Below, the dyadic grid is presented, containing the absolute value of the discrete wavelet coefficients of the above signal. The high coefficients values are in red and the low values in blue. In Fig. 9.12, the second level of decomposition, related to low frequencies, contains high absolute coefficients values on the complete signal. The fifth scale contains mid-range value coefficients in the last part of the signal. Finally, the last scale allows to visualize the high frequency content appearing at the beginning and at the end of the signal.

9.4.2 Signal Energy

Wavelet decomposition can also be used to calculate the energy of a signal for each level of decomposition. Thus, the energy e_j^2 of the signal X in the scale j is given by:

$$e_j^2 = \sum_{k=1}^{2^{j-1}} x_{j,k}^2, \forall j \in \{1, \ldots 2^{j-1}\} \tag{9.3}$$

In other words, from the dyadic grid, the energy associated with the scale j (decomposition level j) is equal to the sum of the squares of the coefficients of the line j. The use of signal leads to a loss of the temporality information. It is also possible to obtain this result using a Fourier transform; however, the DWT provides more opportunities for further work. For example, the wavelet decomposition could be useful if the temporal evolution of the frequency content of signals is investigated in a future work.

9.5 Examples of Feature Extraction

9.5.1 Slope Criterion

For a given participant $i (i = 1, \ldots, 13)$ in a given state (normal or relaxed), each electrode $m (m = 1, \ldots, 58)$ provides a signal X_m. A discrete dyadic wavelet decomposition is performed on this signal by considering 15 scales ($15 = \lfloor \log_2(46{,}000 \rfloor)$, where 46,000 is the number of points in each 3 min EEG signals and where $\lfloor . \rfloor$ is the integer part). From the coefficients obtained, the energy of the signal is calculated for each scale. Figure 9.13 presents these energies as a function of frequency.

The Alpha waves are between 8 and 12 Hz. Thus, according to the literature, only the energies calculated for 4, 8, and 16 Hz are used (black circles in Fig. 9.13). Then, a simple regression is performed (dotted line in Fig. 9.13), and the slope is retained. This coefficient is representative of the evolution of signal energy in the frequency considered. By repeating this process for each electrode, a feature of 58 coefficients (one per electrode) is obtained for an individual in a given state. Thus, a matrix of size 26 × 58 is obtained, representing the slope criterion.

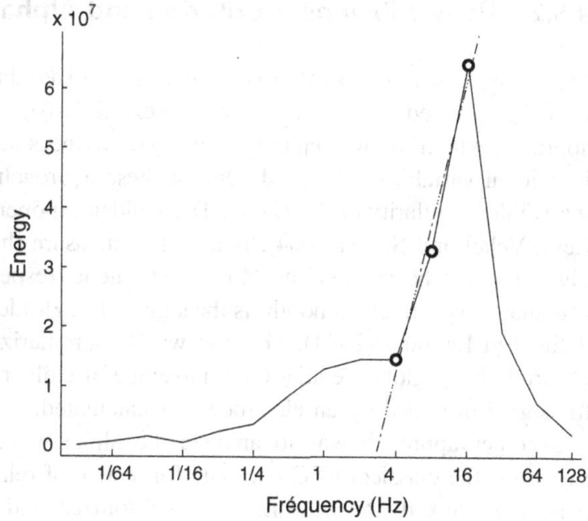

Fig. 9.13 Representation of the energy of signal X_m obtained using a discrete dyadic wavelet decomposition as a function of frequency. To calculate the slope criterion, a simple regression is performed (*dotted line*) on the energies calculated for 4, 8, and 16 Hz (*circles*)

Figure 9.14 gives a representation of the data matrix after a dimension reduction. On the left, the data obtained after a discrete wavelet transform. There are still three dimensions: one for the participants, one for the 15 frequencies, and one for the electrodes. Compared to Fig. 9.9, we switched between time (46,000 points) and frequencies (15 scales). On the right, after the calculus of the slope coefficient, only two dimensions are remaining: one for the participants and one for the electrodes.

To construct a model able to predict the alertness state, some usual classification tools (classification and regression trees or k nearest neighbors for example) will be applied on this matrix in 2 dimensions.

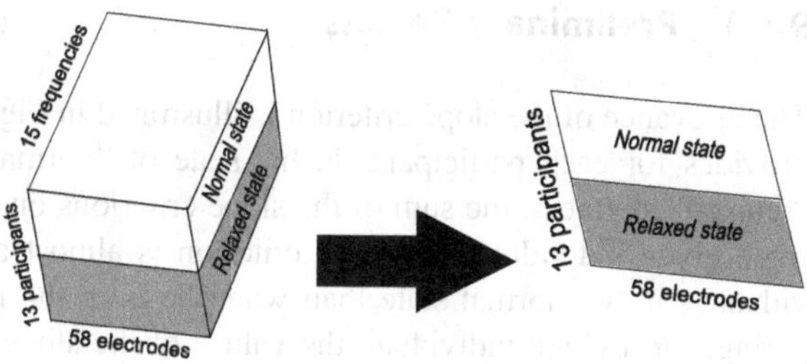

Fig. 9.14 Representation of the data matrix after a dimension reduction. On the *left*, the data obtained after a discrete wavelet transform. There are still three dimensions: one for the participants, one for the 15 frequencies, and one for the electrodes. On the *right*, after the calculus of the slope coefficient, only two dimensions are remaining: one for the participants, one for the electrodes

9.5.2 Hölder Exponent Criterion and Alpha Criterion

Previously, other approaches to obtain a summarized data matrix in two dimensions have been tested on similar signals (Vézard 2010). The goal was to obtain an approach which allows separating the two alertness states and reducing the inter-individual variability observed. One of these approaches was based on the use of the Hölder regularity of the signal. The Hölder exponent, (Jaffard and Meyer 1996, Levy Vehel and Seuret 2004), is a tool to measure the regularity of a signal at a given point. The smaller the Hölder exponent (respectively, large) is, the more irregular (respectively, smooth) is the signal. The Hölder exponent was estimated as defined in Legrand (2004). The aim was to summarize the signal recorded by an electrode in its global regularity. An average of Hölder exponents for each point of the signal provided by an electrode was calculated.

Another approach was to analyze the alpha wave content in signals. Alpha rhythm is the classical EEG correlate for a state of relaxed wakefulness. When the person is relaxed, the neurons are synchronized and operate at a particular and identical rhythm. This rhythm appears to be responsible for the more pronounced appearance of Alpha waves (Niedermeyer and Lopes da Silva 2005). When the person is forced to perform a task that can break the relaxed state, the functioning of neurons vary widely. They seem to act by groups which do not work at a similar rhythm. Alpha waves are then masked by the more pronounced appearance of other waves (such as Beta waves). Thus, the idea was to measure the proportion of alpha waves in the signal (alpha waves divided by the sum of all waves: alpha, beta, theta, and delta).

These two approaches gave a data matrix in two dimensions like that obtained with the slope criterion. However, they did not seem to work as well as the matrix of slopes to discriminate the two states of vigilance (Vézard 2010). Therefore, the slope criterion is investigated in this book chapter.

9.5.3 Preliminary Results

The relevance of the slope criterion is illustrated in Figs. 9.15 and 9.16. Figure 9.15 provides for each participant, in his state of "normal" alertness and his state of "relaxed" alertness, the sum of the slope criterions on all electrodes. It appears that for a given individual, the slope criterion is almost always lower when the individual is in the normal state than when he is in the relaxed state. Thus, by comparing, for a given individual, the values of the slope criterion for the normal and relaxed states it is possible to effectively distinguish the two states. However, for a new individual, a single record is known and the problem remains unsolved. Figure 9.16 shows for each electrode the sum of the slopes of the participants in a "normal" alertness state and participants in a "relaxed" state. The previous observation is also true at the electrode level. In fact, for a given electrode, the slope criterion is higher when considering the record obtained by this electrode after the

Fig. 9.15 Slope criterion summed over all electrodes for each of 13 participants

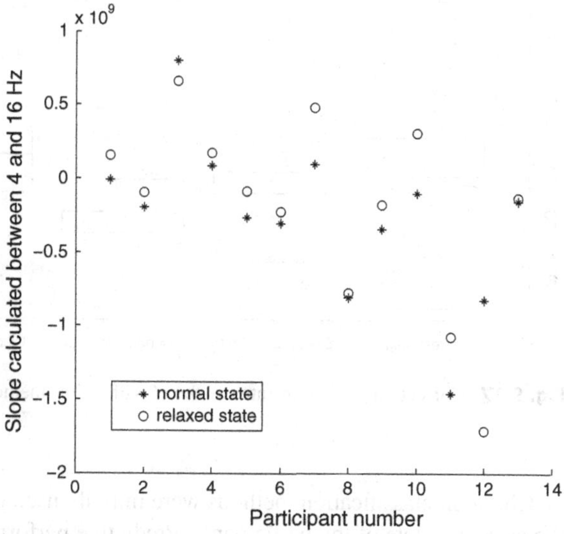

Fig. 9.16 Slope criterion summed over all participants for each of 58 electrodes

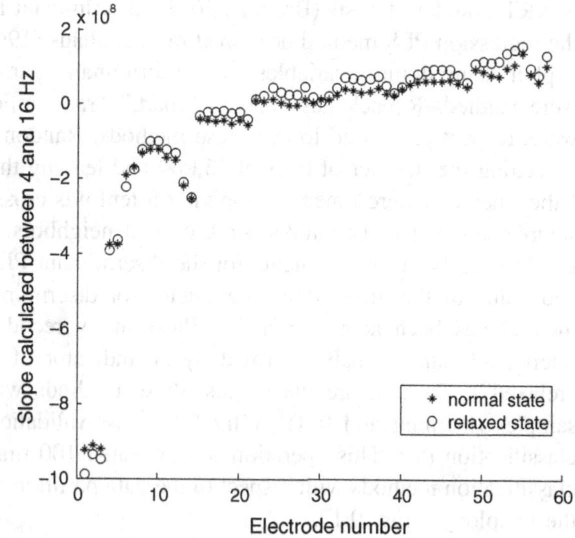

relaxation. Thus, the slope criterion can effectively discriminate the two states of alertness for an individual. However, a strong inter-individual variability can be observed in Fig. 9.15. Because of this strong individual variability, we cannot plot a line on Fig. 9.15 which separates the two alertness states (represented by cross and circles). Then, for a given subject with two EEG records, the slope criterion allows determining which record corresponds to the record done in the relaxed state. However, when only one record is known (new subject), we cannot classify it effectively.

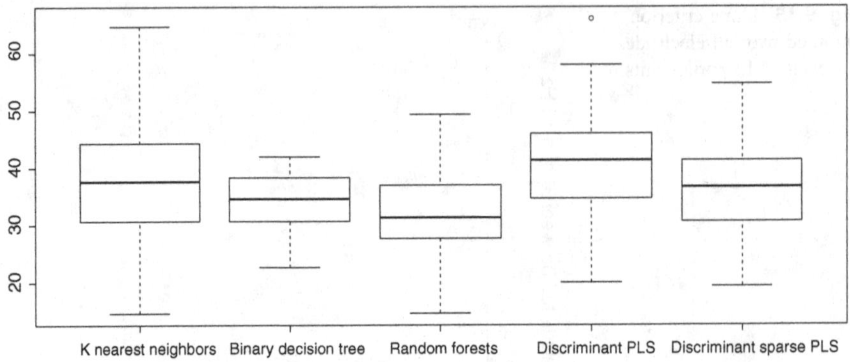

Fig. 9.17 Correct classification rate for the classification methods on the slope criterion

Common classification methods were initially used on the slope matrix to predict the alertness state of the participants. Predictive performance of *k* nearest neighbors [presented in Hastie et al. (2009)], binary decision tree (Breiman et al. 1984) (CART), random forests (Breiman 2001), discriminant PLS [by direct extension of the regression PLS method described in Tenenhaus (1998) recoding the variable to explain using dummy variables], and discriminant sparse PLS (Lé Cao et al. 2008) were studied. R packages "class," "rpart," "randomForest," "pls," and "SPLS" were, respectively, used to test these methods. Random forests have been applied by setting the number of trees at 15,000 and leaving the other settings by default. Other methods were tuned by applying a tenfolds cross-validation on the training sample (number of neighbors for *k* nearest neighbors, complexity of the tree for CART, number of components for the discriminant PLS, number of components, and value of the thresholding parameter for discriminant sparse PLS). The PLS method has been adapted for classification by recoding the variable to predict (alertness) using a matrix formed by an indicator of the modality ("normal" or "relaxed"). To compare the results, these methods were evaluated on the same samples (learning and test). A fivefold cross-validation was used to calculate a classification rate. This operation was repeated 100 times to study the stability of classification methods with respect to the data partitioning. The results are given by the boxplots in Fig. 9.17.

It appears that the median correct classification rate (CCR) is very disappointing. It does not exceed 40 % for most methods. Table 9.1 summarizes the means and standard deviations obtained using classification methods on the slope criterion. Large standard deviations reflect the influence of the data partitioning on the results. In the case of a binary prediction, these results cannot be satisfactory. It is likely that the inter-individual variability observed in Fig. 9.15 has affected the performance of the classification methods. This inter-individual variability is very difficult to include in the classification methods with the available data for this study. Therefore, the preprocessing has been refined to obtain improved classification

Table 9.1 Means and standard deviations of correct classification rate for the classification methods on the slope criterion

	K nearest neighbors	Binary decision tree	Random forests	Discriminant PLS	Sparse discriminant PLS
Mean	37.28	33.98	32.03	40.63	36.25
Standard deviation	10.47	5.15	6.46	8.55	7.96

rates. Specifically, a GA has been used as a feature selection process, to determine the electrode and the frequencies that provide the best discrimination for the slope criterion.

9.6 Feature Selection with a Genetic Algorithm

In this section, a GA is used to improve the slope criterion. So far, previous work in the field, which suggested to focus on the alpha waves, was used. For this reason, the regression was done using frequencies between 4 and 16 Hz. Given the results, this approach will be refined. The algorithm searches for the best range of frequencies (not necessarily adjacent) to perform the regression. Similarly, so far all electrodes were kept. However, one objective of this work is to remove some electrodes to reduce the time required for the installation of the cap. Thus, the best combination electrode/frequencies based on the quality of the prediction is searched for. In this work, 58 electrodes and 15 decomposition levels are available. Then, $58 * 2^{15} = 1{,}900{,}544$ ways exist to choose an electrode and a frequency range. To avoid an exhaustive search, the proposed approach is to use a GA to perform a feature selection (Broadhursta et al. 1997; Cavill et al. 2009).

9.6.1 General Principle of a Genetic Algorithm

These optimization algorithms (De Jong 1975; Holland 1975) are based on a simplified abstraction of Darwinian evolution theory. The general idea is that a population of potential solutions will improve its characteristics over time, through a series of basic genetic operations called selection, mutation and genetic recombination or crossing. From an algorithmic point of view, the general principle is depicted in Fig. 9.18.

The purpose of these algorithms is to optimize a function (fitness) within a given search space of candidate solutions. Solutions (called individuals) correspond to points within the search space, a random set of which are generated, this seeds the algorithm with an initial Population (set of individuals). They are represented by the genomes (binary codes or reals, with a fixed or variable size). All individuals are

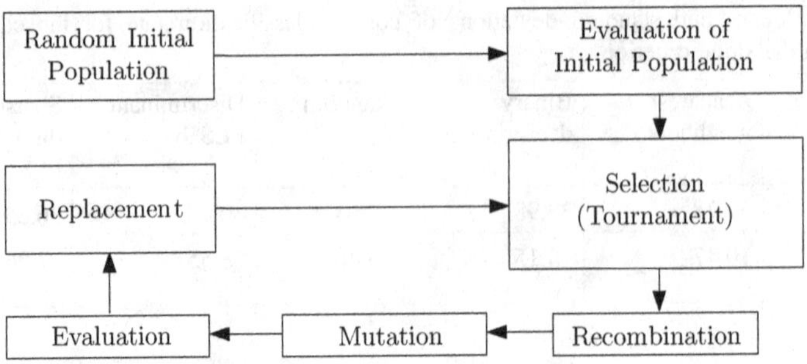

Fig. 9.18 Evolutionary loop of a basic GA

evaluated using a problem-specific objective function called fitness. Individuals are selected based on their fitness (using a series of tournaments), these selected individuals are called Parents. These parents are used to generate new individuals using two basic genetic (search) operations, recombination (random recombination of two or more individuals), and mutation (random modification of a single individual). These newly generated individuals are called Offspring, since they share (genetic) similarities with the Parents used to generate them. Finally, the best individuals (among Parents and Offspring) are selected and replace the initial population. The algorithm is iterated until a stop criterion is reached; for instance, when all individuals are identical (convergence of the algorithm) or after a pre-specified number of iterations.

9.6.2 Algorithmic Choices

In this work, the genome is composed of 16 variables: the first, an integer ranging from 1 to 58, characterizes the number of the electrode selected, the 15 others are binary and correspond to the inclusion (or not) of each frequency to compute the slope criterion. An example of a genome is given in Fig. 9.19. Each genome defines the electrode and the frequencies on which to perform the regression as illustrated in Fig. 9.20.

9.6.2.1 Genetic Operators

The main search operators used with the GA are crossover (recombination) and mutation; both are described in detail next.

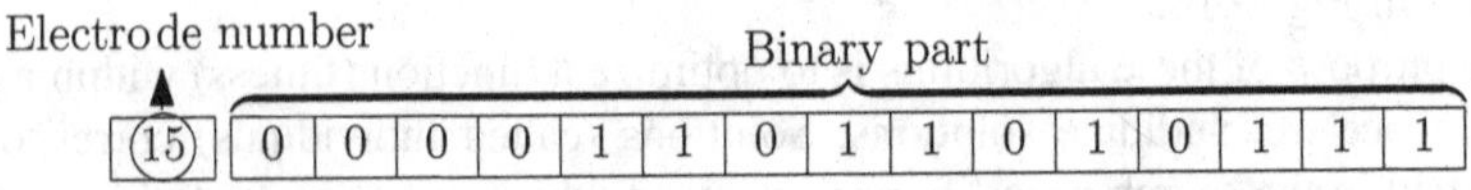

Fig. 9.19 Example of a genome in the GA

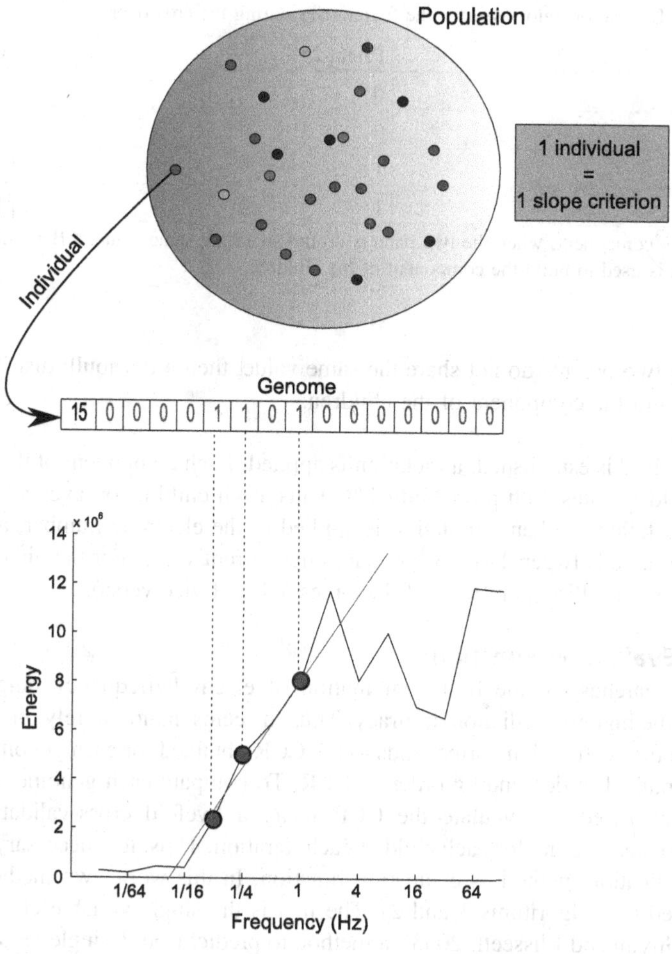

Fig. 9.20 Relationship between the genome and the slope criterion

Crossover

To create a child, two parents are required. A tournament is used to select candidate solutions, keeping the best individual from a randomly selected group (the one with the highest rating based on fitness). The tournament size is set to 2, this keeps selection pressure low and helps maintain a high diversity in the population. The tournament selection is repeated twice to select two parents (tournament "winners"), which are used by the crossover operator to create a single child. The electrode specified by the resulting child is located halfway between the electrodes specified by both parents. The frequency information is combined using the logical operator given in Table 9.2. This crossover is used to balance the production of 1 and 0 s. For a given frequency, when the two parents share the same value (1 for a selected frequency or 0 for a non-selected frequency), the child inherits this value.

Table 9.2 Logical operator used for the frequencies during the crossover

Parent 1	Parent 2	Child
0	0	0
1	0	Bern(1/2)
0	1	Bern(1/2)
1	1	1

For a binary component, when the two parents do not share the same value, a Bernoulli distribution is used to build the component of the children

When the two parents do not share the same value, then a Bernoulli distribution is used to build the component of the children.

Mutation

Once the child is established, a mutation is applied. Each component of the genome of the child mutates with probability 1/8. Thus, each child is, on average, affected by two mutations. When a mutation is applied to the electrode number, a random number (drawn between 1 and 58) replaces the current value. For the binary part, a mutation is a bit-flip operation (a 0 becomes a 1 and vice versa).

9.6.2.2 Evaluation Functions

The GA searches for the best combination of electrode/frequency range which achieves the highest prediction accuracy. Thus, it seems natural to rely on the CCR. Then, the fitness function corresponds to the CCR obtained for each genome. These are then ranked in descending order of CCR. To compare each genome, the same samples are used to calculate the CCR using a fivefold cross-validation. The evaluation step is done for each child at each iteration. Thus, it is necessary to use a fast classification method as evaluation function. In this work, two methods have been tested (see algorithms 1 and 2). The first is the single variable classification (SVC) (Guyon and Elisseeff, 2003), a method to predict from a single variable. The average for each modality (normal or relaxed) is calculated on the individuals in the training set for the variable (feature). Individuals of the test sample are then assigned to the class corresponding to the nearest average. The prediction is compared to ground truth which gives a CCR. The second method is the binary decision tree (CART) (Breiman et al. 1984). Here, the algorithm is used with a single variable which guarantees fast calculation. Then, the fitness function for each genome x is written as:

$$f(x) = \frac{\#\text{well classified participants of the test set}}{\#\text{participants in the test set}} \tag{9.4}$$

The GA searches for the genome which maximizes f.

Algorithm 1 Single Variable Classifier algorithm (SVC)

Require:
$\mathbf{x}_{app}$: Value of the variable for the training set examples,
$\mathbf{x}_{test}$: Value of the variable for the test set examples,
$\mathbf{y}_{app}$: True labels of the training set examples,
$\mathbf{y}_{test}$: True labels of the test set examples,
nb_{test}: Number of individuals in the test set

Ensure: Calculate a threshold T using $\mathbf{x}_{app}$ and a correct classification rate (CCR)
Calculus of the threshold T
$(G_1, G_2) \leftarrow \text{Average}(\mathbf{x}_{app}, \mathbf{y}_{app})$
It calculates the mean of $\mathbf{x}_{app}$ *for the individuals of the first modality* $G1$ *and the second modality* G_2.
$T \leftarrow \frac{G_1 + G_2}{2}$

Prediction and Correct classification rate
For i= 1 to nb_{test} **do**
$Pred^i \leftarrow \text{Predict}(T, \mathbf{x}^i_{test})$
It predicts the class of the ith *individual of the test set using the threshold* T.
end for
$CCR \leftarrow CalculateCCR(Pred, \mathbf{y}_{test})$
It calculates the correct classification rate by comparing the prediction and true labels.
return T, CCR

9.6.2.3 Stop Criterion

The algorithm stops if one of the following three conditions is satisfied:

- The number of iterations exceeds 1,000.
- Parents are the same for 10 generations.
- The number of differences among the parents is less than 3.

To calculate the number of differences for a given population, denoted D, the genomes of the population at iteration i are stored in a matrix, denoted by P^i. Let P^i_j be the column j of the matrix P^i (where $j = 1, \ldots, 16$). Then $D = D_b + D_{elec}$ where:

- D_b is the number of differences for the binary part of P^i_j (columns 2–16). The number of differences for column P^i_j (where $j = 2, \ldots, 16$) is $min(\text{number of 0 in } P^i_j, \text{number of 1 in } P^i_j)$.
- D_{elec} is the number of differences in P^i_1 (column corresponding to the electrode component). Then, D_{elec} is the number of individuals who have a electrode which is different from the electrode most selected in the population.

9.6.3 Results

The algorithm, programmed using Matlab, is run 100 times for each evaluation method with 300 parents and 150 children. The training and test sets are different for two different runs. Figure 9.21 gives CCR values for each run of the GA with CART (stars) and SVC (circles).

For each run, the algorithm is launched two times (one time with CART and the other time with SVC). During a run, CART, and SVC use the same training and test sets in order to obtain comparable results. The correct classification rate obtained by CART (mean of 86.68 % and standard deviation of 1.87 %) exceed significantly (Mann–Whitney paired test with a p-value $= 5.57 \times 10^{-14}$) those obtained by SVC (mean of 83.49 % and standard deviation of 2.37 %), as mentioned in Table 9.3. At the end of the algorithm, some of the best genomes have the same evaluation (due to the low number of individuals and the evaluation method). It is therefore necessary to choose a genome (BEST) among those who have the same score. Thus, the best genomes at the end of each run of the algorithm are stored. The genome that appears most often is considered as the BEST for the evaluation method considered. The two BEST (for CART and SVC) get a correct classification rate equal to 89.33 %. For CART, the BEST is obtained by performing regression between 1/8, 1/4, 2, 4, and 64 Hz on electrode F4 (right frontal area on Fig. 9.1). For SVC, the BEST is obtained from electrode F2 (right frontal area) and the regression between 1/32, 1/16, 2, 4, 8, 64, and 128 Hz (see Table 9.4). Frequencies chosen for these genomes are more extensive than those used in the preliminary study.

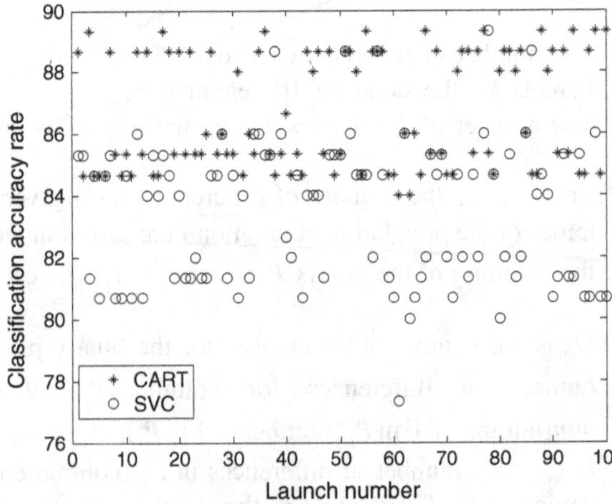

Fig. 9.21 Correct classification rates calculated with CART (*stars*) and SVC (*circles*) for each run of the GA with 300 parents and 150 children

Table 9.3 CCR for the two evaluation methods

Evaluation methods	CCR	
	Mean	Standard deviation
CART	86.68	1.87
SVC	83.49	2.37

Algorithm 2 Classification And Regression Tree algorithm (CART) for an unique variable

Require:

$\mathbf{x}_{app}$: Value of the variable for training set examples,
$\mathbf{x}_{val}$: Value of the variable for validation set examples,
$\mathbf{x}_{test}$: Value of the variable for test set examples,
$\mathbf{y}_{app}$: True labels of the training set examples,
$\mathbf{y}_{test}$: True labels of the test set examples,
$\mathbf{y}_{val}$: True labels of the validation set examples,
nb_{test}: Number of individuals in the test set

Ensure: Create a binary tree T and calculate a correct classification rate (CCR).

Initialisation step

$T \leftarrow \emptyset$
The tree is initialized to the empty set.
$Continue \leftarrow True$
$j \leftarrow 0$

Tree growing step

while Continue **do**
If Current Node is terminal **Then**
$T \leftarrow AssignNode(T, \mathbf{x}_{app}, \mathbf{y}_{app})$
It assigns a modality to each leaf of T using a majority vote.
$Continue \leftarrow False$
, **Else**
$v_j \leftarrow FindThreshold(\mathbf{x}_{app}, \mathbf{y}_{app})$
FindThreshold finds the threshold on the variable $\mathbf{x}_{app}$ *that best separates individuals from the two conditions.*
$t_j \leftarrow ConstructNode(T, v_j)$
It constructs the node using the threshold value v_j*. Individuals of the training sample are split by comparing* $\mathbf{x}^i_{app}$ *and* v_i*.*
End if
$j \leftarrow j+1$
$T \leftarrow t_j$
end while
$n \leftarrow j$

Tree pruning step

$[e_1, e_2, ..., e_n] \leftarrow CalculateError([t_1, t_2, ..., t_n], \mathbf{x}_{val}, \mathbf{y}_{val})$
It compute the error of classification for each subtree using individuals from the validation sample.
$T \leftarrow Pruning([e_1, e_2, ..., e_n], T)$
It prunes the tree T by keeping the subtree that gives the lower classification error e_i*.*
$T \leftarrow AssignNodes(T, \mathbf{x}_{app}, \mathbf{y}_{app})$

Prediction and Correct classification rate

For i= 1 to nb_{test} **do**
$Pred^i \leftarrow$ Predict$(T, \mathbf{x}^i_{test})$
It predicts the class of the ith *individual of the test set using the tree* T*.*
end for
$CCR \leftarrow CalculateCCR(Pred, \mathbf{y}_{test})$
It calculates the correct classification rate by comparing the prediction and true labels.
return T, CCR

Table 9.4 Summary table of results for best genomes

Evaluation methods	BEST genome		
	Electrode selected	Frequency selected (Hz)	CCR (%)
CART	F4	1/8, 1/4, 2, 4 et 64	89.33
SVC	F2	1/32, 1/16, 2, 4, 8, 64 et 128	89.33

Figure 9.22 gives the occurrence of the electrodes in the best genome over the 100 runs. When some genomes have the same CCR at the end of the run, we select the electrode chosen most often among the genomes with equal CCR. The algorithm running with CART selects the electrodes around the number 10 (FZ in Fig. 9.1), 17 (FC1), or 30 (T4). With the SVC method, the electrodes around the 2 (FPZ), the 11 (F2), or the 48 (T6) are mostly chosen. Finally, on average, the population of the evolutionary algorithm converges in less than 50 iterations for both methods. Figure 9.23 gives the number of differences among parents for one run of the algorithm. It shows that the number of differences among parents decreases very rapidly and falls below the threshold of 3 differences in less than 40 iterations. Then, one of the three stop conditions is satisfied and the algorithm stops.

Tables 9.3 and 9.4 summarize the CCR obtained by the GA, which are better than those obtained (see Fig. 9.17) with the criterion of the slopes calculated for frequencies between 4 and 16 Hz (alpha waves). Moreover, Table 9.5 shows that the GA allows for a dimension reduction. SVC classifier cannot be used with more than one variable. Then, Table 9.5 only shows a comparison between the results obtained in Sect. 9.5.3 and those obtained with the GA for the CART classifier.

It also appears that it is more appropriate to use a regression on frequencies of 1/8, 1/4, 2, 4, and 64 Hz for the signal of electrode *F*4 and the CART classifier. Then, this work allows to accurately predict the state of alertness of a new individual. In fact, this electrode and this range of frequencies will be used to calculate the slope criterion for this individual. The CART decision tree, built on the sample formed by the 26 signals (13 study participants in both states of alertness) will be used as a classifier to predict his state of alertness.

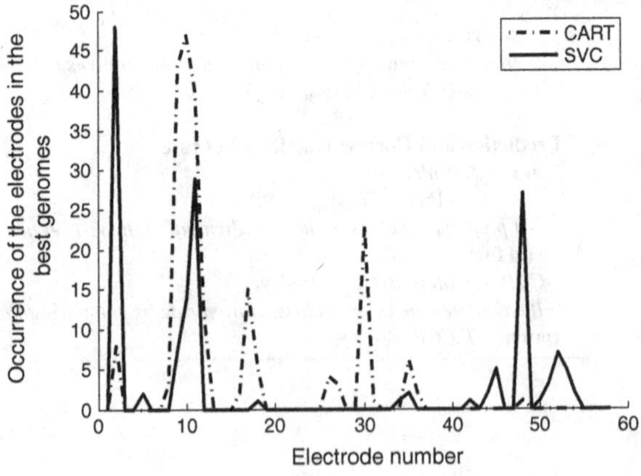

Fig. 9.22 Occurrence of the electrodes in the best genomes for each electrodes during the 100 runs of the GA with 300 parents, 150 children, and CART (*dash-dotted curve*) or SVC (*solid curve*)

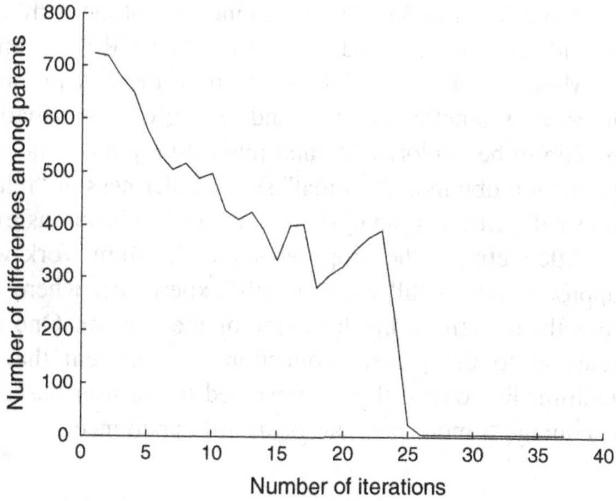

Fig. 9.23 Number of differences among parents for a run of the GA with 300 parents, 150 children, and SVC

Table 9.5 Comparison between CCR obtained in the preliminary study (1st row) and CCR obtained with the genetic algorithm (2nd row)

Evaluation methods	Number of electrodes in the predictive model	CCR	
		Mean	Standard deviation
CART	58	33.98	5.15
CART	1	86.68	1.87

9.7 Conclusions

This chapter presents a system for the automatic detection of human mental states of alertness using EEG data and wavelet decomposition. This contribution is also coupled with a complete protocol of data acquisition, a data validation procedure and a feature selection strategy. Initially, we proposed a criterion to obtain a summarized data matrix in two dimensions. Given the disappointing results obtained by classifying all of the available data, a GA was used as a feature selection step to refine it. This allowed obtaining a reliable classification model that achieves average of classification accuracy equal to 86.68 % with a standard deviation of 1.87 %. The algorithm also selects only a single electrode from the 58 that were initially available; this greatly enhances the possibility of applying the proposed system in real-world scenarios.

An exchange with neurobiologists now seems necessary to link the results obtained by the GA to human physiology. A new campaign to collect EEG data and increase the number of participants included in the study has been undertaken. Increasing the number of data should allow us to improve the precision of the estimate of CCR and thus reduce the number of solutions that have the same score at the end of the GA execution. In addition, an increase of the number of participants allows us to provide an external validation set for the CCR at the end of the GA execution.

Moreover, it is possible to improve the GA proposed in this chapter. In fact, improving genetic operators and testing other evaluation criteria are all paths that remain to be explored. A final interesting point concerns the transformation of the prediction obtained ("normal" state of alertness or "relaxed") to a probability using linear discriminant analysis or logistic regression as evaluation functions.

After refining the proposed method, future work will consider integrating this approach into a full user-friendly experience, where the mental state of the user directly influences the behavior of the system. One example application, that is relevant to the present collection, is a system that automatically modifies the multimedia content that is presented to the user based on his/hers mental state, to encourage a more pleasant or useful experience.

9.8 Questions

1. What is the shape of the raw data?
2. Can we use directly the raw data to classify them?
3. In general, what are the benefits to use a wavelet transform instead of a Fourier transform?
4. In the work presented in this chapter, could we use a Fourier Transform?
5. How can we be sure that the data we use for learning is relevant?
6. Could you summarize in a few lines the behavior of a GA?
7. Why do we use a GA for the optimization of the frequencies and to select the best electrode?
8. Why do we use the slope criterion as a feature?
9. Could this method be used to classify other types of mental states?
10. Suppose that you work on data from individuals in two modalities described by a single variable. Draw a scheme explaining the behavior of the Single Value Classifier (SVC) algorithm and implement this algorithm.

Acknowledgments The authors wish to thank Vérane Faure, Julien Clauzel, and Mathieu Carpentier, who collaborated as interns in the research team during the development of parts of this work.

References

Anderson C, Sijercic Z (1996) Classification of EEG signals from four subjects during five mental tasks. In: Proceedings of the conference on engineering applications in neural networks, London, United Kingdom, pp 407–414

Ben Khalifa K, Bédoui M, Dogui M, Alexandre F (2005) Alertness states classification by SOM and LVQ neural networks. Int J Inf Technol 1:131–134

Breiman L (2001) Random forests. Mach Learn 45:5–32

Breiman L, Friedman J, Olshen R, Stone C (1984) Classification and regression trees., Wadsworth advanced books and softwareCRC Press, Boca Raton

Broadhursta D, Goodacrea R, Ah Jonesa A, Rowlandb JJ, Kelp DB (1997) Genetic algorithms as a method for variable selection in multiple linear regression and partial least squares regression, with applications to pyrolysis mass spectrometry. Anal Chim Acta 348:71–86

Cavill R, Keun HC, Holmes E, Lindon JC, Nicholson JK, Ebbels TM (2009) Genetic algorithms for simultaneous variable and sample selection in metabonomics. Bioinformatics 25:112–118

Cecotti H, Graeser A (2008) Convolutional neural network with embedded fourier transform for EEG classification. In: International conference on pattern recognition, Tampa, Florida, pp 1–4

Daubechies I (1992) Ten lectures on wavelets. SIAM

De Jong KA (1975) An analysis of the behavior of a class of genetic adaptive systems. PhD thesis, University of Michigan

Guyon I, Elisseeff A (2003) An introduction to variable and feature selection. J Mach Learn Res 3:1157–1182

Hastie T, Tibshirani R, Friedman J (2009) The elements of statistical learning: data mining, inference, and prediction, 2nd edn. Springer, Berlin

Hazarika N, Chen J, Tsoi C, Sergejew A (1997) Classification of EEG signals using the wavelet transform. Sig Process 59:61–72

Holland JH (1975) Adaptation in natural and artificial systems. University of Michigan Press, Ann Arbor

Jacobson E (1974) Biologie des motions. Les bases thoriques de la relaxation

Jaffard S, Meyer Y (1996) Wavelet methods for pointwise regularity and local oscillations of functions. Mem Amer Math Soc 123(587)

Jasper HH (1958) Report of the committee on methods of clinical examination in electroencephalography. Electroencephalogr Clin Neurophysiol 10:1–370

Lé Cao K-A, Rossouw D, Robert-Granié C, Besse P (2008) Sparse PLS: variable selection when integrating omics data. Stat Appl Genet Mol Biol 7(Article 35)

Legrand P (2004) Débruitage et interpolation par analyse de la régularité Höldérienne. Application à la modélisation du frottement pneumatique-chaussée. PhD thesis, École Centrale de Nantes et Université de Nantes

Levy Vehel J, Legrand P (2004) Signal and image processing with FracLab. In: Proceedings of 8th international multidisciplinary conference on complexity and fractals in nature

Levy Vehel J, Seuret S (2004) The 2-microlocal formalism. Fractal geometry and applications: a jubilee of benoit mandelbrot. In: Proceedings of symposia in pure mathematics, PSPUM, vol 72, pp 153–215

Lin Y-P, Wang C-H, Jung T-P, Wu T-L, Jeng S-K, Duann J-R, Chen J-H (2010) Eeg-based emotion recognition in music listening. IEEE Trans Biomed Eng 57(7):1798–1806

Mallat S (2008) A wavelet tour of signal processing, 3rd edn. Academic Press

Naitoh P, Johnson LC, Lubin A (1971) Modification of surface negative slow potential (CNV) in the human brain after total sleep loss. Electroencephalogr Clin Neurophysiol 30:17–22

Niedermeyer E, Lopes da Silva F (2005) Electroencephalography, basic principles, clinical applications and related fields, 5th edn.

Obermaier B, Guger C, Neuper C, Pfurtscheller G (2001) Hidden markov models for online classification of single trial EEG data. Pattern Recogn Lett 22:1299–1309

Rosenblith W (1959) Some quantifiable aspects of the electrical activity of the nervous system (with emphasis upon responses to sensory stimuli). Rev Mod Physics 31:532–545

Schultz JH (1958) Le training autogne. PUF

Subasi A, Akin M, Kiymik K, Erogul O (2005) Automatic recognition of vigilance state by using a wavelet-based artificial neural network. Neural Comput Appl 14:45–55

Tecce JJ (1979) A CNV rebound effect. Electroencephalogr Clin Neurophysiol 46:546–551

Tenenhaus M (1998) La régression PLS, Théorie et Pratique

Timsit-Berthier M, Gerono A, Mantanus H (1981) Inversion de polarité de la variation contingente négative au cours d'état d'endormissement. EEG Neurophysiol 11:82–88

Vézard L (2010) Réduction de dimension en apprentissage supervisé. applications à l'étude de l'activité cérébrale. Master's thesis, INSA de Toulouse. Available at the following URL http://www.math.u-bordeaux1.fr/archives/stages/laurentvezard2011.pdf

Vézard L, Legrand P, Chavent M, Faïta Aïnseba F, Clauzel J, Trujillo L (2014) Classification of EEG signals by evolutionary algorithm. Adv Knowl Discov Manage 4:133–153

Vuckovic A, Radivojevic V, Chen A, Popovic D (2002) Automatic recognition of alertness and drowsiness from EEG by an artificial neural network. Med Eng Phys 24:349–360

Walter WG, Cooper R, Aldridge V, McCallum WC, Winter A (1964) Contingent negative variation: an electric sign of sensorimotor association and expectancy in the human brain. Nature 203:380–384

Yeo M, Li X, Shen K, Wilder-Smith E (2009) Can SVM be used for automatic EEG detection of drowsiness? Saf Sci 47:115–124

On Mapping EEG Information into Music

10

Joel Eaton and Eduardo Reck Miranda

Abstract

With the rise of ever-more affordable EEG equipment available to musicians, artists and researchers, designing and building a brain–computer music interface (BCMI) system has recently become a realistic achievement. This chapter discusses previous research in the fields of mapping, sonification and musification in the context of designing a BCMI system and will be of particular interest to those who seek to develop their own. Design of a BCMI requires unique considerations due to the characteristics of the EEG as a human interface device (HID). This chapter analyses traditional strategies for mapping control from brainwaves alongside previous research in biofeedback musical systems. Advances in music technology have helped provide more complex approaches with regard to how music can be affected and controlled by brainwaves. This, paralleled with developments in our understanding of brainwave activity has helped push brain–computer music interfacing into innovative realms of real-time musical performance, composition and applications for music therapy.

J. Eaton (✉) · E.R. Miranda
Interdisciplinary Centre for Computer Music Research (ICCMR), Plymouth University, Plymouth, UK
e-mail: joel.eaton@postgrad.plymouth.ac.uk

E.R. Miranda
e-mail: eduardo.miranda@plymouth.ac.uk

E.R. Miranda and J. Castet (eds.), *Guide to Brain-Computer Music Interfacing*,
DOI 10.1007/978-1-4471-6584-2_10

10.1 Introduction

Articles on brain–computer music interfacing (BCMI) research often open with a sentiment on how far away we are from the science fiction like dreams of thought explicitly controlling computers. However, the ongoing progress in this field in the last decade alone indicates that this is becoming reality; we are not as far away from such dreams as people tend to think.

In a climate where science and technology have the ability to translate primitive emotional states of the brain, develop brain–computer interfacing (BCI) for precise control of machinery and allow for non-speaking persons to communicate by means of brain signals—or brainwaves—mediated by brain scanning technology, it is easy to become enthused about the potentials within neuroscience, especially when applied to the arts (Miranda 2006).

The possibility of BCI for direct communication and control was first seriously investigated in the early 1970s, and the notion of making music with brainwaves (turning BCI into BCMI) is not new. Musicians and composers have been using brainwaves in music for almost the last 50 years. Instrumental in this were a number of highly innovative people, the work of which is discussed in this chapter. This period reflected a significant trend towards interdisciplinary practices within the arts influenced by experimental and avant-garde artists of the time and a growing engagement with eastern music and philosophies by those in this field. It is fair to say that brainwaves in music were initially explored by experimental composers, and the area has been pioneered by a number of notable non-traditional composers and technologists since, and this is reflected in the wide range of applications and research that has been undertaken over the last decade and a half.

Over the last twenty or so years, the world of computer music has been waiting for technology to interpret brainwave information in order to develop BCMI systems. Equipment costs, portability, signal analysis techniques and computing power has rapidly improved over recent times, alongside a deeper understanding of how the brain functions. Now that the line between these two areas is narrowing the playing field is becoming much larger enabling the two to flourish together. Brainwaves have long been considered to be one of the most challenging of biological signals from the human body (known as bio-signals) to harness, and beginning to understand them through music and sound offers clinical as well as creative rewards; for instance, BCMI systems are bound to benefit music therapy.

This chapter focuses on the pressing problem of mapping EEG information into sonic and musical forms. That is, on how to use EEG to control algorithms for synthesising sound or to produce music. A number of mapping methods that have been devised to date are introduced. As we shall see further on, there are a number of different approaches to making music with EEG and the choice of which to use is dependent on the overall objectives of the system.

10.2 Mapping and Digital Musical Interfaces

The pursuit of control within musical systems controlled by the brain has been at the forefront of research ever since it was viable. Control has been a key driver in BCMI research as within it is the ability to convey expression and communication through music. Mapping can be likened to a key that unlocks the creative potentials of control. Mapping allows us to translate an input signal so that it can be understood and used by a musical system. Put simply, mapping is the connection of input controls (via EEG) to an output, which in the case of a BCMI is a musical engine. In the pursuit of enhancing user interactivity in BCMIs, mapping plays a key role in designing creative and practical applications. Even Alvin Lucier, the first composer to perform using EEG signals, had a desire for more comprehensive mappings within his system to allow for greater musical control (Lucier 1976).

Research into mappings and digital instruments has largely focused on gestural control and physical interaction (Miranda and Wanderley 2006). Goudeseune (2002) presents a comprehensive framework of mapping techniques for digital instrument design, building on the proviso that performers can think of mappings as containing the *feel* of an instrument; how it responds to the physical control. Garnett and Goudeseune (1999) refer to the results of mapping as providing 'consistency, continuity and coherence', key factors in the design of musical control systems. Clearly, different strategies for mapping in instruments driven without gestural input, known as *integral interfaces*, are needed to develop BCMI systems (Knapp and Cook 2005).

Mappings can be defined based on the number of connections between the input and output parameters; one-to-one, one-to-many and many-to-many (combinations of one-to-one and one-to-many) (Hunt et al. 2000). Although this framework is useful for evaluating system design, it does not take into account the relationship of the input control to the mapping or any codependencies or rules a mapping may rely on. Goudeseune (2002) recognises the intricacy involved in mapping design, coining the term high-dimensional interpolation (HDI) to define mapping a large number of parameters to a small number of inputs where controls can be interpolated and connected using a variety of rules and techniques.

The investigation of sophisticated mappings in BCMIs, in comparison with other contemporary digital musical instruments and interfaces, has until recently been stifled by the difficulties in eliciting control from EEG information. On the one hand, simple mappings that exemplify EEG control have been favoured as they suit this purpose well. Simple mappings, such as a linear control to modulate a synthesiser's pitch, have been designed to be very effective to facilitate performing and composing with BCMIs for non-musicians (Miranda et al. 2011). On the other hand, new methods of EEG acquisition provide much more accurate real-time control than was previously available, and as a result can accommodate far more advanced mapping techniques leading to complex compositional approaches. Eaton's *The Warren*, a performance BCMI piece that will be discussed later in this chapter, provides a useful example of complex mapping strategies.

As technologies for monitoring brainwave information have advanced so too has the field of computational music. This correlated evolution of technologies and understanding of EEG has shaped the direction of brainwave-controlled music. Both fields have produced knock-on effects in this area, from the introduction of MIDI that led to new applications of brainwaves with music to the advancement of BCI, allowing BCMI research to shift towards its engagement with cognitive control of EEG.

In order to elicit control over EEG, it is essential to be able to decipher meaning within EEG data that directly correlate with the subjective decisions (control choices) of a user, be it a mental state or a cognitive task. This quest for accurate meaning in EEG information has long been at the forefront of BCMI research, as through precision in generating data comes accurate control. Note that the term meaning here refers to understanding the correlation between a user's mental process and an associated brainwave response. Meaning in this manner does not refer embedded or implied thought patterns within brainwaves (unless otherwise stated later on). Mappings are not necessarily dependant on control, as generative mappings that interpret unknown EEG information can produce interesting music, but the two can feed off of each other in terms of complexity. When control is explicit, the ability to introduce complex mapping strategies for more advanced musical control arises.

In this chapter, we use the term *secondary* mappings to refer to a mapping as an aside of an input's primary connection. A secondary mapping may not necessarily be directly presented to a user, it may be used for time-based data harvesting for algorithmic rule-based mapping, or it may just not take precedence over a primary mapping.

10.3 Mapping and Approaches to BCMI

The BCMI systems presented in this chapter differ in terms of application, cost, equipment type and signal processing, data handling and indeed mappings, but all can be said to consist of the following elements (Fig. 10.1):

- **Stimuli** This element is optional and in some cases where it is present provides the feedback link with the system, being part of or being affected by the musical system.
- **EEG Input** Electrodes placed on the scalp, either in the form of a brain cap or a headband to fit them.
- **Signal Processing** Amplification of electrical activity and data extraction to isolate meaningful information. Filtering and further data processing/analysis/classification are applied depending on the EEG technique used.
- **Transformation Algorithm** Transforming the EEG information into parameters within a musical system. This is where mapping of non-musical information to the music engine occurs. This can take various forms from a patch cable from an

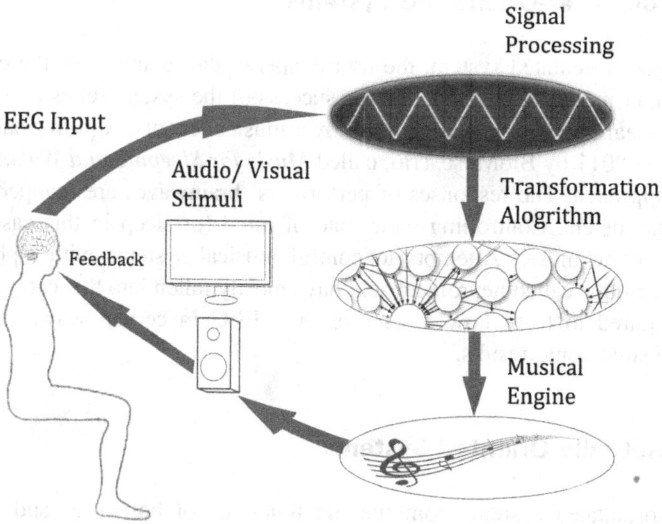

Fig. 10.1 The make-up of a typical BCMI system

EEG amplifier into an analogue synthesiser to a generative software program that triggers musical events.

- **Musical Engine** The musical system receiving commands from the transformation algorithm. This may be external to the algorithm (e.g. a MIDI instrument) or built into it with the appropriate software.

Miranda et al. (2003) identify three types of BCI systems, based on how they interact with a user. BCMIs can also be observed using this categorisation as systems have been developed within all three areas: *user-orientated, computer-orientated* and *mutuallyorientated.*

10.3.1 User-Oriented Systems

A user-orientated type of system is programmed to understand the meaning of user input with in an attempt to adapt to its behaviour in order to achieve control. For the piece *In Tune*, Richard Teitelbaum adapts his system in response to a performer's alpha waves as well as injecting his own musical directions (Teitelbaum 1976). Building user-orientated BCMIs pose difficulties with understanding meaning within EEG. When relying on interpretation, control can be harnessed far better in mutually orientated systems where this problem is addressed two way.

10.3.2 Computer-Orientated Systems

In a computer-orientated system, the user adapts to the functions of the computer. The computer model stays fixed, and the success of the system relies on the ability of a user to learn how to perform control over musical events. A performance piece conceived in 2011 by BioMuse Trio, called *Music for Sleeping and Waking Minds*, uses this approach. The responses of performers' brainwaves are mapped to fixed musical parameters. Controlling their state of mind (or sleep in this case) affects control over the music. Attempts to control musical systems with alpha waves using, a technique called neurofeedback, have mostly fallen into this category as the user is required to learn how to control their EEG in certain ways in order to produce desired sonic results.

10.3.3 Mutually Oriented Systems

Mutually orientated systems combine the functions of both user and computer orientation whereby the two elements adapt to each other. This was the approach used in Eaton's *The Warren.* Here, the system requires the user to learn how to generate specific commands and features mappings that adapt depending on the behaviour of the user.

The majority of BCMIs fall into the category of computer-orientated systems. This allows for fixed parameters to be built that respond to known user brain responses. The use of mutually orientated systems allows for two useful things. Firstly, more sophisticated algorithms derived from EEG behaviour can be mapped onto music. As the system learns the EEG behaviour of a subject over time, this information can be used in series with primary mappings and in parallel through embedding deeper secondary mappings. Secondly, a system where user and computer adapt together increases the likelihood of obtaining accurate EEG as both elements are effectively calibrated to optimise the system performance.

10.3.4 Brainwave Data for BCMI

There are two types of EEG data used in the systems discussed in this chapter: event-related potentials (ERPs) and spontaneous EEG. ERPs are fluctuations of EEG measured in response to events triggered by external stimuli. ERP data are time locked to stimulus and are recognised as positive or negative amplitude deflections. ERPs are categorised by their response time post-stimuli and are associated with brain processing of event expectation and perception.

Systems monitoring spontaneous EEG look at ongoing EEG data, often across multiple frequencies for patterns or trends that correspond to specific brain activities. This can also be time locked to external stimuli, and if so, windows of corresponding data are captured for analysis.

Significant work in using brainwaves for music has been developed with other forms of measurement of brain activity. For instance, fMRI (functional Magnetic Resonance Imaging) has been used to translate brain data as input to offline musical compositions, one example of which is discussed in Chap. 12 in this volume. However, fMRI is currently impractical for developing a BCMI: it is expensive, not portable and has poorer time resolution than EEG, to cite but three encumbering factors.

10.3.5 Methods of Music Generation with Brainwaves

When looking back on research into music and brainwaves, we can separate systems into three categories: ones for *EEG sonification*, ones for *EEG musification* and ones for *BCI control*. EEG sonification is the translation of EEG information into sound, for non-musical and predominantly medical purposes. EEG musification is the mapping of EEG information to musical parameters; however, the EEG data are arbitrary and when possible can offer only loose forms of control. BCI control is inherent in systems where direct cognitive real-time control of music is achievable. In some systems, more than one of these approaches can be found, and in others where one approach has been adopted for investigation of the technique, the application could well be applied to another approach as a result.

It should also be noted that the mapping approaches discussed in this chapter are not wholly comparative, as it charts development in a relatively infantile field, where, as previously mentioned, progress is heavily reliant on the advances within neuroscience. Where considered useful, areas are touched upon that draw parallels between systems as a way of directing the reader through the different approaches and ideas.

Although this chapter does not attempt to explicitly categorise the accuracy of each system, due to the wide range of disparaging technologies and individuals incorporated, it should be carefully acknowledged that accuracy plays a very important part in the derivation of *meaning* within EEG data, and this is considered of high importance.

The sonification of data offers an interesting way of to listening to the sounds of non-musical sources of information. Data harvesting allows us to sonify a world of unlikely information, such as the stock market or even the weather. In sonification, we are concerned with the *sound* of the information relative to itself, and it is a passive process and a way of hearing numerical or graphical data.

Sound has long been used as a way of interpreting biological information, from the use of the stethoscope to the steady beeping of the heart rate monitor. Both of these are methods of *hearing* the body, which when used in real time to help affect control over the signal is known as biofeedback. The visual complexities of EEG have given reason to sonifying its information as a method for understanding activity through the simplification and the natural intuition of discernably listening to multiple elements contained within sounds. As such, the mappings for direct data sonification should be straightforward in order to provide an intuitive correlation between brain activity and sound. Control of EEG in sonification (and some musification) systems is largely passive, whereby the user has no direct control over their EEG. EEG may be influenced external factors, such as tiredness or mood, but in situations where brainwave control is not achieved by explicit choice.

In contrast to sonification, to *musify* data is to map the data into organised musical form. This is rather different from sonification as one is not attempting to understand the data through sonification per se, but rather attaching it to a musical system. Therefore, musical structures are connected to the EEG information based on the patterns or variables apparent within the data. For example, if the EEG delivers five distinguishable data, then these can be directly mapped to five parameters within a pre-designed musical piece. A common factor within EEG musification is the use of generative musical approaches. In musification, BCMI systems a passive approach to EEG control are generally used. EEG data are generally limited in its meaning, and the shift in focus lies heavily on mappings using advanced techniques of interpreting data in useful ways to grant musical success. In summary, the difference between sonification and musification are as follows: (a) sonification produces sounds from EEG data, and the system would normally control a sound synthesiser; (b) sonification is not, in principle, intended for an artistic purpose, but rather as some sort of scientific auditory display of the EEG behaviour.

Both sonification and musification afford no explicit control of the sound of music, and as such, strictly speaking, they could be regarded outside of the realms of BCI research. This is because BCI research is based on the premise that a BCI system allows for the *active* control of a device and/or software by the explicit thought of the command, and the results of the mental activity are fed back to the user in real time (Wolpaw and Birbaumer 2006). This definition of BCI has been harnessed within BCMI to the extent that subjective control over systems is now a realisation. Here is where the challenge of being unable to translate musical thought into direct action has been bypassed through embedding meaning into cognitive processes. For example, where reading the explicit thought of '*play the note D#*', is not feasible, using learnt cognitive processes where a user understands the outcomes may lead to a dedicated brainwave response that can be mapped to play the note D#.

10.4 Observations on Musifying EEG

Musifying brainwave activity without a need for control can offer interesting possibilities with regard to mapping data to music. Although musification is not really BCI, it is nevertheless a valid approach for BCMI for artistic purposes. For instance, Miranda and Soucaret (2008) reported on a mapping method they developed to produce melodies from the 'topological' behaviour of the EEG across a configuration of electrodes on the scalp or *montage*. In this case, the EEG signal

Table 10.1 The montage of 14 electrodes used in EEG melodies

Electrode number	Electrode name
1	Fp1
2	Fp2
3	F7
4	F5
5	F4
6	F8
7	T3
8	T4
9	T5
10	P3
11	P4
12	T6
13	O1
14	O2

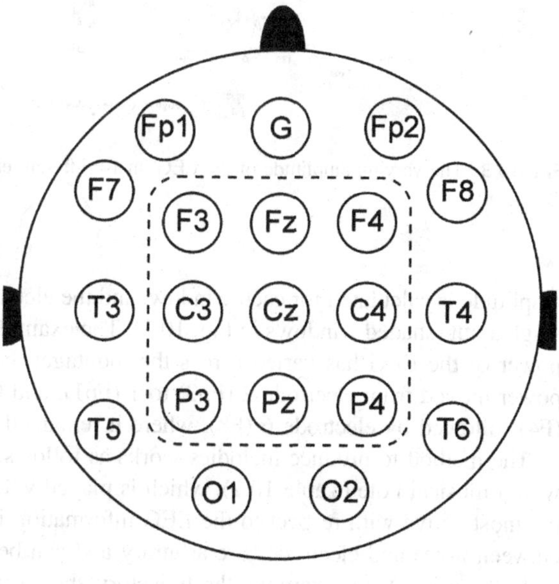

Fig. 10.2 The 10–20 electrode placement scheme recommended by the International Federation of Societies for EEG and clinical neurophysiology

of each individual electrode was analysed individually in order to infer possible trajectories of specific types of EEG information across a montage of 14 electrodes, as listed in Table 10.1; see Fig. 10.2 for placement scheme with labels suggested by the International Federation of Societies for EEG and Clinical Neurophysiology.

As an example, let us assume that we are interested in tracking the behaviour of the overall EEG amplitude. Figure 10.3 plots the amplitude of the EEG on each electrode for approximately 190 s. Each plot is divided into 5 windows of approximately 38 s each; the size of this window is arbitrary. The average

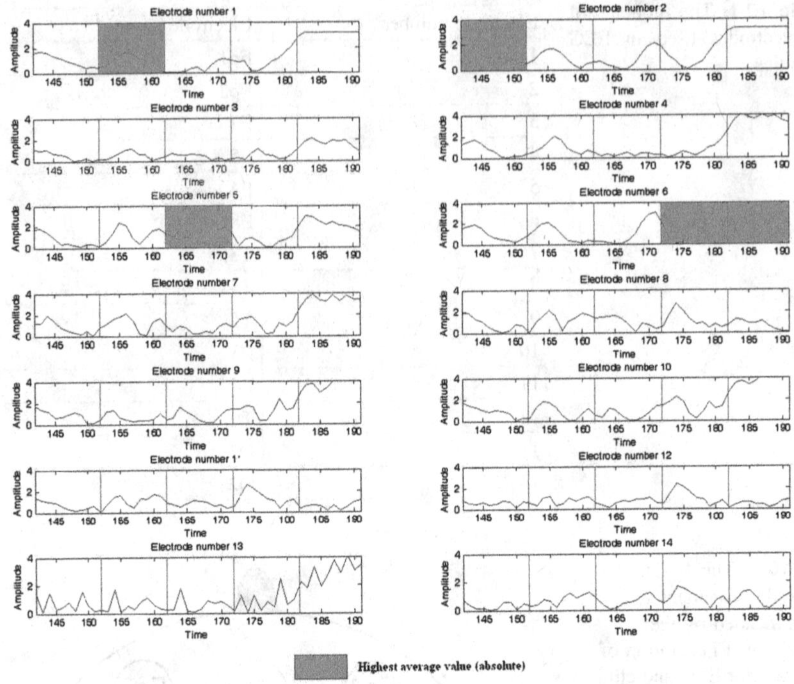

Fig. 10.3 The varying amplitude of the EEG on 14 different electrodes for approximately 190 s

amplitude is calculated for each window, and the electrode with the highest value is singled out (shaded windows in Fig. 10.3). The example in Fig. 10.4 shows how the power of the EEG has varied across the montage: the area with the highest EEG power moved from electrode 2 (Fp2) to 1 (Fp1), and then, it moved to electrode 5 (F4) followed by electrode 6 (F8), where it remained for two windows.

The method to produce melodies works as follows: we associate each electrode with a musical note (Table 10.2), which is played when the respective electrode is the most active with respect to the EEG information in question. The associations between notes and electrodes are arbitrary and can be customised at will.

In the case of our example, the trajectory shown in Fig. 10.4 would have generated the melody shown in Fig. 10.5. (Rhythm is allocated by means of a Gaussian distribution function, which is not relevant for discussion here.)

The authors reported that it was possible to produce interesting pleasant music with the system by forging crafty associations of electrodes and notes, combined with careful generation of rhythmic figures.

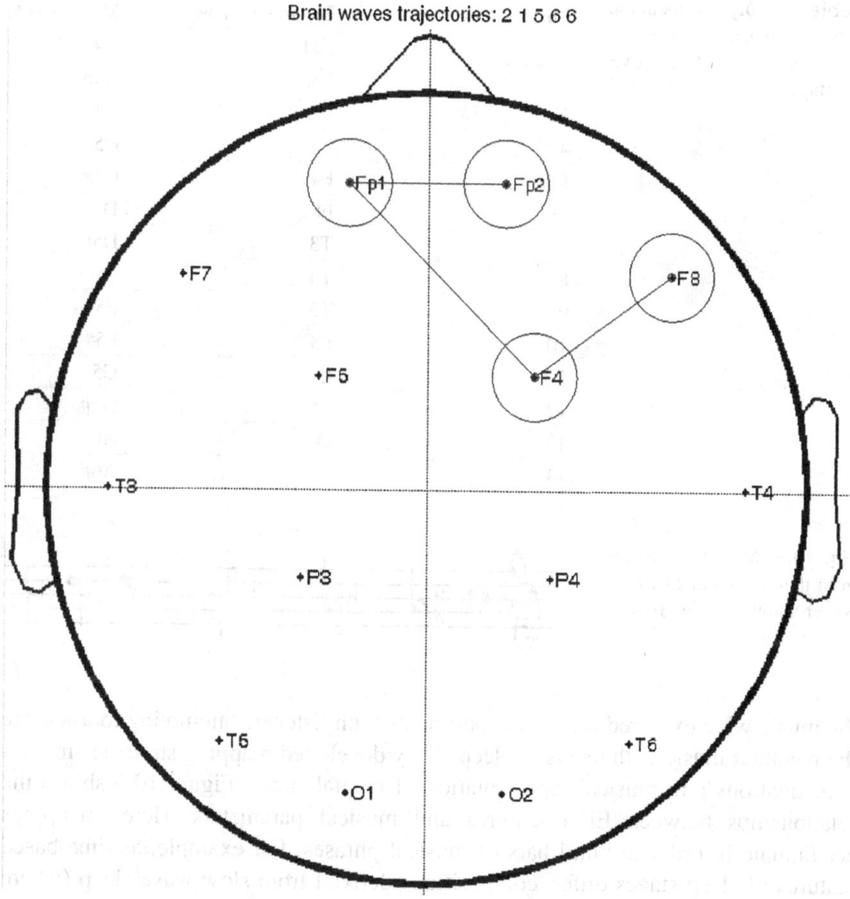

Fig. 10.4 Tracking the behaviour of the amplitude of the EEG signal across a montage of electrodes. In this example, the area with the highest EEG power moved from electrode 2 (Fp2) to 1 (Fp1), and then it moved to electrode 5 (F4), followed by electrode 6 (F8), where it remained for two windows

A number of analyses can be performed in order to track the behaviour of other types of EEG information. For instance, they generated two concurrent melodies by tracking the trajectory of alpha rhythms and beta rhythms simultaneously. They also generated polyphonic music by tracking other types of EEG information simultaneously, such as correlation between electrodes or sets of them, synchronisation between one or more electrodes, and so on.

Another example of musification was reported by Wu and colleagues. They harnessed EEG data generated by variations in sleep to compose music (Wu et al. 2009). The pitch and duration of notes were derived from formulas that mapped each EEG wave to a determinate pitch and its period to duration. Characteristics of

Table 10.2 Associations between musical notes and the electrodes of a given montage

Electrode number	Electrode name	Musical note
1	Fp1	A4
2	Fp2	A4#
3	F7	B4
4	F5	C5
5	F4	C5#
6	F8	D5
7	T3	D5#
8	T4	E5
9	T5	F5
10	P3	F5#
11	P4	G5
12	T6	G5#
13	O1	A6
14	O2	A6#

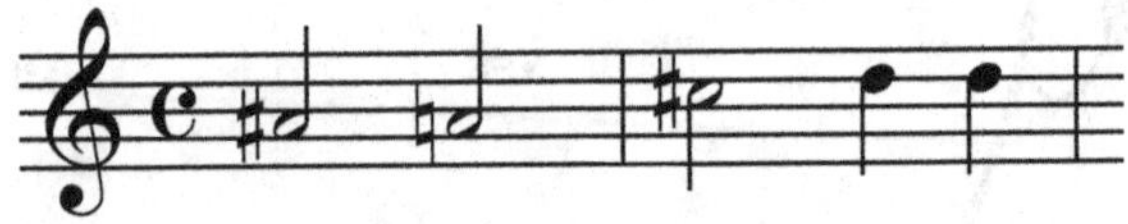

Fig. 10.5 Melody generated from the behaviour of EEG power shown in Fig. 10.4

the music were explored through experiments with listeners attempting to associate the resultant music with levels of sleep. They developed mapping strategies in their investigations into musical representation of mental states. Figure 10.6 shows the relationships between EEG features and musical parameters. Here, mappings accumulate in order to build bars of musical phrases. For example, as time-based features of sleep stages differ, compositions derived from slow wave sleep (where activity is high in low-frequency delta and theta rhythms; see Chaps. 1, 2, 7 and 9 for more on EEG rhythms), are higher in amplitude and lower in pitch than compositions generated from rapid eye movement EEG (where alpha activity is more prominent, albeit with low amplitudes) (Wu et al. 2010). This ability to directly map time-based features, such as the prominent frequency and amplitude, gives way for direct musical evocations of the mind's state, allowing a listener to hear, through music, brain states of arousal and relaxation.

10.5 Early Research into Biofeedback and Music

In 1965, Alvin Lucier performed a piece for live percussion and brainwaves titled *Music for Solo Performer*. The piece was inspired by Luciers' experiments, with the physicist Edmond Dewan, into controlling bursts of alpha activity with meditative states. Brainwaves mapped to sounds, in real time, created a neurofeedback loop, allowing Lucier to affect sonic changes based on the feedback of the previous

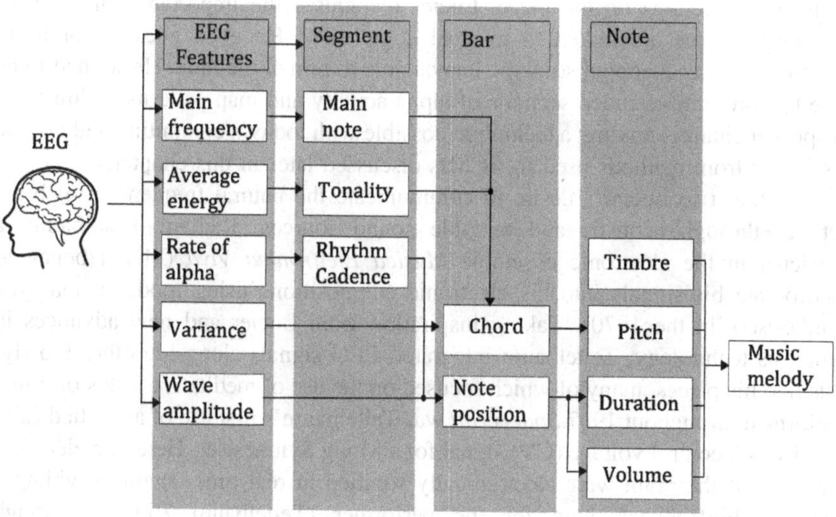

Fig. 10.6 Mapping diagram for musification of EEG proposed by Wu et al. (2010)

brainwave states as he heard them. Alpha waves, or alpha rhythms, are the term given to describe brain activity within the range of 8 and 13 Hz and are commonly associated with relaxed states of attentiveness (Cahn and Polich 2006).

During the performance, Lucier amplified his alpha waves, read from two electrodes positioned on his forehead, through a series of loudspeakers. As the frequencies contained in alpha waves are below the threshold of human hearing, the loudspeakers were coupled with resonant percussive instruments including cymbals, gongs, bass drums and timpani as a way of musifying brainwave activity (Lucier 1976).

This simple method of directly mapping brainwave intensity to instrument resonance was the first attempt of its kind to interpret brainwave activity in real time into a form of experimental music. The theatrical dramaturgy of a man on a darkened stage with wires on his head and his brain generating music was surely impressive enough, but Lucier was considerate in his approach applying deeper mapping considerations to increase the sonic possibilities. The input to the system was alpha rhythms produced in phrases of varying duration, and this one limited parameter from the brain was carefully utilised. The amplitude was operated by a manual control (either by an assistant or by Lucier himself) and mixed between individual speaker channels. The known behaviour of these three parameters (duration, volume and channel mixing) in response to alpha activity was used to design the output stages of the system, or the musical engine, instrument type, speaker placement, and the involvement of extra materials, such as cardboard boxes or metal bins. Additionally, a threshold switch was used for alpha above a certain amplitude level to trigger pre-recorded tape loops of alpha activity, transposed upwards into the audible realm for the audience to hear.

In his reflections on the piece, Lucier recognises the importance of how his mapping choices are linked to musical complexity. He even goes as far as to identify a further mapping strategy, unavailable to him at the time. He wished to be able to store time-encoded sections of alpha activity and map patterns within them to speaker channel mixing; a technique possible with today's computing and not too dissimilar from methods used in BCMIs discussed later in this chapter.

In contrast to Lucier's desire to communicate the natural frequencies of brain activity through acoustic and tangible sound sources, Richard Teitelbaum, a musician in the electronic ensemble *Musica Elettronica Viva* (MEV) began to incorporate bio-signals into his electronic compositions using modular analogue synthesisers in the 1970s. Taking inspiration from Lucier and new advances in synthesis technology, Teitelbaum integrated EEG signals alongside other bio-signals into his pieces, many of which focused on the use of meditative states of mind. Performed throughout 1967 *Spacecraft* was Teitelbaum's first use of amplified EEG activity as a control voltage (CV) signal for a Moog Synthesiser. Here, the electrical activities of the brain were electronically sonified in real time, again providing a real-time biofeedback loop for the performer (Teitelbaum 2006). Although *Spacecraft* was a wholly improvised composition, it provided a foundation for his later uses of brainwaves that sought to investigate elements of control and musical interaction.

In Tune, perhaps Teitelbaum's most popular work, was first performed in Rome, 1967. What stands out in later versions of the piece (referred to by the composer as the expanded version of the piece) is the introduction of a second performer's EEG within his system. Alongside other bio-signals, including heartbeat and amplified breathe, alpha activity was measured and then split into two paths within a modular system comprised of analogue synthesis modules, a mixer and audio effects. Before any audio processing took place, a threshold gate was set to allow only alpha signals generated with eyes closed to pass; the amplitude of alpha rhythms is markedly increased by closing one's eyes. This provided a simple control switch for performers; system ON with eyes shut and system OFF with eyes open. Precise control within an ON state of the system's parameters was largely unattainable beyond basic changes of alpha amplitude increase and attenuation. With the gate open, the alpha of a performer was split from an envelope follower into two directions within the system to provide a one-to-many mapping. The first path allowed for a direct DC signal to be mapped to two voltage-controlled oscillators, thus modulating a preset centre pitch for each. The second path sent the EEG signal to an envelope generator, which allowed for variable control of a voltage-controlled amplifier (VCA) and voltage-controlled filter (VCF). This parallel mapping of one EEG signal allowed for real-time modification of pitch, rhythm and amplitude of the synthesised waveforms coupled with magnetic tape recordings being played back through the same VCA and VCF. Again, these mapping choices were not arbitrary but were in keeping with Teitelbaum's artistic aims for the composition. The heavy breathing and sexualised moaning sounds played back from one tape machine being rhythmically enveloped by the alpha were designed to play alongside the live breath and vocal sounds from a throat microphone (Teitelbaum 1976).

The method for signal processing was repeated for the second performer whose alpha controlled a third and fourth oscillator via a second envelope generator for their amplification and that of a secondary tape machine (but no subsequent filter in this path).

With two performers generating biological signals, Teitelbaum performed the role of conductor. He manually played the system controls (synthesis, reverb and mixing parameters) in response to the performer's alpha alongside injecting his own musical intuition. Alongside, its use of brainwave information as a control input to an electronic musical system *In Tune* introduces the use of brainwaves as a collaborative musical tool for performers and raises interesting questions regarding the potential influences of biofeedback between individuals in shared musical environments not just of brainwaves but from other bio-signals.

The fields of biofeedback and aesthetic experience became increasingly popular in the late 1960s and early 1970s. During his time at the Laboratory of Experimental Aesthetics, part of the Aesthetic Research Center of Canada, David Rosenboom conducted a thorough body of research into biofeedback and the arts, definitively recorded in his 1990 writing *Extended Musical Interface with the Human Nervous System* (Rosenboom 1990).

Other artists at this time were also experimenting with alpha, such as Finnish artist Erkki Kurenniemi's instrument *Dimi-T*, where EEG was used to control the pitch of an oscillator (Ojanen et al. 2007). Manfred Eaton's ideas for an adaptive biofeedback instrument presented in his book *Bio-Music* (Eaton 1971) presented his concept of a musical brain system powered by visual and auditory stimuli. What is significant in his idea is that the images or sounds that are presented as stimulus for generating brainwave activity can be semantically removed from the music as long as the corresponding brain activity is one desired by the composer. This concept is now a common tool in contemporary BCMI design, where stimuli are used to generate specific brainwave information or *meaning*, but is unrelated to the musical outcomes; this will be discussed in more detail further on.

The study of alpha rhythms in music offered a rich time of creative practice. Ultimately, musical and artistic works were restricted by the limits of control that came with generating and analysing alpha. In order to use the brain for more advanced musical applications, new methods of harnessing and interpreting brain information were required. Yet the work undertaken in using alpha waves to control music was an important landmark in the field of BCMI, as it suggests that the notion of music controlled by thought was actually achievable.

In 1995, Roslaie Pratt and colleagues at the Biofeedback Research Laboratory in Brigham Young University reported on experiments where children with ADD and ADHD used neurofeedback training with the aid of music containing discernible rhythms, to increase focused behaviour through the reduction of theta activity (Pratt et al. 1995). These experiments provided benefits that were still discernible 6 months later. Years later, sound and music were the focus in Hinterberger and Baier's body of work in providing aural elements to an slow cortical potential (SCP)-driven communicative tools, such as rewarding musical jingles linked to successful EEG control, and in their system POSER, short for Parametric Orchestral

Sonification of EEG Rhythms (Hinterberger and Baier 2004). Spurred on by research indicating the superiority of audio over visual feedback in a system with multiple inputs (Fitch and Kramer 1994), POSER applied musical mappings to assist real-time analysis of EEG information. In initial implementations of POSER, features of multiple brainwave rhythms were mapped to MIDI instruments and presented to users. Continuous sounds were modulated in pitch and volume according to changes within the bandwidth of a corresponding rhythm. Reports showed that users were able to evoke control over individual EEG rhythms, as successfully as 85 % during trials, using musical notes as real-time feedback for simultaneous EEG data. This approach is later adopted in a system that screens EEG for dynamic characteristics (Baier et al. 2007), such as those prominent in diseases including epilepsy and Alzheimer's (Jeong 2002). Here, events of interest within EEG are mapped to digital synthesis parameters in Csound music software (Boulanger 2000), to aid in the distinction between normal and abnormal rhythms in patients. By connecting expected EEG artefacts to synthesis features such as amplitude modulation and harmonic content, a sonic real-time interpretation of meaningful data is available. In another system, the use of sound localisation via an array of speakers is used to reflect the horizontal location, across the scalp, of the current activity. Further work into these sonification techniques also addressed interaction and user acceptance issues (de Campo et al. 2007).

10.6 Computer Music and the Brain

The mappings in early experiments with music and brainwaves were built into the hardware that was used. They were pre-determined by the equipment available, they were fixed and they were difficult to change or undo. BioMuse, a hardware and software system developed by Benjamin Knapp and Hugh Lusted in the 1990s, introduced a major departure from this, with the use of real-time digital computing to process EEG data (Knapp and Lusted 1990).

BioMuse provided a portable kit for digitally processing bio-signals, but what was ground breaking was that it was able to convert these signals into MIDI data. Thus, creating a MIDI controller based on bodily responses, BioMuse also measured eye movements, muscle movements and sound from a microphone input. This use of the MIDI protocol allowed for an EEG signal to be mapped to the input of MIDI-enabled equipment, such as a synthesiser, a drum machine or a sequencer. Furthermore, the technology allowed for fine-tuning of input data. An input threshold switch and a channel sensitivity control meant that the system could be calibrated for different users and different applications. Adjusting the threshold allowed for amplitudes over a specified level to trigger a specified MIDI command, and increasing the channel sensitivity increased the number of MIDI values in a corresponding range. A demonstration of BioMuse presented at the New Music Seminar 1990 in New York City showcased this method of mapping multiple bio-signals to MIDI parameters.

The BioMuse software provided the ability to manipulate bio-signal to MIDI mappings. With the large number of MIDI commands available, this feature allowed alpha waves to be mapped to note-specific MIDI commands, such as Note On or Note Off, or to affect sounds triggered by other bio-signals, such as Control Change messages. From 1987, bursts of alpha activity were sonified via a MIDI synthesiser (Lusted and Knapp 1996), and again the use of opening and closing the eyes was incorporated into compositions to generate significant differences in alpha activity.

Earlier we mentioned the piece *Music for Sleeping and Waking Minds*, which is a more recent work using updated versions of these tools. This is an 8-h-long composition intended for night-time listening. Four performers wearing EEG sensors affect properties of tones using simple direct mappings, in order to project basic changes in their brainwave activity to an audience. Alongside alpha activity, delta rhythms and spindles are also measured and mapped to parameters of audio. The contrast in input parameters is reflected through the resulting sound. Where alpha rhythms are prominent during modes of light sleep and through closing of the eyes, delta rhythms, waves between approximately 0–4 Hz, are associated with deepest levels of sleep. A spindle is recorded as a spike in activity between 11–16 Hz with a duration ≥0.5 s and combines with muscle twitches during periods before deep sleep (Babadi et al. 2012). These three classes of brain activity associated with different stages of sleep are mapped to three musical parameters. Within the composition are sixteen tones of differing spectra. Each performer controls parameters relating to four of these tones. An increase in alpha activity applies a tremolo effect to the tones, prominent delta waves change the timbre of the tones, and spindles trigger enveloped tones through a delay effect with feedback (Ouzounian et al. 2011). Whereas delta activity and spindles are not wholly controllable, these three elements of brain activity are effectively communicated through the act of watching the performers sleep as well as listening to the resulting audio.

10.7 Event-Related Potentials and Auditory Stimuli

Research into using brainwave activity for musical purposes has not been limited to translating alpha and other rhythms related to meditative states. Studies into ERPs have led to BCMIs designed to measure brain activity as a direct result of sensory, cognitive or motor responses. The ability to actively generate brain activity using ERPs has led to BCMI systems whereby a user has full control over the musical outcomes.

ERPs are electrophysiological brain responses produced by perception to stimuli that is presented to a subject. They are locked in time to the event of the stimuli, and they are sources of controlled and visible variability in brain activity (Donchin et al. 1978). ERPs highlight the role of anticipation within brain processing as they can be elicited by deviation from expected events provided by, on the whole repetitive, stimuli.

In 1990, Risto Näätänen reported on a number of experiments in measuring brain activity relating to attention using auditory stimuli. Even though attention research involving ERPs had been going on for over 50 years at the time, Näätänen was keen to distinguish between the brain's automatic responses to stimuli and responses derived from someone's attention and their interpretation of the heard stimuli (Näätänen 1990). The idea of a subject being able to shift their attention at will to auditory stimuli opened up possibilities of BCMI systems controlled by a user's attention to elements of what they are hearing.

Research into attention and sound has long been investigated even before the use of EEG, and earlier research observed a phenomenon known as dichotic listening in regard to how we focus our hearing attention. Dichotic listening is the process of paying attention to sound from one ear whilst ignoring sound from the opposite ear. When asked to focus on speech arriving at one ear, subjects were often unable to recall speech of the same volume from the opposite ear (Cherry 1953). In Näätänen's experiments, he found that the brain reacts to deviations from repetitive sounds automatically, even when a listener focuses their attention away from what they are hearing. This was measured with a P300 EEG response, where the potential begins with a positive deflection and peaks at around 300 ms after the onset of the stimuli. This 'oddball paradigm' implied that when presented with recurring audio information, the brain reacts automatically, and predictably, to deviations in audio patterns.

Throughout the 1990s and early 2000s, further research into how the brain responds to auditory stimuli shed light on how the brain processes our perceptions of music. A key area in this field is the study of meaning held within ERPs, building upon previous research into how the brain processes language (Besson and Macar 1987). Here, the term meaning has more depth than mere EEG association to input. Besson and Faïta (1995) demonstrated how different responses within ERPs are elicited when subjects listen to musical phrases that end either congruently or incongruently in pitch or rhythm. The results also show how differences between musicians and non-musicians indicate that musical expertise can influence aspects of music processing, aside from mere perception.

In 2003, Besson and Schön reported that the P600 ERP response (a positive deflection peaking at around 600 ms post-stimuli) is associated with syntactic violations in language and music such as grammatical errors and incongruously ending musical phrases. Whereas increases in the N400 (negative deflection around 400 ms) ERP are associated with unexpected semantic violations in language, such as 'The pizza is too hot to cry' (Besson and Schön 2003). The amplitude of the ERP is relative to the degree of the violation; the more abstract the meaning results in a potential with higher amplitude.

This research indicates that there is a separate mechanism in the brain for processing music, and although the P600 is a slower response than the N400, it nonetheless provided a basis for further research into applying auditory perception into controlling music. A difficulty in using ERPs as a control source in BCMIs is the issue of identifying potentials amongst non-related EEG information. To address this, epochs of ERPs are summed and averaged from many presentations of the same stimuli in order to gauge whether the response is positive or not. This extra time adds a delay to the signal processing, distancing control away from real-time musical influence.

10.8 EEG Classification and Auditory Stimuli

By the early 2000s, there were several headband-based systems that could play music from EEG data (Miranda 2001). The majority of these provided only two electrodes and very limited tools for interpreting the raw EEG data. Moreover, the quality of the EEG obtained with these less costly systems did not match the minimum standards required to implement reliable BCI system. Nevertheless, in 2001, Alexander Duncan, then a PhD candidate working under the guidance of Eduardo Miranda and Ken Sharman at the University of Glasgow, proposed a BCMI system based on musical focusing through performing mental tasks whilst listening to music, alongside EEG pattern classification (Duncan 2001). Duncan proposed a number of data classification methods for collecting a subject's EEG profile to create an offline neural network classifier, which is used for comparative analysis of EEG readings. This system could effectively be trained to understand the brain signals of a user so that in practice there was a built-in model to apply 'best-fit' rules to derive the meaning within the EEG. Here, EEG was extracted through power spectrum analysis, instead of ERPs. Power spectrum analysis uses Fourier transformations to observe the amplitudes of EEG frequencies. In this set-up, EEG generated from external stimuli was analysed by a computer to create classifications of patterns over multiple trials. Building such a classification systems used artificial intelligence to create models of expected users responses. A model is built from the averages of many practice tests of an individual's response to stimuli, which in effect trains the system. When the system is then engaged in an experiment, it reads an incoming EEG signal and classifies it against the artificial neural network stored within its memory.

Researchers based at the Interdisciplinary Centre for Computer Music Research (ICCMR), University of Plymouth implemented this approach in experiments that combined auditory attention with data classification to analyse features within a short epoch of post-stimuli EEG. In 2003, Miranda and colleagues reported on three experiments that investigate methods of producing meaningful EEG, two of which were deemed suitable for practical musical control. The first of the two uses the technique of *active listening*, and the second uses *musical focusing*.

In the first experiment, small epochs of EEG measured across 128 electrodes were analysed to determine any difference between the acts of *active listening* (replaying a song in the *minds ear*) and *passive listening* (listening without focus). Trials were multiplied and looped to build a portfolio of EEG readings. Musical stimuli consisted of melodic phrases being played over rhythmic patterns. In different trials during a break between melodies, subjects were asked to do three different things. In the first trial to replay the tune in their heads, in a second to try relax their minds without focusing on anything in particular, and in a third to count. Trials were carried out in a number of orders for greater disparity, and a mental counting exercise was factored in as a test of whether musical concentration through active and passive listening was extrinsic to standard methods of mental concentration focusing (Miranda et al. 2003).

The second experiment set to determine whether EEG could identify if a subject was engaged in *musical focusing* (paying particular attention to an element of music being heard) or *holistic listening* (listening to music without any effort). During the *musical focusing* experiments, subjects were asked to focus attention to an instrument within the music that was positioned either in the left or right stereo field.

These tests suggested that it might be possible to accurately measure EEG differentiation between someone engaged in mentally focusing on music and holistic listening. The second test suggested that it might be possible, although to a lesser degree, to record whether a subject is focusing on sound arriving in the left ear or the right ear, whilst in both experiments, the counting exercise provided a different response in the EEG indicating that musical focus uses different brain processing mechanisms that other forms of concentration.

The experiments were conducted in blocks of multiple trials, and the results were derived offline. However, their outcomes led to two initial concepts for BCMIs. *b-soloist* is a BCMI system designed to detect active and passive listening. A continuous rhythm is presented to a subject with regular melodic phrases overlaid. Straight after the melody is played the system looks for either an EEG reading of active or passive listening. If the reading shows active listening has occurred, then the next melody line will be a variation of the last. If the reading shows passive listening occurred, then the next melody played will be exactly the same as the last (see also Chap. 1). *b-conductor* was designed to use musical focusing to affect changes in either left or right channels of music (Fig. 10.4). When presented with music in both channels, a user selects a channel through attentively focusing on the instrumentation it contains. At regular intervals, the system detects the channel of attention in the EEG, and this recognition is mapped to the music, turning up the volume of the focused channel. After a change is made, the volume then returns to a default value until the next command to change is received.

In 2004, Miranda and colleagues reported on a further experiment that investigates EEG derived from auditory imagery. In this, they further the search for distinctions between mental tasks looking for any distinguishable differences between *active listening* and tasks based on *motor imagery* and *spatial navigation*, whereby a subject focus their attention to a physical movement whilst remaining still (Miranda et al. 2004). Tests again used power spectrum analysis but with three pairs of electrodes (7 in total with a reference electrode) to determine a classification system through building a neural network. The three extra tasks assigned were for a subject to imagine opening and closing the right or left hand (motor) and to imagine scanning the rooms of their home (spatial). A separate pair of electrodes read EEG data corresponding to each task, and the voltage difference between the pairs was derived. It was observed which pair produced EEG readings that could be most easily discriminated against another. Again, results were very positive with the largest distinction recorded between auditory imagery and spatial imagery.

Not only did this latter test minimise the number of electrodes for accurately reading overall EEG, thus likely reducing interference and preparation time, but it also narrowed the gap between BCMIs and EEG techniques within other BCI fields such as assistive technologies, where patients already accustomed to motor imagery would need less training.

Importantly, these experiments indicated that subjective choices can elicit expected brain responses. Unlike the previous experiments with auditory stimuli, they do not rely on the subject's expectation or perception of stimuli but allow for a user to impose a subjective decision that has the possibility of becoming separate from the meaning within the music being used. This is a crucial step in the leap towards BCI control of music through neurofeedback.

This element of subjective control aside, the systems discussed in this section rely on an intrinsic link between the stimuli and resultant music. They are in effect one and the same, creating the ultimate feedback loop. Attempting to implement such a BCMI as an interoperable interface with musical systems outside brain-related activity becomes extremely difficult when using auditory stimuli as the driver for generating EEG. Issues of attention become prominent when a user is required to focus on specific sounds to generate EEG, which then have a separate effect as they produce or affect unrelated music as the result. BCMIs designed specifically for utilising these features, such as the *b-soloist* and *b-conductor* ideas, rely on the use of the stimuli as the driver and the receiver of neurofeedback. However, to design any systems outside such a tight link, the element of neurofeedback can become confused and even lost, as the cause is disengaged from the effect. To counter this, a compromise in neurofeedback loss is made, heavy user training is required to reassign unrelated mappings through decision making, or as noted by Miranda et al. (2003), higher levels of intelligence are imparted in compositional algorithms detracting from cognitive musical control.

10.9 Towards BCI Control of Music

Currently, there are a number of systems offering EEG detection linked to musical functions commercially available, e.g. WaveRider, g.tec, Emotiv, to name but three. These systems provide various methods of processing raw EEG that can be mapped to musical engines, in effect providing the hardware for a BCMI system. At the time of publication, there are few systems that allow for mapping EEG directly to musical programs without direct access to APIs and designing bespoke tools; however, the Emotiv system offers the ability to map raw EEG into open sound control (OSC) data, and software such as Brainbay and WaveRider provides tools for mapping EEG to MIDI. We note however that the prices of EEG equipment can differ enormously. The reader should exercise caution here because cheaper equipment does not always match the quality of more expensive ones; EEG requires good quality electrodes and decent amplifiers.

To develop sophisticated systems of BCI, control relevant stimuli are required, and unless using in-the-box methods of analysis and data processing, the appropriate means of data acquisition and methods of mapping to a musical engine are necessary, and this requires expertise.

In 2005, Miranda adopted the approach of designing the musical engine of a BCMI with sufficient artificial intelligence in order to create sophisticated meaning from simpler EEG readings. Here, he applied a process known as Hjorth analysis, a second method of extracting EEG alongside power spectrum analysis. Hjorth analysis is the extrapolation and measure of time-based features within short windows of EEG information. These are referred to as the activity, mobility and complexity within the reading, and measures of each are produced involuntarily as they lie within overall EEG data. Using these techniques, the *BCMI-Piano* attempts to guess the mental state of the user and performs real-time generative piano music in response, with features based on the techniques of composers such as Beethoven and Schumann, as discussed in Chap. 3.

The P300 oddball paradigm, earlier mentioned in relation to auditory stimuli research, was used by Grierson (2008) for a BCMI controlled by focusing visual attention to stimuli displayed on a computer screen (See also Chap. 3). The P300 potential was found to contain information relative to visual attention of repetitive stimuli. In the same manner, as deviations in auditory stimuli were found to trigger P300 responses (Näätänen 1990) as an automatic response, the P300 could also be elicited by an unexpected interruption within a repetitive visual pattern. In the case of P300 spelling devices that allow a user to select letters to form words and sentences, the deviant information contains the letter the user desires, and as such is injected with the meaning that a BCI system can knowingly respond to. In the first incarnation of his BCMI, Grierson replaces letters for musical notes for a user to select via a visual interface.

Over the course of trials, Grierson recorded that four out of five subjects were able to perform subjective decision making, with regard to specific note selection and with no training, that were understood by the system 75 % of the time. As ERPs are difficult to detect within EEG, conducting multiple trials improves the reliability of the system to detect these choices and increases the percentage of success. The downside is the time lapse introduced from the initial cognitive decision being made to the end of the trials and the subsequent data processing. Grierson recognises this factor opting for a minimal trial approach in an attempt to link control as close to cognition as possible. The stimuli in this system presented the names of note values over three octaves. Each note name was displayed for approximately 50 ms then removed for up to 1,800 ms, in a quasi-random order. A subject was asked to select a specific note and count each time it was displayed, generating the associated ERP information in synchronisation with each display. Experiments recorded time delays of approximately 12 s, with one subject successfully initiating control over approximately 7 s with less trials, where total time = flash time × choices × trials, e.g. 50 ms × 36 × 7 = 12.7 s.

Although these times are lengthy in comparison to EEG response times in other BCMI devices, what (Grierson et al. 2011) accomplished with his system was the ability to widen choice to a range of values. Instead of a 'one or the other' decision, the meaning within the stimuli was designed to visually represent many more choices, up to 36 in this case example. Grierson and colleagues have since developed a suite of BCMI applications based upon the NeuroSky bluetooth headset (Grierson et al. 2011).

The research into ERPs also went as far as to indicate that BCMI control may not need to rely on a subject training their brain to act accordingly to the intelligence of a BCMI. By relying on the ability of the brain to respond to the focus of attention in a multi-variable environment, no training was necessary as long as the user had the ability to recognise visual events and perform the counting task. As a result of these factors, this method for eliciting P300 for control was subsequently utilised by the neurotechnology company g.tec in their commercial BCI system.

As previously mentioned, the ERP response to a single event is problematic to detect on a single trial basis, as it becomes lost in the noise of ongoing brain activity. However, if a user is subjected to repeated visual stimulation at short intervals (at rates approximately between 5 and 30 Hz), then before the signal has had a chance to return back to its unexcited state, the rapid introduction of the next flashing onset elicits another response. Further successive flashes induce what is known as the steady-state response in the brain's visual cortex, a continuously evoked amplification of the brainwave (Regan 1989). This removes a need for performing numerous delayed trials as the repeated visuals are consistently providing the stimuli required for a constant potential, translated as a consistent increased amplitude level in the associated EEG frequency.

This technique, steady-state visual-evoked potential (SSVEP), was adopted in a BCMI system designed for a patient with locked in syndrome (Miranda et al. 2011) as a tool for providing recreational music making. Here, four flashing icons were presented on a screen, their flashing frequencies correlating to the frequencies of corresponding brainwaves measured in the visual cortex. The user selects an icon simply by gazing at it, and the amplitude of the corresponding brainwave frequency increases. Whilst EEG data are analysed constantly, the system looks for amplitude changes within the four frequencies. The icons represent four choices, always available to the user at the same time. These controls are in turn mapped to commands within a musical engine, as well as being feedback into the display screen to provide visual feedback to the user. The instantaneous speed of the EEG response to the stimuli finally brought real-time explicit control to a BCMI, which required no user or system training beyond the task of visual focusing. Please refer to Chap. 1 for more information on this system.

As well as the selection of commands, a second dimension of control was gathered through the level of focused gazing. This elicited a relative linear response within the amplitude of the corresponding brainwave. This allows users to employ proportional control methods akin to intrinsically analogue tasks such as pushing a fader or turning a dial. This differs from previous selective, more digital tasks in BCMIs, such as a switch or a toggle function. In this system, Miranda and colleagues utilised this control to trigger a series of defined notes within a scale (Miranda et al. 2011).

The SSVEP-based BCMI by Miranda and colleagues broke new ground in BCMI research. This is the first instance of a system whereby a user can precisely control note-specific commands with real-time neurofeedback. It is interesting to refer back to the BCI definition of Wolpaw and Birbaumer (2006) who may well define such systems as outside the realm of true BCI as it relies on the EEG interpretation of eye position and not pure thought processes. That said, in the pursuit of real-time control of brainwaves so far SSVEP, in comparison with motor imagery and P300 BCIs, has been found to offer the quickest and most accurate EEG response and with the least amount of training (Guger et al. 2011). Also, the advantages of these types of systems over previous BCMIs are clear to see. One of the outcomes of this initial SSVEP research was the use of BCMIs in collaborative musical applications. In terms of music used as a real-time communicative tool between people, this system allows a user to play along with a musician, or potentially, with another BCMI user. This was recognised as an important breakthrough for the potential BCMI systems in therapeutic situations and for potentially launching the BCMI into a wider field of collaborative musical applications.

In 2012, we reported on further mapping and compositional techniques using SSVEP within a BCMI (Eaton and Miranda 2012) for the composition of *The Warren*, a multichannel electronic performance piece designed to explore the boundaries of mapping strategies in a BCMI system to generate real-time compositional rules (Eaton 2011). Control of EEG performs generative functions that control macro-level musical commands, such as shifts in arrangement, tempo and

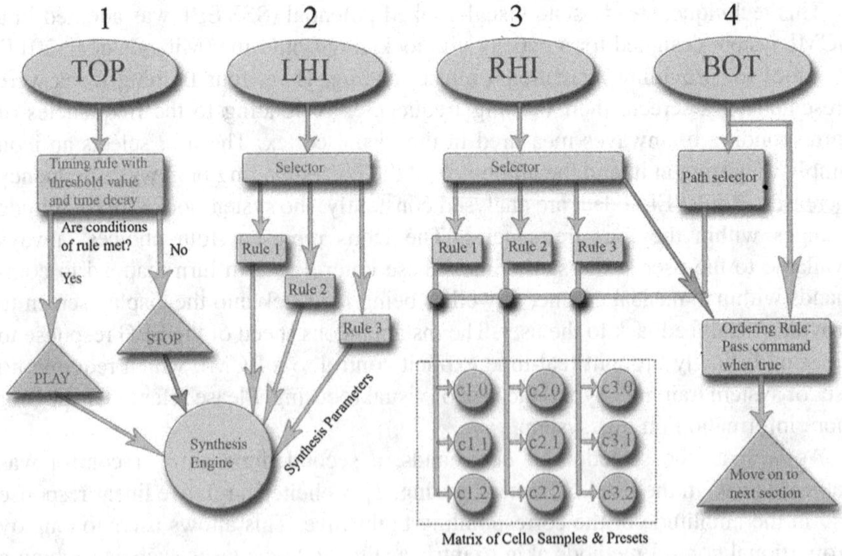

Fig. 10.7 A mapping diagram from a short section of *The Warren*. Here, each icon (1–4) is assigned to a number of commands based on the requirements of the composition

effects over the master channel, alongside control of micro-level functions, such as control over individual pitches or synthesis parameters (Fig. 10.7). This approach provided a framework for addressing performance considerations often associated with more mainstream digital interfaces. The piece was engineered to communicate expressive musical control and to provide a loose framework of musical elements for the performer to navigate through selecting areas for precise manipulation. An important feature of the design was to emulate the unpredictable nature of performing with acoustic instruments, so often safeguarded in performances with electronic instruments. Slight deviations of learnt control patterns or miscalculations when navigating through the piece could result in the *wrong* result, such as bringing the composition to an abrupt end or injecting unwanted silences or dissonance into the piece. This approach forced the concentration of the performer, underlying the importance of successfully interpreting the meaning within the control EEG. To achieve the desired complexities and nuances, mapping rules were designed to suit the musical functions, a break away from previous systems where compositional mappings were intrinsic to the meaning of EEG. Here, the meaning of EEG was designed through the use of the stimuli and therefore learnt or understood by the user. With such an abundant amount of meaningful data, *The Warren* also makes musical use of non-meaningful data to provide deeper complexity through secondary mappings. For example, ordering rules were applied to control-specific musical parameters through monitoring the performer's control behaviour. The order in which icons were selected over x amount of control changes would result in different generative rules being applied, the results unbeknown to the performer who would be concentrating on the current primary task. This harks back to Miranda's technique and the integration of Hjorth analysis, adding intelligent feedback to the system as part of the compositional process and making the system learning between performer and computer mutually exclusive.

When designing mappings for a structured performance piece, as opposed to an instrument or an improvised piece, the mappings need to adapt to the arrangement of the composition and the functions. This reverse engineering method of mapping design based on musical function and necessity provides an interesting arena for creativity. As a result, the mappings explored in *The Warren* vary widely depending on the compositional choices, the sonic intentions of composer (and performer) and the limitations of the input controls. Instead of summarising these mappings solely in numerical terms, the nature of how the control is governed can be presented in parallel with Dean and Wellman's (2010) proportional-integral-derivative (PID) model. This approach defines control as the 'effect' of the input signal onto the output's value, regardless of the number of parameters connected. Proportional control dictates that output values are relative to input; the output is value X because the input is X. Integral control provides an output value based solely upon the history of the input, whereas derivative control gives an output value relative to the rate of change of the input signal.

These principles are adopted in a number of ways in *The Warren*, and the inclusion of conditional rules and variations allows for an abundance of creative implementations. For example, in the first movement, a cello sound can be played using the derivative measurement of the increment and decrement of one of the four EEG input channels. Alongside this a second input channel has an integral control to regulate a modulation index of the cello sound processing, an example of interpolating two different primary controls to manipulate one sound. To add further control within these selection-based mappings, mapping rules were applied to the four incoming EEG data streams and used at various times during the piece depending on the required function. Here, we look at three of these rules, *threshold values, timing* and *ordering*.

10.9.1 Threshold Values

All four of the brainwave control signals can act as a single selector using mappings for when the amplitude is high or low. Beyond this, each input signal can be assigned to elicit a number of commands. In this technique, user control of brainwave amplitude (of a specific channel) was mapped to a series of functions across a range of evenly spaced threshold values scaled according to the input range. When the input signal passes a threshold value, a control command is triggered. For example, an input range of 1-25 could be treated with the following rules:

if input == 5 play note C2
if input == 10 play note D2
if input == 15 play note E2
if input == 20 play note F2

Without further consideration, an input signal rising and falling through this range would excite all of the notes on the way up and on the way down. The use of timing rules (below) provided the performer with the ability to make specific selections whilst avoiding triggering unwanted commands.

10.9.2 Timing

The majority of the mappings within *The Warren* are led by timing rules. Calculating the time, a user takes to complete cognitive tasks allows for an added dimension of control. Expanding the simplified threshold example shown above the speed at which a brainwave increases towards a threshold value would dictate whether the in between mapping rules are accepted or not. This allows the performer to choose how many of the threshold values within one input range to select during any one command. For example, if the time between initial excitation and input signal reaching a value of 15 is greater than X, then only note E2 would be played. If the time taken was less than X and more than 1/2X, then notes D2 and E2 would be played. Less than 1/2X then C2, D2 and E2, and so forth. In practice, the timing rules were used like this alongside the threshold values and also separately on their own. They were mapped to parameters ranging from audio effects settings including delay, filter and distortion parameters and audio playback sample chopping controls such as playback position, pan position, and texture blending.

Further complexity was added through exploiting the features of using timers. A hold-and-release function allowed for a change in control to occur at the point of release. The time between the hold command and the release command being received provided selectable options. When an input value increases, a timer begins until the value decreases. Upon this decrease, the value of time is compared against a series of rules. In practice, the accuracy of brainwave control can vary due to a range of factors such as tiredness, environment, mood and electrical interference. To accommodate this instability when attempting to sustain brainwave amplitude through SSVEP, a further time delay rule monitors the EEG. For example, if we define a threshold input value of 5, so that when the input value increases above 5 a hold command is activated. If the input stays above 5, then the hold command stays on, and if the value decreases below 5, it is released. To add some flexibility to this simple hold-and-release function, a time delay of 3 s is added to the hold function. Therefore, if the input decreases below 5 for less than 3 s and then increases to above 5, the hold command remains on. If the input decreases to below 5 for longer than 3 s, then the release function is activated. This technique creates a rule whereby an icon needs to be fixated on constantly to generate a command sent to the performance system, akin to the constant attention required to play a sustained note on an acoustic instrument. Deviation from this attention is allowed for a time span of up to 3 s, allowing for the performer to utilise other input commands to manipulate the sound via different parameters or to control other aspects of the music. This flexibility can able help combat irregularities in the input signal. To help performer calculate times during a performance, a digital clock display was built into the visual interface.

10.9.3 Ordering

The mappings and structure of the *The Warren* were designed to allow loose periods for the performer to 'play' the system. Within this it was unlikely that the exact manner in which the controls were used would ever be the same twice. To add a layer of surprise and quasi-randomness to the piece, as well as to further engage the concentration of the performer to adapt to the system secondary mappings were dominated by applying rules to the order in which icons were selected and which commands were triggered. At times, these rules were mapped to stochastic musical parameters ensuring a controlled level of unpredictability.

The level of depth attained within these mapping strategies requires a high level of mental concentration and awareness of time, external and in relation to the music within a performance. Here, the mappings had to be tested, learned, practiced and optimised for system performance and user ability.

The Warren demonstrated that BCMI technology could be used in place of more traditional digital controllers as well as in a live performance setting. In 2013s *Flex* (Eaton and Miranda 2013a), this idea was taken a step further through a BCMI built using affordable hardware and open-source software. In an effort to make music making with brainwaves more accessible, *Flex* used SSVEP with an EEG headset by Emotiv and two laptop computers, one providing the visual interface, EEG signal processing and transformation algorithms and the other the musical engine. The gap between compositional and mapping design used in *The Warren* was disregarded here as the two elements were intertwined. *Flex* is designed to be between approximately 10–15 min long depending on the how the controls are found and used. The composition combines sound sources recorded in surround sound, ranging from fairground ambience to bell chimes, with synthesised and heavily processed sounds. A key aim of the performance is to convey the narrative of the composition whilst attempting to engage an audience with the control tasks being undertaken by the performer.

Flex uses the idea of control as a key theme. Instead of merely providing control, *Flex* hides control and moves it around forcing the performer to adapt to the system and learn control before it is taken away again. In effect, the controls corresponding to the icons are randomised; different elements of the composition are presented without any way of the performer knowing in advance. Built-in rules allow for the presence of mappings corresponding to the current sounds being played, but the choice of parameters is selected at random from an array of predetermined functions. Performed in quadraphonic sound, mapping rules mix the sounds across the four channels as well as control the arrangement of the piece. Additional mapping rules control micro-level functions such as audio sample playback and audio effects (Fig. 10.8).

Indeed, there are more mappings available that can be used in any one performance, which helps make every performance different. The line between active and passive control becomes somewhat blurred here due to the manner in which control is attained. Control in *Flex* can be difficult to establish, and this brings elements of the unexpected and even the undesired into a performance. Again, hidden secondary mappings are also built into add elements of surprise, in effect further flexing the rigidity of control throughout the piece. Overall, the mapping system is designed for control to be manageable, and where control becomes lost, it is relatively easy to recover. As such, certain safety features are implemented in order to prevent complete chaos. Performing *Flex* becomes a musical game, where the aim and reward are control (although the success rate of control is not a primary concern) and where the audience is rewarded with the resulting music and performance.

One of the main issues with performing with a BCMI system is what could be considered as a lack of obvious 'performance'. EEG measurement requires

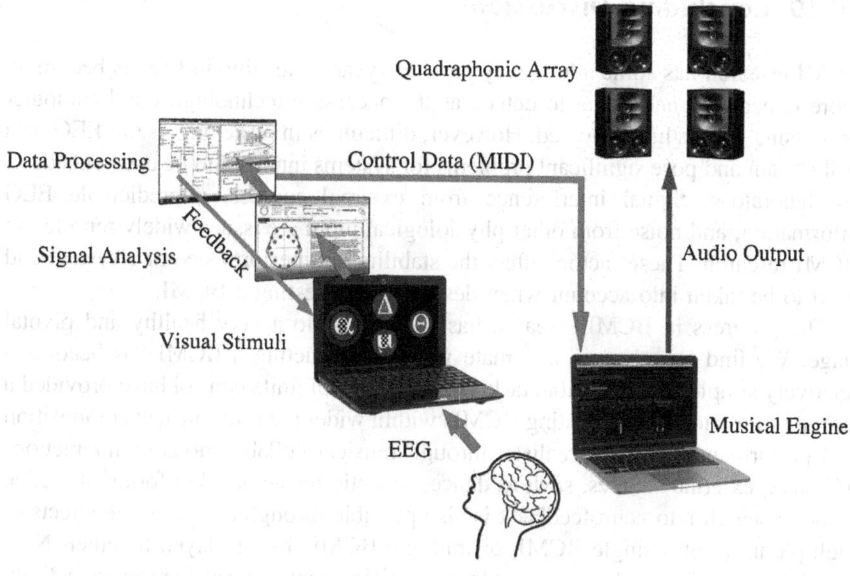

Fig. 10.8 The system components for *Flex* built using consumer grade EEG hardware, two laptop computers and open-source software

motionless concentration, and the use of visual stimuli requires staring at a computer screen, both of which offer a rather disengaging spectacle for viewing however sophisticated the underlying processes are. We are fortunate to now be at a stage where the technology is no longer the only focus of attention, yet we need to be mindful of how we communicate the practices of BCMI to an audience and the aesthetic effects of the tools we choose. This has led recent work to move away from brainwave control over electronic sounds towards integrating brainwave control and external musical bodies, including acoustic instruments and musicians.

In 2013, Eaton and Miranda reported on *Mind Trio*, a proof-of-concept BCMI that allowed a user to control a musical score in real time, choosing from short pre-composed musical phrases (Eaton and Miranda 2013b). SSVEP provides choice over fours phrases during a window of time. These windows are synchronised with a metronome and dynamic score presented to a musician via a computer monitor. Within each window, the user selects the phrase that is displayed at the next sync time. The musician is presented with the current phrase and is shown the next phrase shortly before it becomes active. This extension of brainwave control designed to accommodate the involvement of a third party is the basis of *Activating Memory*, an experimental composition for a string quartet and a BCMI quartet. It uses the same principle as *Mind Trio* for users to choose phrases of music that a corresponding musician then performs. All four systems are synchronised via a master clock across two movements. *Activating Memory*'s debut performance was at the 2014 Peninsula Arts Contemporary Music Festival, Plymouth, UK.

10.10 Concluding Discussion

BCMI research has come a long way in recent years. Meaning in EEG is becoming more understood and easier to detect, as the necessary technologies and computer processing speeds have allowed. However, difficulties in retrieving useful EEG data still remain and pose significant problems for systems intended to be used outside of the laboratory. Signal interference from external sources, unpredictable EEG information, and noise from other physiological input are issues widely reported in BCMI research. These factors affect the stability and performance of a system and need to be taken into account when designing and testing a BCMI.

The progress in BCMI research has brought us to a very healthy and pivotal stage. We find ourselves in a climate where constructing a BCMI has become a relatively simple and affordable task. New systems of finite control have provided a strong foundation for integrating BCMIs within wider areas of musical composition and performance, perhaps realised through musical collaborations or interactions with live, external sources, such as dance, acoustic music or other forms of media. Wider research into neurofeedback is also possible through assessing the affects of multiple users of a single BCMI, or multiple BCMIs being played together. Now that the appropriate tools are available, we anticipate an increase in research activity across a wider playing field, with a particular emphasis on compositional integration. We are slowly beginning to see brainwave control creep into everyday tech culture, and as in all successful interdisciplinary areas, we expect it to be prominent in all of the clinical, therapeutic and recreational interpretations of what a BCMI is.

Events bringing researchers and practitioners together have produced fruitful experiments in the past, as evident in programs such as eNTERFACE (Arslan et al. 2006; Benovoy et al. 2007). In the current climate of expansion in BCMI research, the dissemination of ideas and collaboration between practitioners linking BCMI research and related areas together is an opportunity to be embraced to further accelerate work in this field and should not be ignored.

It can still be argued that more meaning within EEG is needed, not only in BCMI research but also in our overall understanding of the brain. As we have seen, meaning leads to control and in turn complexity, and advances in this offer exciting prospects. One area of research that promises to widen the scope of interpreting meaning in EEG is the study of emotional responses in brain activity, and evolving research in this field is already uncovering very direct links with emotional responses and music (Crowley et al. 2010; Kirke and Miranda 2011).

The use of modern BCMI systems for performance in concert settings has marked the arrival of more accessible, responsive and sophisticated platforms for designing and building successful BCMI systems, bringing brainwave control and music full circle. In place of Lucier's percussive instruments are dynamic scores and complex musical engines. And instead of bursts of alpha activity there are layers of sophisticated EEG control on offer. The importance of considering mapping strategies in the development of BCMI systems can be traced all the way back to Alvin Lucier and his *Music for Solo Performer*, an interface that offered such a unique and tangible interaction with brainwaves, from such limited input. With this, and the availability of today's tools, in mind, we hope to see a rise in the creative applications of brainwaves in music coming from composers as well as researchers through approaches applying the complexity in compositional and mapping strategies that have now become a reality.

10.11 Questions

1. What were the first type of brainwaves used for musical control and how were they controlled by a subject?
2. What is the difference between the sonification and the musification of brain-wave signals?
3. What is the function of the transformation algorithm in a BCMI system?
4. With todays technology in mind, consider an approach to modernising the mappings in Alvin Lucier's *Music for Solo Performer*. How could the piece be reworked?
5. What features of ERPs make them useful for mapping to musical functions?
6. Design a concept for a BCMI that uses two techniques of EEG extraction as input signal. Do the techniques you have chosen fit the concept well? Could other techniques be used instead?
7. What are the benefits of a user-orientated BCMI over a computer-orientated BCMI?
8. Compare the mappings of the two pieces *Music for Sleeping and Waking Minds* and *In Tune*. How would each piece differ if the systems were swapped for both performances?
9. What are the main differences between the P300 and SSVEP techniques and how do they affect musical control?
10. Consider a musical extension of a BCMI. How could integrate BCMI into another type of musical interface of your choice?

References

Arslan B, Brouse A, Castet J, Le´hembre R, Simon C, Filatriau J, Noirhomme QA (2006) Real time music synthesis environment driven with biological signals. In: IEEE international conference on acoustics, speech, and signal processing, Toulouse, France

Babadi B, McKinney SM, Tarokh V, Ellenbogen JM (2012) DiBa: a data-driven bayesian algorithm for sleep spindle detection. Biomed Eng 59(2):483–493

Baier G, Hermann T, Stephani U (2007) Multi-channel sonification of human EEG. Paper presented at the 13th international conference on auditory display, Montréal, Canada, 26–29 June 2007

Benovoy M, Brouse A, Corcoran TG, Drayson H, Erkhurt C, Filiatriau J, Frisson C, Gundogdu U, Knapp B, Lehembre R, Mühl C, Ortiz Pe´rez MA, Sayin A, Soleymani M, Tahiroglu K (2007)

Audiovisual content generation controlled by physiological signals for clinical and artistic applications. In: Enterface'07, Istanbul, Turkey

Besson M, Fai'ta F (1995) An event-related potential (ERP) study of musical expectancy: comparison of musicians with nonmusicians. J Exp Psychol 21(6):1278–1296. doi:http://dx.doi.org/10.1037/0096-1523.21.6.1278

Besson M, Macar F (1987) An event-related potential analysis of incongruity in music and other non-linguistic contexts. Psychophysiology 24(1):14–25. doi:10.1111/j.1469-8986.1987.tb01853.x

Besson M, Schön D (2003) Comparison between language and music. In: Peretz IaZ RJ (ed) The cognitive neuroscience of music. Oxford University Press, New York

Boulanger R (2000) The Csound book. The MIT Press, Cambridge

Cahn BR, Polich J (2006) Meditation states and traits: EEG, ERP, and neuroimaging studies. Psychol Bull 132(2):180–211. doi:10.1037/0033-2909.132.2.180

Cherry EC (1953) Some experiments on the recognition of speech with one and with two ears. J Acoust Soc Am 25:975–979

Crowley K, Sliney A, Pitt I, Murphy D (2010) Evaluating a brain-computer interface to categorise human emotional response. In: IEEE 10th international conference on advanced learning technologies (ICALT), 5–7 July 2010, pp 276–278. doi:10.1109/ICALT.2010.81

Dean TL, Wellman MP (2010) Planning and control. Morgan Kaufmann Publishers, San Mateo

de Campo A, Höldrich R, Wallisch A, Eckel G (2007) New sonification tools for EEG data screening and monitoring. In: 13th International conference on auditory display, Montreal, Canada

Donchin E, Ritter W, McCallum WC (1978) Cognitive psychophysiology: The endogenous components of the ERP. In: Callaway E, Tueting P, Koslow SH (eds) Event-related brain potentials in man. Academic Press, New York

Duncan A (2001) EEG pattern classification for the brain-computer music interface. Glasgow University

Eaton J (2011) The Warren. http://vimeo.com/25986554. Accessed 3 Jan 2014

Eaton J, Miranda E (2012) New approaches in brain-computer music interfacing: mapping EEG for real-time musical control. In: Music, mind, and invention workshop, New Jersey, USA

Eaton J, Miranda, E (2013a) Real-time brainwave control for live performance. in: 10th International symposium on computer music multidisciplinary research (CMMR): sound, music and motion, Marseille, France

Eaton J, Miranda, E (2013b) Real-time notation using brainwave control. In: Sound and music computing conference (SMC) 2013, Stockholm, Sweden

Eaton M (1971) Bio-music: biological feedback, experiential music system. Orcus, Kansas City

Fitch WT, Kramer G (1994) Sonifying the body electric: superiority of an auditory display over a visual display in a complex, multivariate system. In: Kramer G (ed) Auditory display—sonification, audification, and auditory interfaces. Addison-Wesley, Reading, p 307

Garnett G, Goudeseune C (1999) Performance factors in control of high-dimensional spaces. In: International computer music conference (ICMC 1999), Beijing, China, pp 268–271

Goudeseune C (2002) Interpolated mappings for musical instruments. Organised Sound 7 (02):85–96. doi:10.1017/s1355771802002029

Grierson M (2008) Composing with brainwaves: minimal trial P300 recognition as an indication of subjective preference for the control of a musical instrument. In: ICMC, Belfast, Ireland

Grierson M, Kiefer C, Yee-King M (2011) Progress report on the EAVI BCI toolkit for music: musical applications of algorithms for use with consumer brain computer interfaces. In: ICMC, Huddersfield, UK

Guger C, Edlinger G, Krausz G (2011) Hardware/software components and applications of BCIs. In: Fazel-Rezai R (ed) Recent advances in brain-computer interface systems. Rjeka, Croatia, pp 1–24

Hinterberger T, Baier G (2004) POSER: parametric orchestral sonification of EEG in real-time for the self-regulation of brain states. In: International workshop on interactive sonification, bielefeld

Hunt A, Wanderley MM, Kirke R (2000) Towards a model for instrumental mapping in expert musical interaction. In: International computer music conference, San Francisco, USA. International Computer Music Association

Jeong J (2002) Nonlinear dynamics of EEG in Alzheimer's disease. Drug Dev Res 56(2):57–66. doi:10.1002/ddr.10061

Kirke A, Miranda E (2011) Combining EEG frontal asymmetry studies with affective algorithmic composition and expressive performance models. In: International computer music conference (ICMC 2011), Huddersfield, UK

Knapp RB, Cook PR (2005) The integral music controller: introducing a direct emotional interface to gestural control of sound synthesis. Paper presented at the international computer music conference, Barcelona, Spain

Knapp RB, Lusted H (1990) A bioelectric controller for computer music applications. Comput Music J 14(1):42–47

Lucier A (1976) Statement on: music for solo performer (1971). In: Rosenboom D (ed) Biofeedback and the arts: results of early experiments. Aesthetic Research Center of Canada, Vancouver

Lusted H, Knapp RB (1996) Controlling computers with neural signals. Sci Am 275(4):82–87

Miranda E, Sharman K, Kilborn K, Duncan A (2003) On harnessing the electroencephalogram for the musical braincap. Comput Music J 27(2):80–102

Miranda ER, Roberts S, Stokes M (2004) On generating EEG for controlling musical systems. Biomed Tech 49(1):75–76

Miranda ER (2001) Composing music with computers. Focal Press, Oxford

Miranda ER (2006) Brain-computer music interface for composition and performance. Int J Disabil Human Dev 5(2):119–125

Miranda ER, Magee WL, Wilson JJ, Eaton J, Palaniappan R (2011) Brain-computer music interfacing (BCMI): from basic research to the real world of special needs. Music Med 3 (3):134–140. doi:10.1177/1943862111399290

Miranda ER, Soucaret V (2008). Mix-it-yourself with a brain-computer music interface. In: 7th ICDVRAT with ArtAbilitation, Maia/Porto, Portugal

Miranda ER, Wanderley MM (2006) New digital musical instruments: control and interaction beyond the keyboard. Computer music and digital audio series. A-R Editions, Madison

Näätänen R (1990) The role of attention in auditory information processing as revealed by event-related potentials and other brain measures of cognitive function. Behav Brain Sci 13 (2):201–288

Ojanen M, Suominen J, Kallio T, Lassfolk K (2007) Design principles and user interfaces of Erkki Kurenniemi's electronic musical instruments of the 1960s and 1970s. In: New interfaces for musical expression (NIME07), New York, USA

Ouzounian G, Knapp B, Lyon E (2011) Music for sleeping and waking minds. Concert notes, Science Gallery Dublin and The League of Consciousness Providers, Dublin, Ireland

Pratt RR, Abel HH, Skidmore J (1995) The effects of neurofeedback training with background music on EEG patterns of ADD and ADHD children. Int J Arts Med 4(1):24–31

Regan D (1989) Human brain electrophysiology: evoked potentials and evoked magnetic fields science and medicine. Elsevier, New York; London

Rosenboom D (1990) Extended musical interface with the human nervous system. Leonardo MonoGraph series

Teitelbaum R (1976) In Tune: some early experiments in biofeedback music (1966–1974). In: Rosenboom D (ed) Biofeedback and the arts: results of early experiments. Aesthetic Research Centre of Canada Publications, Vancouver

Teitelbaum R (2006) Improvisation, computers and the unconscious mind. Contemp Music Rev 25 (5–6):497–508. doi:10.1080/07494460600990026

Wolpaw JR, Birbaumer N (2006) Brain–computer interfaces for communication and control. In: Selzer M, Clarke S, Cohen L, Duncan P, Gage F (eds) Textbook of neural repair and rehabilitation, vol 1, Neural repair and plasticity, vol 1. Cambridge University Press, UK, pp 602–614

Wu D, Li C, Yin Y, Zhou C, Yao D (2010) Music composition from the brain signal: representing the mental state by music. Comput Intell Neurosci, 267671. doi:10.1155/2010/267671

Wu D, Li CY, Yao DZ (2009) Scale-free music of the brain. PLoS ONE 4(6):e5915

Retroaction Between Music and Physiology: An Approach from the Point of View of Emotions

11

Pierre-Henri Vulliard, Joseph Larralde and Myriam Desainte-Catherine

Abstract

It is a well-known fact that listening to music produces particular physiological reactions for the auditor, and the study of these relationships remains a wide unexplored field of study. When one starts analyzing physiological signals measured on a person listening to music, one has to firstly define models to know what information could be observed with these signals. Conversely, when one starts trying to generate some music from physiological data, in fact, it is an attempt to create the inverse relationship of the one happening naturally, and in order to do that, one also has to define models enabling the control of all the parameters of a generative music system from the few physiological information available, and in a coherent way. The notion of emotion, aside from looking particularly appropriate in the context, reveals itself to be a central concept allowing the articulation between musical and physiological models. We suggest in this article an experimental real-time system aiming at studying the interactions and retroactions between music and physiology, based on the paradigm of emotions.

P.-H. Vulliard (✉) · J. Larralde · M. Desainte-Catherine
University of Bordeaux, LaBRI, UMR 5800, Bordeaux, Talence, France
e-mail: pierre-henri.vulliard@labri.fr

J. Larralde
e-mail: larralde@labri.fr

M. Desainte-Catherine
CNRS, LaBRI, UMR 5800, Bordeaux, Talence, France
e-mail: myriam.desainte-catherine@labri.fr

E.R. Miranda and J. Castet (eds.), *Guide to Brain-Computer Music Interfacing*,
DOI 10.1007/978-1-4471-6584-2_11

11.1 Introduction

The experimental software setup presented in this article, called MuZICO (french acronym for Zen Music and Brain Computer Interface), allows the approach of interactions between music and physiology from two points of view: The first, the low-level one, is based on the sonification of physiological signals produced by the autonomous nervous system, thus creating a direct link between the unconscious manifestation of a physiological state and the production of acoustic vibrations. The second point of view, the high-level one, introduces an abstraction layer dealing with the recognition of vigilance states (vigilance, relaxation, sleep) and/or emotional states of the listener and allows to control the generative music system parameters in real time according to these states.

Until now MuZICO has been tested with two different devices for physiological signals capture. The first is the Thought Technologies ProComp5 infinity biofeedback system, a five inputs box for various compatible sensors. The following Thought Technologies sensors have already been used with MuZICO:

- abdominal/thoracic breath sensor
- skin conductivity sensor
- electrocardiograph
- electroencephalograph

The other system we are mainly using is an Emotiv EPOC headset, which is a sixteen electrodes wireless EEG headset. However, the signals coming from this headset can be analyzed either as EEG signals, or as muscular signals.

In the following sections, we describe the evolution of our work on sonification and interaction between vigilance states and musical synthesis then we propose a generalization of vigilance states to emotional states and its validation in the context of a collaboration with experts on emotions expressed by body postures. Finally, we inspect prospects offered by the synthesis of these experiences for the study of interactions and retroactions between music, perceived emotions, and physiology.

11.2 Processing of Physiological Signals from the Energetics Approach

11.2.1 Sonification: Direct Mapping

In our case, sonification is used to catch the listener's attention and to help him understand the link between the manifestations of his autonomous nervous system, usually unintentional, and the music they generate. The energy deployed by the internal organs and sampled by the sensors is directly translated into sound energy, thus leading to a synesthesic perception phenomenon.

In specific terms, we compute in real time the variation speeds of the data flows coming from all sensors, except the signals coming from the EEG electrodes. These variation speeds are translated into energy by a direct control of the volume, spectral richness, or pitch of a sound. As regards the heart and breath sensors, the frequency of the cycle is evaluated (heart rate or inhalation-exhalation period), and the variations of this frequency are also translated into volume, spectral richness, or pitch variations. Concerning the signals from the EEG electrodes, the evolution of energy ratios between wave bands reflecting typically vigilance states is sonified in the same way.

11.2.2 Interaction Between Vigilance and Musical Synthesis

In addition to the production of sound directly from physiological signals, the energetics approach also involves a higher level of abstraction by analyzing the listener's arousal in order to generate a musical flow, using harmonic, rhythmic, and spectral rules. The idea here is to control the dynamism of this musical flow according to the listener's arousal.

All the physiological data are therefore integrated into one continuous parameter representing the listener's arousal, and this parameter is then redistributed over the whole set of musical generation parameters. It is a $M \rightarrow 1 \rightarrow N$ mapping.

A study carried out by Pierrick Legrand and Frédérique Faïta Ainseba (Vezard et al. 2011) shows the link between specific musical composition parameters, the energy evoked by the resulting generated music, and the arousal level induced by its listening.

These parameters, as they are used in the MuZICO environment, have either discrete or continuous values and are either defined by adjectives (classes) or by numerical values. For example, the spectrum of a sound can take values from muted to bright, which corresponds to specific energy values that are measurable in the signal and are linked to its spectral centroid. The other parameters related to energy in music are as follows: the tempo of the music, the attack of the notes' dynamic envelope, the pitch of the notes, the number of instruments playing simultaneously, the more or less "natural" sound of the instruments, the percussive or sustained aspect of sounds, as well as the number of simultaneous notes (Livingstone et al. 2010).

All these parameters are characterized by bounded intervals used by MuZICO to generate music.

Let be $V_E$min and $V_E$max the bounds of the intervals of values taken by musical parameters, corresponding to low and high energies evoked by the music, respectively (see Table 11.1). The parameters that have undefined units are translated by MuZICO into various units according to the algorithms used to produce the sounds (synthesis, sample playing, audio-effect control). For example, one can translate the brightness of a sound by a particular setting of the modulation index in a FM

Table 11.1 An excerpt of MuZICO's musical parameters intervals

Parameter (unit)	$V_E$min	$V_E$max
Tempo (bpm)	[0.1; 60]	[144; 300]
Spectrum	Muted	Bright
Notes attack slope (ms)	[100; 1000]	[1; 30]
Notes pitch (Hz)	[110; 220]	[440; 880]
Instrumental sounds	Natural	Synthetic
Nature of sounds	Percussive	Sustained
Number of voices	1	Many
Notes density	Low	High

synthesis algorithm, or by the control of the cutoff frequency of a low-pass filter applied to a sample player.

11.2.3 Applications

The MuZICO system has already been presented to various audiences in artistic (use in concert as a generative musical accompaniment) and scientific mediation (presentation during workshops and as an interactive station) contexts. Its applications can therefore be artistic, educational, or therapeutic.

In the artistic context, MuZICO has been used in two configurations: One, strongly implying one of the authors as a saxophone player, consists in generating the accompaniment from the breath, heart rate, or EEG sensors, as an improvisation support for the musician. Concretely, this configuration has been used during electroacoustic music concerts organized by the SCRIME [at the Marché de Lerme in Bordeaux (see Fig. 11.1), and at the Hôtel de Région d'Aquitaine], or during performances (at the Cap Sciences scientific museum and at the i-boat club in Bordeaux). The other has been set up in the context of the ANR Care project and allows a dancer (Gaël Domenger) to generate the music supporting his dance improvisation thanks to a motion-capture suit (see Fig. 11.2). A public show involving this configuration took place at the Casino de Biarritz in March 2011.

In a more educational context, various sessions introducing the MuZICO system have been presented to the public on an ad hoc basis: Workshops were held during the Eurêka days of the Fête de la Science 2011 at the Hôtel de Région d'Aquitaine, and during the Semaine du Son 2012 at Cap Sciences. In a more sustained way, an interactive station allowing to test the system was set up in Cap Sciences during the exhibition Numériquement Vôtre (from march to December 2011). The public could interact with MuZICO thanks to a skin conductivity sensor that was integrated to the station.

Concerning the therapeutic domain, the potential usefulness of music in the field of relaxation techniques has already been discussed (Ganidagli et al. 2005; Loewy et al. 2005; Zhang et al. 2005; Peretz et al. 1998; Sammler et al. 2007; Trainor and

Fig. 11.1 Pierre-Henri Vulliard at the Marché de Lerme

Fig. 11.2 Gaël Domenger using eMotion and MuZICO softwares

Schmidt 2003). Some exploratory experiments have been carried out from the start of the project by one of the authors, involving professionals such as masseurs and hypnotists.

11.3 Interaction Between Emotions and Synthesis of Music

11.3.1 Representation of Emotions

We propose here a generalization of vigilance states to emotional states in order to increase the expressive potential of the musical generation, by adding a second

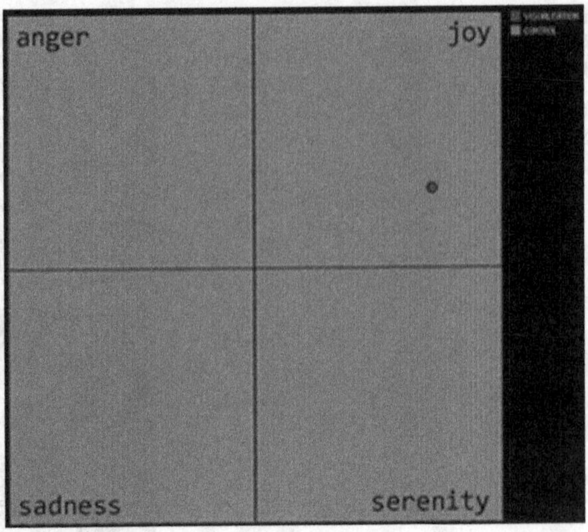

Fig. 11.3 Ekman's primary emotions placed in the bidimensional space of emotions

dimension to the first dimension of vigilance. There are many definitions and representations of emotions in the psychological field of study, but nowadays, few of them are used in computer science.

We chose to use the same model as in Livingstone et al. (2010) and Wallis et al. (2008), commonly used in psychology under the name of Plutchik's emotions wheel (Russel 1980). It is a bidimensional space defined by two axes corresponding to excitation (which we consider to be the same as vigilance in the context of music generation), and to valence (pleasure/unpleasure), in which we can place Ekman's primary emotions (Ekman 1992) (see Fig. 11.3).

11.3.2 Care Project—eMotion Software

Our collaboration in the context of the Care project (Clay et al. 2011) with a team studying motion-based emotions recognition allowed us to test the generation of music using the bidimensional space of emotions.

For the interactive dance show that was presented as the conclusion of this project, we generated all the music and sounds in real time with MuZICO.

11.3.3 Validation of the System by the Audience

The audience that came to see the second representation of the Care project's end show was given questionnaires in order for the scientific team to collect feedback from them (Clay et al. 2012): The second representation of the show was presented in front of an audience of about 120 people. Upon their arrival, we distributed them questionnaires for collecting feedback about the show. We collected 97 questionnaires (i.e. roughly an 80 % return rate). For a non-researcher, it is quite difficult to apprehend the scientific world. This show presented scientific results in a both artistic and playful way; scientists were also directly in front of the audience, which the audience appreciated. All those factors made the audience enjoy the show. Then, to answer the question "What did you like?" 90 % of the people who gave a feedback cited the innovative or even magical aspects of the show. The audience found, however, that the generated music expressed too frankly the emotion being portrayed (75 % of the audience) and that the music should have taken more distance with the expressed emotions and leave more interpretation from the spectator. This criticism is interesting as it validates the parameters we chose for music generation, but pinpoint the fact that more artistic choices would have been more relevant. Finally, when asked about the potential improvements that could be made, 98 % of the people stated that they wished to witness the evolution of the show, should it be in the scientific content, the choreography, the sound generation, and/or the graphical choices. The audience hence had a very strong interest in following this collaboration between art and science; this interest went further, as it sparkled reaction from medias in the form of articles and an interview on local television.

11.3.4 Valence and Musical Complexity

According to Livingstone et al. (2010), the musical parameters related to the valence of emotions evoked by music are tonality and complexity. We propose here harmonic, rhythmic and melodic generative models, as well as a description of their parameters from the point of view of musical complexity.

The way these models (implemented as modules in MuZICO) communicate between each other is explained by Fig. 11.4.

11.3.4.1 Pitch Scales

We suggest a model of generative pitch scales base on the following sequence:

$$U_{n+1} = \left(U_n \times \omega^{\frac{V_{n \bmod (m+1)}}{\theta}} - U_0 \right) \times \omega^{-p}$$
$$\text{where } U_0 = f_f, \quad \text{and} \quad p \in \mathbb{N}^+ \text{such that } U_0 < U_{n+1} < \omega \times U_0 \tag{11.1}$$

where ω is the octave ratio, θ the number by which the octave is divided to get the temperament, f_f the root frequency of the generated scale, V a m-dimensional vector, and $V_{i,i \in [1..m]}$ the ith component of V.

We consider that the complexity of a pitch scale depends on the number of iterations needed to generate it, and on the values of ω, θ, and V.

For example, let us consider the case where $\omega = 2$, $\theta = 12$ (tonal occidental music).

Let be $V = (7)$ (generation of the pitch scale by iterating through steps of fifths), and N the number of iterations. This gives:

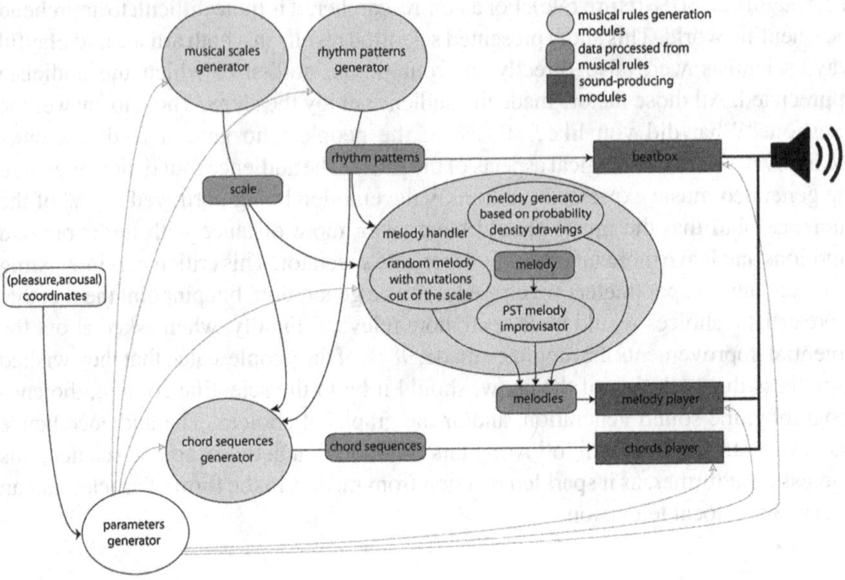

Fig. 11.4 Communication between MuZICO modules

- the pentatonic scale or the Maj 6/9 chord for $N = 4$
- the diatonic scale for $N = 6$
- the chromatic scale for $N = 11$

These scales have an increasing harmonic complexity. Scales of different complexities can also be generated by varying V: Let $V = (4)$ and $N = 2$, we then generate a pitch scale corresponding to an augmented chord. Let $V = (3)$ and $N = 3$, we then generate a pitch scale corresponding to a diminuated chord. Let $V = (3, 4, 4, 3)$ and $N = 6$, we then generate the ascending melodic minor scale.

11.3.4.2 Chords Sequences

In MuZICO, the chords sequences are generated by a grammar that takes the context into account to oversee the transpositions related to the modulations. The complexity of a chord depends on several factors that we identified:

- the number of notes that make it up
- the complexity of the scale on which it is built
- the order of appearance of its notes in the building of this scale by iterations

The complexity of a chords sequence also depends on several factors:

- the complexity of the chords that make it up

- the number of different chords in the sequence
- the harmonic complexity of the chords series, mainly taking into account the cadences and resolutions in the context of modern modal and tonal music, and the modulations in the context of the latter.

11.3.4.3 Rhythmic Patterns

The rhythmic patterns are created by layering various iterative rhythmic patterns, all synchronized to the same underlying pulse, each pattern being defined by an offset from the common initial pulse and by a rhythmic density (number of pulses between two onsets in the pattern). The superimposition is done by a logical OR operation, considering the superimposed patterns to be bit vectors, a value of 1 representing an onset, and a value of 0 no onset.

Let m be the rhythmic patterns length, q the number of superimposed patterns, of_i the offset from the initial pulse of pattern number i, and d_i the rhythmic density of pattern number i.

Let us consider that a bit vector is equivalent to a finite set of integer numbers, each number of the set representing a positive bit at the index corresponding to its value in the vector, a rhythmic pattern R created by superimposition of patterns P_i can thus be defined by:

$$\begin{gathered} R = \cup_{i=0}^{q} P_i \\ \text{where } P_i = \cup_{n=0}^{E((m-of)/d)} (of_i + d_i \times m) \end{gathered} \tag{11.2}$$

As regards rhythmic complexity, we use a model based on Toussaint's complexity measure or other techniques of rhythmic complexity evaluation that assign each pulsation a weight (see Fig. 11.5).

11.3.4.4 Melody Generation

In order to generate melodies, MuZICO has a module taking as input the current pitch scale and several rhythmic patterns generated using the techniques explained in the previous section. Each note has a weight corresponding to its apparition order in the process of building the current pitch scale. For each onset in the rhythmic pattern, a note is chosen randomly using a Gaussian probability density varying according to the corresponding rhythmic weight. An initial melodic profile is thus obtained, which will be subsequently developed in a musical way by various algorithms such as the Probabilistic Suffix Tree (PST). This profile is then modified according to the modulation information sent by the grammar which generates the chords, resulting in the final melody to be plaid.

The complexity of such a melody can be evaluated and depends on the Gaussian probability densities, as well as the PST configuration parameters.

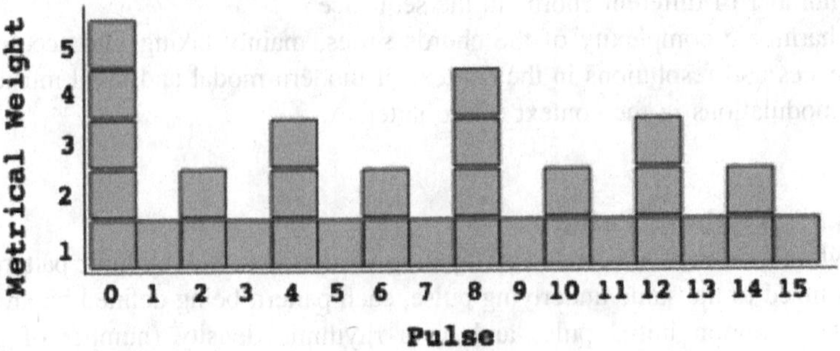

Fig. 11.5 Weight of the pulses in a rhythmic pattern of length 16

11.3.5 Seeking Valence in EEG Signals

Although valence as a musical concept is widely documented and has long been discussed by music theorists, valence as a component of felt emotions is a more ill-defined notion. The latest research activities led by our team mainly focus on this problem. Following a number of discussions with researchers at Numediart Institute in Mons, Belgium, in the context of a collaboration with a movie director on an interactive cinema project, our former intentions of seeking clues for valence estimation in the study of inter-hemispheral EEG activity were confirmed by our discovery of the work of Louis A. Schmidt and Laurel J. Trainor (Schmidt and Trainor 2001):

> …positively valenced musical excerpts elicited greater relative left frontal EEG activity, whereas negatively valenced musical excerpts elicited greater relative right frontal EEG activity… In addition, positively valenced (i.e., joy and happy) musical excerpts elicited significantly less frontal EEG power (i.e., more activity) compared with negatively valenced (i.e., fear and sad) musical excerpts, and there was significantly less frontal EEG power (i.e., more activity) in the left compared with the right hemisphere across valence.

Furthermore, we also consider interpreting the electric signals coming from the headsets as a measure of the face's muscular activity. By applying different filtering and analysis techniques to the same signals, we may define new independent variables which could help characterize the valence, as in standard facial emotion recognition.

11.3.5.1 Experimental Setup

We attempt to characterize and detect valence states and variations in EEG signals, therefore we must use stimuli which will influence the valence of the subjects so that we are able to measure it.

As this is a hot research topic shared by a community of people, standards have already been defined, at least regarding the media allowed to be used in these experiments from a medical point of view. We will of course use the provided databases so that our results remain compatible and comparable with the ones obtained by other teams. We are currently finalizing an experimental setup based on the use of the IAPS (Lang et al. 1999) and IADS (Bradley and Lang 2007) databases. Each file in these databases is annotated with mean and standard deviation values for valence, arousal, and dominance components. These values were obtained by asking subjects to rate each file from the three points of view. Mean and standard deviation were then computed from these ratings for each file.

IAPS and IADS are reference databases on which we can rely to validate our analysis of valence by comparing the results of our estimations to their original valence mean and standard deviation values in these databases.

We also plan to do the same with video excerpts, as video stimuli greatly differ from static pictures stimuli. To the best of our knowledge, no such database has been established so far.

11.3.5.2 Hardware

The first headset we started working with is a ProComp, a driver of which was formerly written by our team. In the context of the interactive cinema project initiated by Marie-Laure Cazin, we also got several wireless Emotiv EPOC headsets and started experimenting with them. These headsets come with a SDK which allows to get a grab on the raw electrodes data.

11.3.5.3 Software

We are currently developing several complementary software parts: One is a client that gets the raw electrodes data from the SDK, analyzes it, and sends both raw and analyzed data to any network via the Open Sound Control (OSC) (Wright 2005) protocol. This is an alternative driver for the headsets, rewritten in order to precisely fit our needs, also integrating the older driver for the ProComp.

Another part is an application allowing to play audio and video stimuli, display pics, and optionally record video footages of the experiences. This software is controlled in real time from other applications, via OSC.

Finally, the central software we use for the experiments is INRIA's OpenVibe (Renard et al. 2010), an open-source project dedicated to the analysis of electroencephalographic signals, which includes a powerful python scripting interface. These scripting capabilities easily allow one to enable bidirectional OSC communication and for example, define scenarios that will control the stimulation software, or drive MuZICO's musical generation algorithms. OpenViBE can keep track of EEG recording sessions into binary files containing the signals and the stimulations sent via OSC. The recorded experiences can then be replayed if desired, and one can focus on particular moments of them afterward for in-depth analysis.

11.3.5.4 Experimental Setup

The experiments we defined in order to validate valence recognition algorithms take place in three phases: During the first one, while simultaneously recording raw and preprocessed EEG signals from the subject, OpenViBEs successively sends control messages to the application that displays IAPS pictures, and for each displayed picture writes a stimulation in the recorded EEG signal stream corresponding to a "positive" or "negative" tag, basing itself on a displaying scenario and the IAPS database.

These tags are used for classification purpose: During the second phase, we feed a linear discriminant analysis (LDA) classifier with the file containing the signals and the tags. The classifier then builds rules to recognize positive or negative valence artifacts in the signals from this information. This occurs "offline," and no EEG signals are recorded during this phase (Fig. 11.6).

The third phase is the validation phase: We do the same as during the first phase with a new picture displaying scenario, but this time, the LDA runs in analysis mode, so it continuously gives an estimation of the current valence from the subject's incoming EEG signals. We can compare this estimation with the corresponding valence values in the IAPS database (Fig. 11.7).

In a previous work, TCTS laboratory had already assessed the quality of Emotiv EPOC headsets, by comparing their performance in the framework of a P300-based brain–computer interface with the performance reached using a medical EEG system [on the basis of the same electrode configuration, of course (Duvinage et al. 2012). Figure 11.8 illustrates that, albeit giving worse results than those obtained with the medical system, Emotiv EPOC headsets are usable for BCI applications.

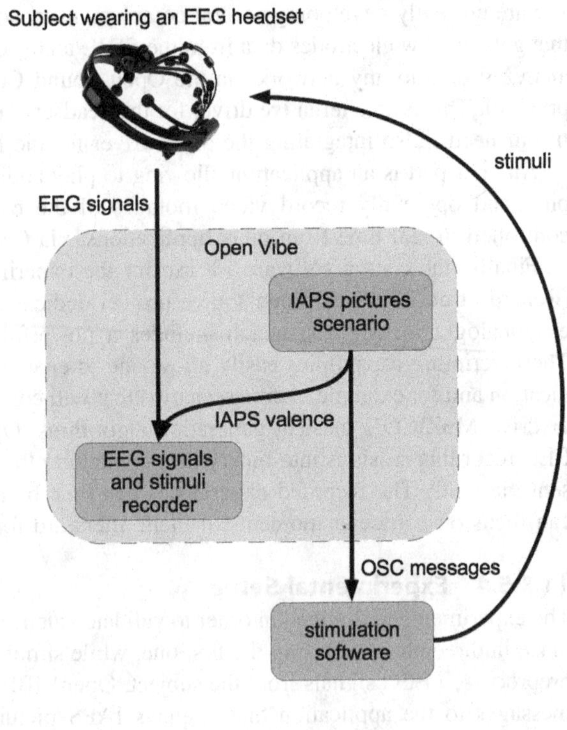

Fig. 11.6 First phase of the IAPS valence experiment

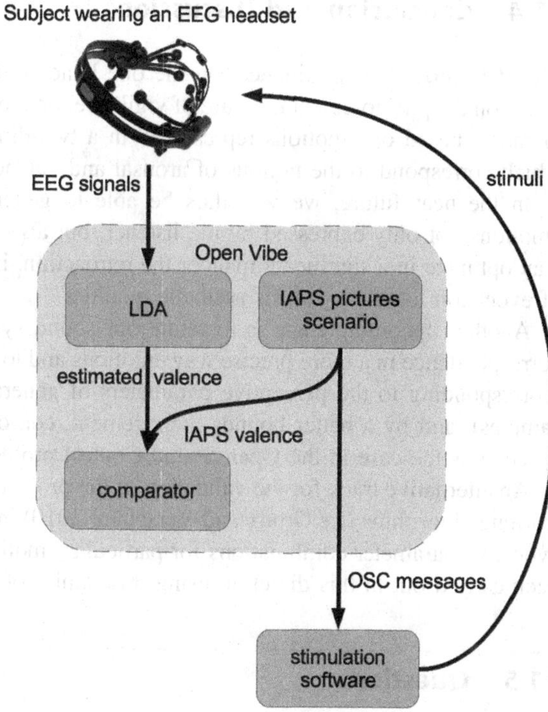

Fig. 11.7 Third phase of the IAPS valence experiment

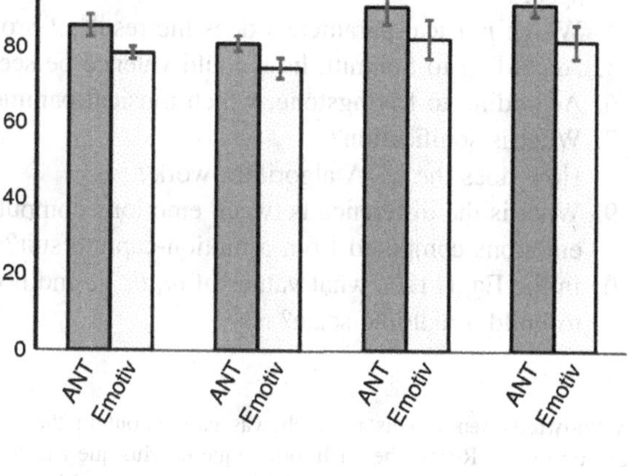

Fig. 11.8 Average and standard error values of classification rates obtained with a medical EEG system (ANT) and the Emotiv EPOC headsets under sitting and walking conditions. The chance level is 25 %

11.4 Conclusion and Discussion

MuZICO proved its pertinence, on the one hand by its ability to generate music based on the physiological measure of vigilance, on the other hand in the generation of music based on emotions represented in a two-dimensional space, the axes of which correspond to the notions of arousal and valence.

In the near future, we will thus be able to generate music as a function of emotions not only expressed by the listener, but also felt by this one, and by this way optimize in a significant manner the retroaction, improving the music's power of evocation together with its aesthetic qualities.

Another upcoming stage is to refine our sound synthesis engine by putting in correspondence in a more precise way emotions and lower level musical parameters (corresponding to the perceptive parameters of generated sounds and used sound samples), and by a better bounds management, conforming to a URL like namespace as is the case in the Open Sound Control protocol.

An alternative track for the validation of the project is to use reinforced machine learning algorithms (Le Groux and Verschure 2010) in order to determine the most evocative parameter combinations for particular emotions. Some experiments have been carried out in this direction using this family of algorithms.

11.5 Questions

1. How does MuZICO process emotions?
2. How is the IAPS system used in the experiments on emotion recognition?
3. What emotional components does MuZICO use as dimensions of its emotional space?
4. Which musical parameters does the result of arousal analysis control?
5. According to Schmitt, how could valence be seen in EEG brainwaves?
6. According to Livingstone, which musical parameters are related to valence?
7. What is sonification?
8. How does the LDA algorithm work?
9. What is the difference between emotions computed from an EEG headset and emotions computed from a motion-capture suit?
10. In the Eq. (11.1), what values of ω, θ, V, and how many iterations are needed to build a diatonic scale?

Acknowledgments This research was carried out in the context of the SCRIME (Studio de Création et de Recherche en Informatique et Musique Electroacoustique, scrime.labri.fr) project which is funded by the DGCA of the French Culture Ministry, the Aquitaine Regional Council. SCRIME project is the result of a cooperation convention between the Conservatoire of Bordeaux, ENSEIRB-Matmeca (school of electronic and computer scientist engineers) and the University of Sciences of Bordeaux. It is composed of electroacoustic music composers and scientific researchers. It is managed by the LaBRI (laboratory of research in computer science of the University of Bordeaux, www.labri.fr). Its main missions are research and creation, diffusion and pedagogy thus extending its influence.We would like to thank Pierre Héricourt, system engineer at LaBRI, for developing the EEG headsets' drivers, allowing us to interface the headsets with all the software parts, and thus to set up our experiments for real.

References

Bradley MM, Lang PJ (2007) The international affective digitized sounds (; IADS-2): affective ratings of sounds and instruction manual. University of Florida, Gainesville, FL, Technical Report B-3

Clay A, Domenger G, Couture N, De La Rivière J, Desainte-Catherine M, et al (2011) Spectacle augmenté: le projet CARE, un processus de recherche. http://hal.inria.fr/hal-00651544/

Clay A, Couture N, Decarsin E, Desainte-Catherine M, Vulliard PH, Larralde J (2012) Movement to emotions to music: using whole body emotional expression as an interaction for electronic music generation. In: Proceedings of NIME'12. University of Michigan, Ann Arbour. pp. 180–186

Duvinage M et al (2012) A P300-based quantitative comparison between the Emotiv Epoc headset and a medical EEG device. In: Proceedings of the 9th IASTED international conference on biomedical engineering, p 2

Ekman P (1992) An argument for basic emotions. Cogn Emot 6(3–4):169–200

Ganidagli S, Cengiz M, Yanik M, Becerik C, Unal B (2005) The effect of music on preoperative sedation and the bispectral index. Anesth Analg 101(1):103–106

Lang PJ, Bradley MM, Cuthbert BN (1999) International affective picture system (IAPS): technical manual and affective ratings. University of Florida, Gainesville, FL

Le Groux S, Verschure P (2010) Towards adaptive music generation by reinforcement learning of musical tension. In: Proceedings of the 6th Sound and Music Computing Conference, pp 134–137

Livingstone S, Muhlberger R, Brown A, Thompson W (2010) Changing musical emotion: a computational rule system for modifying score and performance. Comput Music J 34(1):41–64

Loewy J, Hallan C, Friedman E, Martinez C (2005) Sleep/sedation in children undergoing EEG testing: a comparison of chloral hydrate and music therapy. J Perianesthesia Nurs 20(5):323–332

Peretz I, Gagnon L, Bouchard B (1998) Music and emotion: perceptual determinants, immediacy and isolation after brain damage. Cognition 68(2):111–41

Renard Y, Lotte F, Gibert G, Congedo M, Maby E, Delannoy V, Bertrand O, Lécuyer A (2010) Openvibe: an open-source software platform to design, test, and use brain-computer interfaces in real and virtual environments. Presence Teleoperators Virtual Environ 19(1):35–53

Russel J (1980) A circumplex model of affect. J Personnal Soc Psychol 39(6):1161–1178

Sammler D, Grigutsch M, Fritz T, Koelsch S (2007) Music and emotion: electrophysiological correlates of the processing of pleasant and unpleasant music. Psychophysiology 44:293–304

Schmidt LA, Trainor LJ (2001) Frontal brain electrical activity (EEG) distinguishes valence and intensity of musical emotions. Cogn Emot 15(4):487–500

Trainor LJ, Schmidt LA (2003) Processing emotions induced by music. Oxford University Press, Oxford

Vezard L, Chavent M, Legrand P, Faita-Ainseba F, Clauzel J, et al (2011) Caractérisation d'états psychophysiologiques par classification de signaux EEG. Intégration de ces résultats dans le projet PSI

Wallis I, Ingalls T, Campana E (2008) Computer-generating emotional music: the design of an affective music algorithm. In: International conference on digital audio effects, pp 7–12

Wright M (2005) Open sound control: an enabling technology for musical networking. Organ Sound 10(03):193–200

Zhang XW, Fan Y, Manyande A, Tian YK, Yin P (2005) Effects of music on target? controlled infusion of propofol requirements during combined spinal? epidural anaesthesia. Anaesthesia 60(10):990–994

Creative Music Neurotechnology with Symphony of Minds Listening

12

Eduardo Reck Miranda, Dan Lloyd, Zoran Josipovic and Duncan Williams

Abstract

A better understanding of the musical brain combined with technical advances in biomedical engineering and music technology is pivotal for the development of increasingly more sophisticated brain–computer music interfacing (BCMI) systems. BCMI research has been very much motivated by its potential benefits to the health and medical sectors, as well as to the entertainment industry. However, we advocate that the potential impact on musical creativity of better scientific understanding of the brain, and the development of increasingly sophisticated technology to scan its activity, should not be ignored. In this chapter, we introduce an unprecedented new approach to musical composition, which combines brain imaging technology, musical artificial intelligence and neurophilosophy. We discuss *Symphony of Minds Listening*, an experimental composition for orchestra in three movements, based on the fMRI scans taken from three different people, while they listened to the second movement of Beethoven's *Seventh Symphony*.

E.R. Miranda (✉) · D. Williams
Interdisciplinary Centre for Computer Music Research (ICCMR), Plymouth University, Plymouth PL4 8AA, UK
e-mail: eduardo.miranda@plymouth.ac.uk

D. Williams
e-mail: ducan.williams@plymouth.ac.uk

D. Lloyd
Department of Philosophy, Program in Neuroscience, Trinity College, Hartford CT 06106, USA
e-mail: dan.lloyd@Trincoll.edu

Z. Josipovic
Psychology Department, New York University, 6 Washington Place, Room 482a, New York, NY 10003, USA
e-mail: zoran@nyu.edu

E.R. Miranda and J. Castet (eds.), *Guide to Brain-Computer Music Interfacing*,
DOI 10.1007/978-1-4471-6584-2_12

12.1 Introduction

BCMI research has been very much motivated by its potential benefits to the health and medical sectors, as well as to the entertainment industry. Yet, advances in the field tend to be assessed in terms of medium, rather than content. For instance, let us consider the field of music technology. Much has been said on the improvement of technology for music recording and distribution, from vinyl records and K7 tapes to CDs and the Internet. However, not much is said on the impact of these media to creative processes. Have these media influenced the way in which music is composed? Likewise, not much has been said on the creative potential of BCMI technology. Might it lead to new ways to make music, or to the emergence of new kinds of music?

We believe that the potential impact on musical creativity of better scientific understanding of the brain, and the development of increasingly sophisticated technology to scan its activity can be huge. Musicians have an unprecedented opportunity today to develop new approaches to composition that would have been unthinkable a few years ago.

In this chapter, we introduce an unprecedented new approach to musical composition, which combines brain imaging technology (Bremmer 2005), musical artificial intelligence (AI) (Miranda 2000), and new philosophical thinking emerging from neurophilosophy (Churchland 2007). The first outcome of this approach is *Symphony of Minds Listening*, an experimental composition for orchestra in three movements, based on the fMRI scans taken from three different people, while they listened to the second movement of Beethoven's *Seventh Symphony*: a ballerina, a philosopher (co-author Dan Lloyd), and a composer (co-author Eduardo R. Miranda). In simple terms, we deconstructed the Beethoven movement to its essential elements and stored them with information representing their structural features. Then, we reassembled these elements into a new composition with a twist: the fMRI information influenced the process of reassembling the music.

The chapter begins with a discussion on the philosophical ideas behind the work. Next, before delving into more technical details, it gives an overview of the compositional approach we have been developing. It follows with an introduction to the brain scanning and data analysis methods. Then, it introduces MusEng, the system that we developed to deconstruct and recompose music and demonstrate the core processes behind the composition of *Symphony of Minds Listening*.

12.2 Neurophilosophy of Music

The human brain is allegedly the most complex object known to mankind: it has circa one hundred billion neurones forming a network of an estimated one quadrillion connections between them. The amount of information that circulates through this network is huge. The operation of individual neurones is fairly well understood nowadays, but an important piece of the jigsaw is missing: the way they cooperate in ensembles of millions has been fiendishly difficult to understand. This piece of the puzzle is important because it most probably holds the key to unlock our understanding the origins of the mind.

There has been a school of thought, which considered that the mind is divorced from the brain. What is more, it has been suggested that minds would not even need brains to exist. Although the separation between mind and brain still has currency in some circles, nowadays it is common sense to consider the mind as resulting from the functioning of the brain. However, we do not have a clear understanding of how brain activity actually gives rise to the mind.

Much research is being developed from a number of approaches all over the globe to understand how the brain gives rise to the mind. Our research is looking into establishing a musical approach to understand the brain. We believe that the brain can be viewed as a colossal, extraordinarily large symphonic orchestra and the mind as a correspondingly complicated symphony. The 'mind as music' hypothesis is explored in length in Lloyd (2011).

At Plymouth University's Interdisciplinary Centre for Computer Music Research, we are looking into the relationship between music and a specific aspect of our mind: *emotions*. We hope to be able to determine which aspects of a musical composition elicit specific emotions on listeners. The hypothesis is that if one can predict which musical features are likely to cause the feeling of, say, joy or sadness, then it might be possible to build technology that would allow new music to steer our emotions more effectively. For example, it would be highly beneficial for humankind if physicians could have the option to prescribe a musical composition as part of the treatment to help take a patient out of depression. Not unlike chemists, future musicians could be trained with the skill to compose with specific musical ingredients aimed at inducing particular affect in listeners. Our work is aimed at making this ambitious dream a reality, but the challenges to achieve this are not trivial.

Similar to the fact that we have unique fingerprints, which differ from person to person, our brains are also unique. Indeed, the mechanisms whereby we make sense of music differ from person to person. Even though all human brains share a common basic plan, the detailed neurological circuitry differs from one person to another. Unlike our fingerprints, however, our brain circuits are continually changing and this makes scientific research into unveiling how the brain functions rather challenging. Paradoxically, it seems that the more we study the brain, the more difficult it becomes to draw firm conclusions. A balance needs to be established between identifying the commonalities and acknowledging the differences of our brains. *Symphony of Minds Listening* is inspired by the later: it is an artistic expression of how different brains construct their own unique reality.

12.3 An Overview of the Approach

Functional magnetic resonance imaging (fMRI) is a procedure that measures brain activity by detecting associated changes in blood flow. The measurements can be presented graphically by colour-coding the strength of activation across the brain. Figure 12.1 shows a typical representation of an fMRI scan of a person listening to music, displaying the activity of the brain at a specific window of time. In this case,

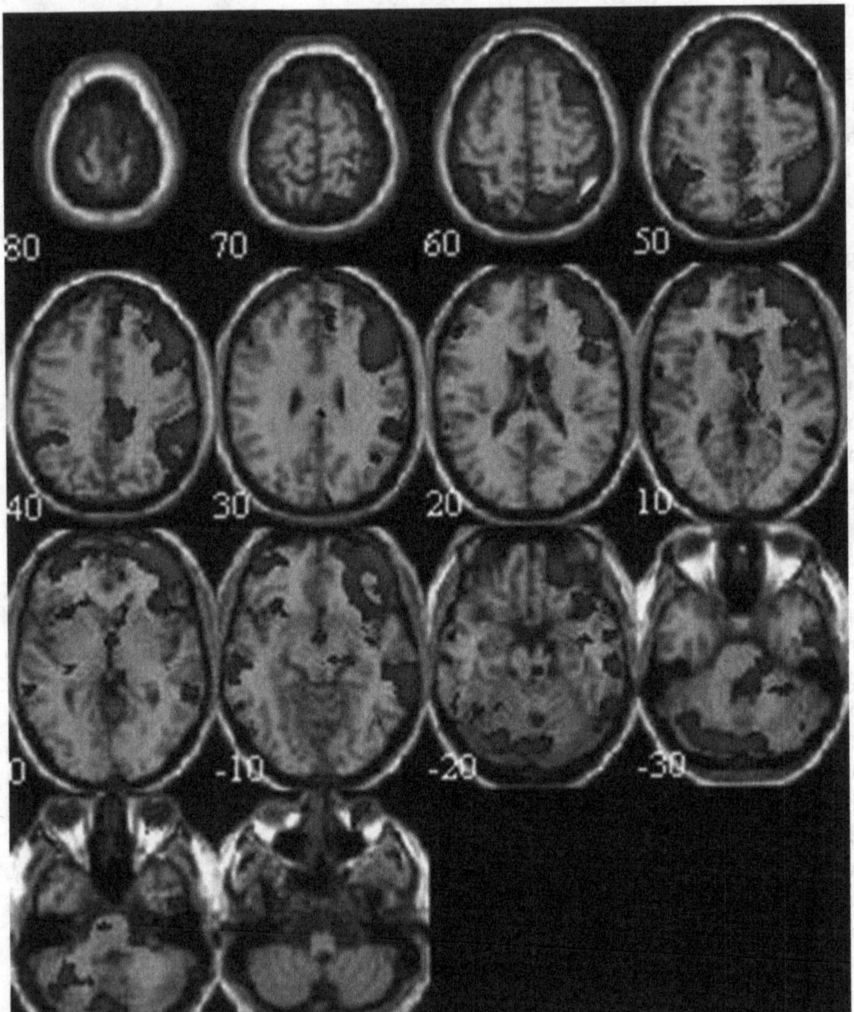

Fig. 12.1 A typical representation of an fMRI scan, showing 14 slices of the brain. The actual scanning for this project comprised 36 slices' snapshots taken every 2 s

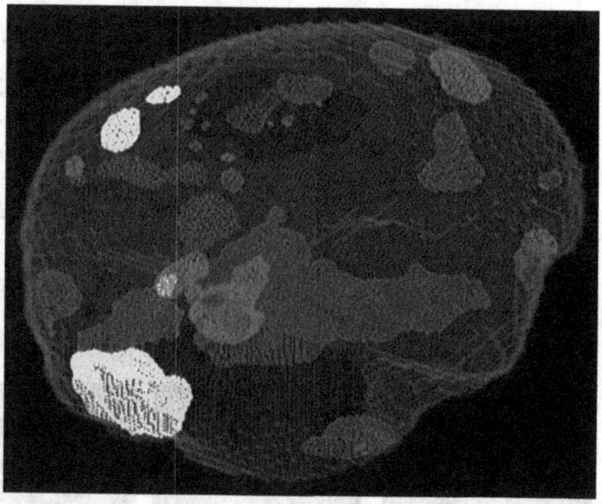

Fig. 12.2 An artistic 3D rendering of an fMRI scan

each time window lasts for two seconds. The figure shows 14 planar surfaces, or slices, from the top to the bottom of the brain, and the respective activity detected in these areas. Figure 12.2 is an example of an artistic 3D rendition of such an fMRI scan. It shows different areas of the brain, represented by different colours (that is, shades of grey), responding in a coordinate manner to the music.

Each scanning session generated sets of fMRI data, each of which associated with a measure of the second movement of Beethoven's seventh symphony. This is shown schematically in Fig. 12.3.

Firstly, the movement was deconstructed by means of MusEng, a piece of software, which extracted information about the structure of the Beethoven piece. Then, we programmed MusEng to use this information and the fMRI data to generate new musical passages.

During the compositional phase, the fMRI information was used on a measure-by-measure basis to influence the composition. This procedure, which is shown schematically in Fig. 12.4, involved diverse modes of data processing and transformation of Beethoven's music. The resulting musical passages bore varied degrees of resemblance to the original.

Not surprisingly, the fMRI scans differed among the three listeners. Therefore, brain activity from three different minds yielded three different movements in the resulting composition that resemble the original in varied ways. The instrumentation is the same as for Beethoven's original instrumentation, and each movement is named after the respective person who was scanned:

- 1st Movement: *Ballerina*
- 2nd Movement: *Philosopher*
- 3rd Movement: *Composer*

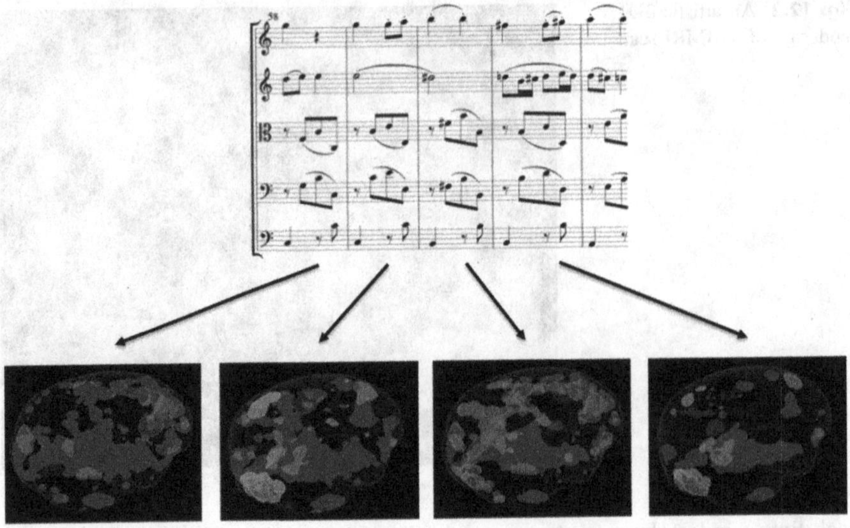

Fig. 12.3 The result of a scanning section is a set of fMRI data for each measure of Beethoven's piece

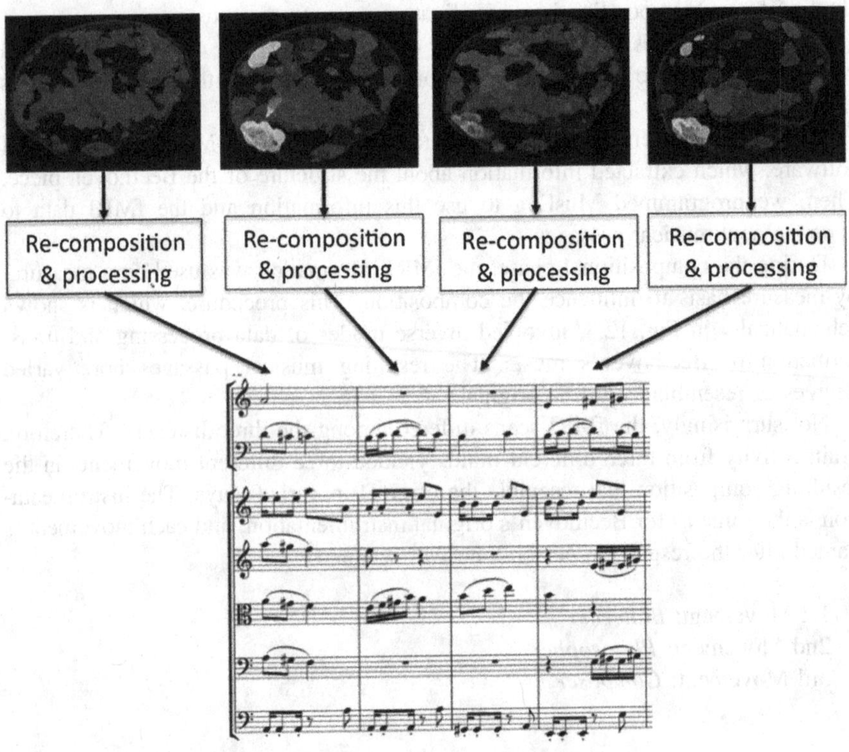

Fig. 12.4 The fMRI data inform the reassemblage of the piece

12.4 Brain Scanning: Materials and Methods

The brain images were collected using equipment and parameters that are typical in cognitive neuroscience. The scanner was a Siemens Allegra 3-T head-only scanner with a head coil. ('T' stands for Tesla, a measure of magnetic field strength. Contemporary scanners range from 1.5 to 7 T.) A T2-sensitive echo planar imaging (EPI) pulse sequence was used to obtain blood oxygenation level-dependent (BOLD) contrasts: TR = 2,000 ms, TE = 30 ms, 36 axial slices, 3 × 3 × 3 mm, 64 × 64 matrix in a 192 × 192 mm FOV. That is, each full-brain image took two seconds to collect, to yield 36 image slices of the brain. Each slice comprised 64 × 64 picture elements, known as voxels or volume pixels. Thus, each image comprised approximated 150,000 continuously varying voxels.

Subjects heard the second movement of Beethoven's *Seventh Symphony* twice. The subjects were instructed to attend to the music with their eyes closed. The fMRI recording began with 30 s without music, then 460 s of Beethoven, then 18 s without music and finally the same 460 s of Beethoven previously heard. Thus, each run generated 484 images.

12.5 fMRI Analysis

The raw fMRI scans were first preprocessed following usual procedures for functional neuroimaging. These included correcting for head motion, morphing the individual brains to conform to a standard anatomical atlas, and spatial smoothing, which is a procedure that reduces random fluctuations by calculating a moving average of each voxel in the context of its spatial neighbours. These preprocessing steps were implemented using Statistical Parametric Mapping software (Ashburner et al. 2013).

Each of the 484 images produced 150,000 voxels, which are very complex for direct analysis. Instead, the image series were further processed with independent component analysis, abbreviated as ICA (Stone 2004). Informally, ICA separates ensembles of voxels that oscillate in unison. These are unified as supervoxels representing temporally coherent networks of brain activity. The coloured patches in Fig. 12.2 are examples of independent components. A total of 25 components were calculated for the three subjects in the experiment.

In order to select which of these components might be musically significant, the activity of each component during the first pass through the Beethoven listening was compared to that same component during the second pass. If these two segments of a component time series were correlated, we hypothesised that the activity was at least partly musically driven, since the stimulus, that is, the music, would be identical at the corresponding time points in the two passes through the music. Although 25 independent component time series were identified, only the strongest 15 were selected to influence the compositional process. The order of strength of the selected 15 ICA components is as follows: 25, 15, 14, 8, 5, 10, 11, 18, 6, 2, 4, 1,

Table 12.1 The values of the strongest 5 ICA components for the first 10 measures of Beethoven's music yielded by the subject 'composer'

Beethoven measure	ICA 25	ICA 15	ICA 14	ICA 8	ICA 5
1	7	5	5	5	2
2	5	5	8	5	8
3	7	3	5	5	6
4	5	8	3	5	2
5	5	7	4	4	4
6	6	6	4	5	3
7	7	8	5	6	3
8	4	6	3	4	3
9	6	6	4	5	4
10	5	7	5	5	3

17, 16 and 13. The time series were normalised to range from 0 to 9. As a last step, the varying components were resampled to match the timing of the Beethoven score measure by measure. Thus, each time point was indexed to a measure of the Beethoven score. The movement comprises 278 measures; therefore, each ICA component comprises a time series of 278 values, ranging from 0 (meaning lowest fMRI intensity) to 9 (highest fMRI intensity). As an example, Table 12.1 shows the values of the first 5 strongest ICA components (that is, 25, 15, 14, 8 and 5) for the first 10 measures of Beethoven's music, yielded by the fMRI of the subject 'composer' during the first listening pass in the scanner.

To accompany the premiere of *Symphony of Minds Listening*, the ICA data were animated on a large screen projection behind the orchestra. The whole brain appeared as a transparent frame, derived from a standard anatomical template. Within this image, each component was assigned a distinct colour, and brightened and faded according to the intensity of component activity at each time point. The animations were created using MATLAB software (MathWorks 2014), using custom-made functions. The remaining of this chapter focuses on the compositional process and the MusEng software.

12.6 The Compositional Process

The actual composition of *Symphony of Minds Listening* is primarily the work of the first co-author and involved a number of creative stages and practices, some which were not systematically documented. That is to say, the compositional process involved manual and computer-automated procedures.

There generally are two approaches to using computer-generated materials in composition, which we refer to as the *purist* and *utilitarian* approaches, respectively. The purist approach to computer-generated music tends to be more concerned with the correct application of the rules programmed in the system, than with the musical results per se. In this case, the output of the computer tends to be considered as the final composition. The composer would not normally modify the music here, as this would meddle with the integrity of the model or system. At the other end of the spectrum is the utilitarian approach, adopted by those composers who consider the output from the computer as raw materials for further work. Here, composers would normally tweak the results to fit their aesthetic preferences, to the extent that the system's output might not even be identifiable in the final composition. Obviously, there is a blurred line dividing these two approaches, as practices combining aspects of both are commonly found. *Symphony of Minds Listening* tends towards the utilitarian approach.

The composition of the piece evolved in tandem with the development of MusEng. MusEng was programmed with artificial intelligence to learn musical information from given examples and use this information to generate new music. Incidentally, a few of MusEng's functionalities were firstly applied manually to compose a section of the piece, before they were implemented in software to aid the composition of other sections. Indeed, a number of compositional processes did not make it into the software on time. The symphony had a deadline to be delivered for its premiere in February 2013, at Peninsula Arts Contemporary Music Festival, in Plymouth, UK. The software development, however, is still in progress. And other pieces are planned, and the compositional approach is being refined as we write this chapter; for instance *Corpus Callosum*, for a chamber group of 25 musicians.

For a discussion on how science and technology can inform and inspire the act of musical composition, the reader is referred to Miranda (2013, 2014). Both references advocate the use of computers as assistants to the creative process, rather than as autonomous composing machines.

For the composition of *Symphony of Minds Listening*, the first step was to deconstruct the score of Beethoven's piece into a set of basic materials for processing. These materials were subsequently given to MusEng as input for a machine learning phase, which will be explained in more detail in the next section of this chapter.

First of all, Beethoven's piece was divided into 13 sections:

- Section 1: from measure 1 to measure 26
- Section 2: from measure 26 to measure 50
- Section 3: from measure 51 to measure 74
- Section 4: from measure 75 to measure 100
- Section 5: from measure 101 to measure 116
- Section 6: from measure 117 to measure 138
- Section 7: from measure 139 to measure 148
- Section 8: from measure 149 to measure 183
- Section 9: from measure 184 to measure 212

- Section 10: from measure 213 to measure 224
- Section 11: from measure 225 to measure 247
- Section 12: from measure 248 to measure 253
- Section 13: from measure 254 to measure 278

The 13 sections informed the overarching form of the 3 movements of the new symphony. This provided a template for the new piece, which preserved the overall form of the original Beethoven movement. Indeed, MusEng did not learn the whole Beethoven piece at once. Rather, it was trained on a section-by-section basis and the musical sequences for the respective new sections of the new movements were generated independently from each other. For instance, Section 1 of the movement *Ballerina* has 26 measures and was composed based on materials from the first 26 measures of Beethoven's music. Next, Section 2 has 24 measures and was composed based on materials from the next 24 measures (26–50) of Beethoven's music, and so on.

A block diagram portraying the compositional procedures is shown in Fig. 12.5. The blocks with thicker lines represent procedures that can be influenced and/or controlled by the fMRI. After the segmentation of the music into 13 sections, the flow of action bifurcates into two possibilities: manual handling of the segments (left-hand side of Fig. 12.5) and computerised handling with MusEng (right-hand side of Fig. 12.5). A discussion of manual handling is beyond the scope of this chapter, but as an example we can show the transformation of Section 1 of Beethoven's original into the opening section of *Ballerina*. Figure 12.6 shows the first 10 measures of Beethoven's music focusing on the parts of the violas, violoncellos and double basses. Figure 12.7 shows how those measures were recomposed to

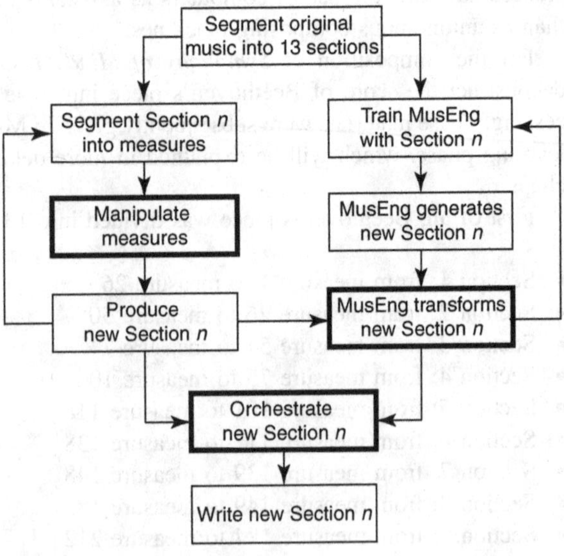

Fig. 12.5 Block diagram of the overall compositional process

Fig. 12.6 The first 10 measures of Section 1 of Beethoven's music, showing the viola, violoncello and double bass parts

Fig. 12.7 Ten measures from the opening of *Ballerina*, the first movement of *Symphony of Minds Listening*, showing the viola, violoncello and double bass parts

form 10 measures for the opening of the first movement of the new symphony. Note the visible rhythmic transformation of measures 4, 6, 8 and 10.

The path on the right-hand side of the block diagram in Fig. 12.5 represents the computer handling of the segments with MusEng. This will be explained in more detail in the next section.

Table 12.2 Mapping ICA activation values onto musical dynamics

Activation	Dynamics
0	*ppp*
1	*ppp*
2	*pp*
3	*p*
4	*mp*
5	*mf*
6	*f*
7	*ff*
8	*fff*
9	*fff*

Finally, once a new segment has been generated, it is orchestrated and appended to the respective score of the new movement accordingly. The fMRI occasionally influenced the instrumentation and the orchestration. For instance, in *Philosopher*, the second movement, different ICA components were associated with groups of instruments of the orchestra (e.g. ICA 25 = violins and violas, ICA 15 = trumpets and horns, ICA 14 = oboes and bassoons and so on); these associations changed from section to section. Then, for example, if the flute is to play in a certain measure *x* of *Philosopher*, the ICA activation value of the respective component in measure *x* of Beethoven's music would define how the flute player should produce the notes. We defined various tables mapping ICA activations to instrumental playing techniques and other musical parameters. For instance, Table 12.2 shows a mapping of ICA activations onto musical dynamics.

As a hypothetical example, let us consider a case where the flutes would play the sequence shown in Fig. 12.8 in measures 5, 6 and 7 of the third movement: *Composer*. If we assume that the flute is associated with ICA 8, then according to the values shown in Table 12.1, the activations for measure 5, 6 and 7 are equal to 4, 5 and 6, respectively. Thus, the dynamics attributed to these 3 measures would be as shown in Fig. 12.9.

Fig. 12.8 Three new measures for the flutes

Fig. 12.9 The measures from Fig. 12.8 with added dynamics informed by fMRI information

12.7 The Musical Engine: MusEng

MusEng has three distinct phases of operation: a *learning phase*, a *generative phase* and a *transformative phase*.

The learning phase takes a musical score and analyses it in order to determine a number of musical features. A dataset comprising these features and rules representing the likelihood of given features appearing in the data is then stored in memory. At the generative phase, these data inform the generation of new sequences, which should ideally resemble the sequences that were used to train the system in the first phase. Finally, at the transformative phase, the outcome from the generative phase is modified according to a number of transformation algorithms. It is in this final phase that the fMRI information is used to influence the resulting music. Note that we are not interested in a system of rules that reproduces an exact copy of the original music. Rather, we are interested in producing new music that resembles the original. Hence, the transformative phase was added to further modify the results from the generative phase. The role of fMRI information is to control the extent of the transformations. Essentially, stronger activity in a given ICA component of the fMRI data results in larger amounts of transformation in the musical outcome.

MusEng reads and outputs musical scores coded in the MIDI format. Musical instrument digital interface (MIDI) is a protocol developed in the 1980s, which allows electronic instruments and other digital musical tools to communicate with one another. Music notation software normally has an option to save and read files in this format. This is useful because it is straightforward to make a MIDI file representing the Beethoven symphony to train the system. MusEng outputs can be loaded into any suitable music notation software for further work and adjustments.

MusEng only processes monophonic musical sequences, that is, sequences of one note at a time. Obviously, Beethoven's movement is a polyphonic orchestral symphony. To circumvent MusEng's monophonic limitation, we developed two approaches to process the music. In the first approach, we train the system with the part of one instrumental voice of the orchestra at a time (violins, violoncellos, etc.), and then, we generate sequences for those respective parts individually. In the second approach, we reduce the orchestral music to one monophonic voice and then generate various monophonic sequences, which are subsequently orchestrated.

12.7.1 Learning Phase

MusEng implements an adapted version of iMe (short for Interactive Musical Environments), a system developed at Plymouth University's Interdisciplinary Centre for Computer Music Research with Marcelo Gimenes (Miranda and Gimenes 2011). MusEng takes a MIDI file as an input and extracts the following 5 features from the encoded music:

Table 12.3 Musicodes for the first two measures of the musical sequence in Fig. 12.10

	1	2	3	4	5	6	7	8	9
Melody direction	0	−1	+1	−1	0	+1	+1	−1	−1
Melody interval	0	5	2	4	0	1	1	1	1
Event duration	120	120	120	120	60	60	120	60	60
Note pitch	E5	G#4	B5	E4	B5	C5	D5	C5	B5
Modality	E Maj A harmonic min				A min C Maj				

The rows correspond to the event number or, in this case, number of notes in the sequence: the first two measures comprise a total of 9 notes

Fig. 12.10 An example of a musical sequence

- Pitches of the notes
- Melody directions between successive notes in a sequence
- Melody intervals, i.e. the amount of change between the pitches of successive notes in a sequence
- Note durations
- Modalities implied by groups of notes in a sequence

These features are stored as event-based vectors, referred to as *musicodes*. Table 12.3 shows the musicodes for the first two measures of the musical excerpt shown in Fig. 12.10.

Melody direction can be −1, 0, or +1, referring to descending, motionless or ascending movement. The current note in a sequence is compared with the previous note; the very first note in a sequence therefore returns a value equal to 0. Melody intervals are represented in terms of half steps, which are also calculated with reference to the current note's distance from the previous note. Again, the first note in the sequence returns a value equal to 0. With note durations, the value 240 is assigned to quarter notes, and other durations are calculated with reference to this value; for example, half notes are equal to 480 and eighth notes are equal to 120. Values for pitches are readily extracted directly from the corresponding MIDI code, for instance, MIDI 21 = Note A0, MIDI 23 = Note B0, MIDI 24 = Note C1 and so on.

In general, the number −2 is used to represent the absence of data in a musicode vector. Thus, the note pitch musicode for a musical rest would be equal to −2. With respect to the implied modality of segments, the system creates a label specifying a tonal pattern and indicates when the estimation is ambiguous. For example, in the first measure of the music shown in Fig. 12.10, the system sees E, G# and B, as an E Major chord, but the G# has also implied A harmonic minor.

As we shall see below, MusEng builds a musical memory in terms of small segments of music. Ideally, the system would segment the music based on perceptual criteria. The original iMe system sported such a method, inspired by Gestalt psychology (Eysenck and Keane 2005). However, for this project, we programmed MusEng to segment the music according to a user-specified number of measures, for instance, every measure, or every two measures, or every three and so on. The rationale for this decision is that we wanted to synchronise the fMRI analysis to the input score by handling the fMRI data on a measure-by-measure basis, as it was shown schematically in Figs. 12.3 and 12.4. Therefore, it made more sense to establish the measure as a reference value to segment the music.

MusEng's memory consists of a series of Feature Tables (FTs), which comprise vectors of musicodes for material that the system has been exposed to. As the musicodes are extracted from incoming measures, the system may or may not create new FTs, depending on whether the respective musicodes have already been seen by the system. If a certain vector of musicodes is identical to one that has been previously seen by the system, then the system updates the relevant FT by increasing a weighting factor, represented by the variable ω (Eq. 12.1). This variable is generated by summing the total number of FTs and then dividing the number of instances of each individual FT by the total. In essence, this becomes a simple moving average. In Eq. (12.1), the value of ω indicates the weighting factor associated with a given FT. The variable x represents the number of instances of a given FT in the series, and n the total number of FTs in the series.

$$\omega_{(x)} = \frac{\sum \mathrm{FT}_{(x)}}{\sum \mathrm{FT}_{(n)}} \tag{12.1}$$

This moving average has the effect of lowering the value of ω for vectors of musicodes that do not appear as often as more frequent ones, in the same way that it raises the value of ω for more commonly used vectors, to a maximum value of 1.0. The value of ω informs the probability of a given musical segment being generated later on by the system. Typically, a decrease in the value of ω causes the system to 'forget' to utilise the corresponding FT entry in the subsequent generative phase.

In order to illustrate how MusEng's memory is built, let us examine a hypothetical run through the sequence previously shown in Fig. 12.10, commencing with an empty memory. The first measure (Fig. 12.11) is analysed, and the respective musicodes are generated. For the sake of clarity, this example will focus on three of the five features: melody direction (dir), melody interval (int) and event duration (dur).

MusEng creates in its memory the first feature table, FT1, with musicodes derived from the first measure of the training sequence (Fig. 12.11) as follows:

Fig. 12.11 The first measure for the example analysis

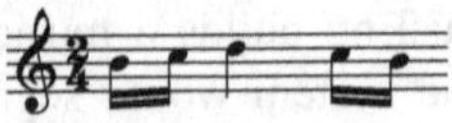

Fig. 12.12 The second measure for the example analysis

Fig. 12.13 The third measure for the example analysis

$$\text{dir} = \{0, -1, +1, -1\}$$
$$\text{int} = \{0, 5, 2, 4\}$$
$$\text{dur} = \{120, 120, 120, 120\}$$
$$\omega = 1/1 \text{ or } 1.0$$

Then, the system creates FT2 with musicodes extracted from the second measure of the training sequence (Fig. 12.12) as follows:

$$\text{dir} = \{0, +1, +1, -1, -1\}$$
$$\text{int} = \{0, 1, 1, 1, 1\}$$
$$\text{dur} = \{60, 60, 120, 60, 60\}$$
$$\omega = 1/2 \text{ or } 0.5$$

Next, MusEng creates FT3, with musicodes from the third measure of the training sequence (Fig. 12.13) as follows:

$$\text{dir} = \{0, +1, 0\}$$
$$\text{int} = \{0, 1, 0\}$$
$$\text{dur} = \{120, 120, 240\}$$
$$\omega = 1/3 \text{ or } 0.33$$

The fourth and fifth measures are processes next, but MusEng does not create new FTs in these cases because they are repetitions of previous measures; that is, their respective musicodes have already been seen by the system. In this case, only the values of ω for the respective FTs are adjusted accordingly. Thus, at this point of the training phase, the ω values for each FT are as shown in Table 12.4.

MusEng's memory after the training phase, complete with 3 FTs, is shown in Table 12.5. It is important to stress that particular FTs gain or lose perceptual

Table 12.4 Values of ω after three FTs have been created and stored in memory, calculated by dividing the number of instances of a given FT by the total number of FTs analysed

	FT1	FT2	FT3
ω	1/5 = 0.2	2/5 = 0.4	2/5 = 0.4

Table 12.5 MusEng's memory after being trained with the musical sequence shown in Fig. 12.5

	dir	int	dur	ω
FT1	0, −1, +1, −1	0, 5, 2, 4	120, 120, 120, 120	0.2
FT2	0, +1, +1, −1, −1	0, 1, 1, 1, 1	60, 60, 120, 60, 60	0.4
FT3	0, +1, 0	0, 1, 0	120, 120, 240	0.4

importance depending on how often the system is exposed to them. Notice, therefore that FT2 and FT3 have higher ω values than that of FT1, because they appeared twice.

12.7.2 Generative Phase

At the generative phase, MusEng generates new FTs by mutating the musicodes of an existing FT towards those of another FT in memory. This process is influenced by the values of ω: FTs with larger ω values are selected more often than FTs with smaller ω values. Note that we wrote 'tend to be selected'. This is because MusEng uses a Gaussian distribution function to make this selection.

The very first measure of a newly generated structure is typically informed by the first FT in memory (FT1). Let us consider this as the source FT for the mutation. A second FT, the target FT, is selected from memory according to the values held in memory for the variable ω, as mentioned above, and FTs with higher ω values tend to be selected as targets more often than FTs with lower ω values.

The generative process is illustrated below by means of a simple example using the memory from the previous learning phase, but considering a mutation on a single musicode only: melodic direction (dir). Therefore, let us assume the memory scenario shown in Table 12.6.

In order to generate a new measure, the dir musicode of the source FT1 will be mutated towards the respective musicode values of a target FT. In this case, both FT2 and FT3 have the same ω so there is an equal chance of FT2 or FT3 being selected as the target FT, and a smaller chance of FT1. Let us assume that FT2 is selected as the target. Thus, FT2's dir musicode is applied to FT1's dir musicode to produce a mutation (represented in bold) as follows:

$$\{0, +1, +1, -1, -1\} + \{0, -1, +1, -1\} = \{0, \mathbf{0}, +1, -1, -1\}.$$

Note that the dir musicode has outlying maximum and minimum values of +1 and −1; hence, only the second value is actually mutated (+1) + (−1) = 0. Therefore, the newly generated FT contains a dir musicode of {0, 0, +1, −1, −1}.

Table 12.6 A memory scenario with three FTs

	FT1	FT2	FT3
dir	0, −1, +1, −1	0, +1, +1, −1, −1	0, +1, 0
ω	0.2	0.4	0.4

Fig. 12.14 The musical rendering of the new FT that was generated by mutating the dir musicode from FT1 and FT2

Mutating other musicodes (melody interval, event duration, note pitch, etc.) would yield more variation. Mutations are possible across all musicodes in a similar manner, with the only exception being mutations in modality. These are accomplished by a process of transformation whereby the intervals between successive absolute pitches in the given FTs are forced to conform to preset intervals for major, minor or diminished modes.

Finally, the new FT is rendered into a musical measure (Fig. 12.14) and saved into a MIDI file.

The above example only showed the generation of a single measure. For longer musical sequences, further FTs are generated by using the next FT in memory as the source FT and mutating it with a target FT that again is selected according to the value of the variable ω of all other FTs stored in memory.

12.7.3 Transformative Phase

The transformative phase comprises a number of transformation algorithms that modify a given musical sequence, three of which will be explained in this section.

Although there are some differences in the specific processing undertaken by each algorithm, the basic signal flow is quite similar for all of them. The generated input signal is modified towards values given by one of the transformation algorithms. With most of the transformation algorithms, the amount of modification is scaled according to the fMRI data. The fMRI data, or more specifically the data extrapolated from the fMRI scans by ICA analysis, are referred to as the fMRI_index. These data are provided to MusEng on a ten-point scale with values between 0 and 9. In order to use the fMRI index as a control signal (CS) for the transformation algorithms, MusEng first scales the data to a range between 0.1 and 1.0. The system applies the following simple scaling process to the value of the fMRI_index (Eq. 12.2).

$$CS = \{(\text{fMRI\_index} + 1) * 0.1\} \tag{12.2}$$

A difference value d between the input and the transformed musicodes is also calculated. This difference is then multiplied by the CS to give a final scaled modifier value (SMV). The SMV is summed with the input signal to directly transform the output. This gives a degree of fMRI-controlled variability in each transformation: a high fMRI_index value will result in larger transformations to the music, whereas a low fMRI_index value will result in smaller transformations.

Below are examples of three of the transformation algorithms, which illustrate the effect of varying the fMRI_index: pitch inversion, pitch scrambling and pitch delta.

12.7.3.1 Pitch Inversion Algorithm

Given an input musical sequence, the pitch inversion algorithm creates a new sequence, which is effectively the input sequence turned upside down. For instance, a sequence rising in pitch would descend in pitch after being passed through this transformation. In order to illustrate this, let us consider the measure produced in generation phase example, as shown in Fig. 12.14. Incidentally, this measure will be used as the starting point for the following two transformation examples as well.

The melody interval musicode for this measure is {0, 0, 3, 2, 1}, and the note pitch musicode is {B4, B4, D5, C5, B4}. In this case, the MIDI values are 71, 71, 74, 72 and 71, respectively; MIDI uses a range of 128 pitch values. There are a variety of ways to accomplish a pitch inversion, including diatonic and chromatic options, or inversions around a specific sounding pitch. MusEng processes pitch inversion simply by subtracting the current MIDI pitch value from 128, and substituting in the resulting natural number as the new pitch value. For instance, the transformed pitch values for our example created using this technique would be as follows: (128 − 71 = 57), (128 − 71 = 57), (128 − 74 = 54), (128 − 72 = 56) and (128 − 71 = 57).

The resulting MIDI values are 57, 57, 54, 56 and 57, yielding the following pitch sequence {A3, A3, F#3, G#3, A3}. Note that the inverted sequence maintains the original melody interval musicode of {0, 0, 3, 2, 1}, while giving an upside down melody, as shown in Fig. 12.15.

The example above assumed a maximal fMRI index value of 9, which once scaled to create a CS gives 1.0. However, as mentioned in the introduction to this section, varied degrees of transformations are also possible by scaling the amount of transformation according to the value of the fMRI_index. The difference between the input and the transformed pitches is multiplied by the CS, before being summed with the input to create the final transformed output value (Eq. 12.3).

$$\begin{aligned}\text{New\_pitch} = \{&\text{Input\_pitch} + ((\text{Input\_pitch} - \text{transf\_pitch}) \\ &*[(\text{fMRI\_index} + 1) * 0.1])\}\end{aligned} \tag{12.3}$$

Let us examine what happens if we assume an fMRI_index equal to 5, which yields a CS equal to 0.6. In this case, we would expect an output approximately halfway between the original pitch and the inversion, in other words an almost neutral set of intervals. First, the difference d between the maximal inversion and the input signal for each of the musicode values needs to be calculated as follows:

Fig. 12.15 Newly inverted sequence, after transformation of measure in Fig. 12.14

Fig. 12.16 Sequence after inversion with fMRI_index = 5, giving a nearly neutral set of pitch intervals

$$d = \{(57-71), (57-71), (54-74), (56-72), (57-71)\}$$
$$d = \{-14, -14, -20, -16, -14\}$$

Then, the scaled modifier values are calculated by multiplying the difference values by the value of CS:

$$\text{SMV} = \{(-14 * 0.6), (-14 * 0.6), (-20 * 0.6), (-16 * 0.6), (-14 * 0.6)\}$$
$$\text{SMV} = \{-8.4, -8.4, -12, -9.6, -8.4\}$$

Finally, the SMV values are summed with the original input to give a transformed set of output values:

$$\text{New\_pitches} = \{(71 - 8.4), (71 - 8.4), (74-12), (72 - 9.6), (71 - 8.4)\}$$
$$\text{New\_pitches} = \{62.6, 62.6, 62, 62, 62.6\}$$

Pitch values are rounded up to the nearest whole number as per the MIDI standard, giving a transformed set of pitch values equal to {63, 63, 62, 62, 63}, which is rendered as {D#4, D#4, D4, D4, D#4}, as shown in Fig. 12.16.

12.7.3.2 Pitch Scrambling Algorithm

In simple terms, the pitch scrambling algorithm orders the pitch values of the input signal into a numerical list, which is then reordered randomly. This provides a stochastic component to the transformation algorithm. Using the same measure as for the previous example (Fig. 12.14) as a starting point, let us examine the result of applying this transformation. The process is as follows:

- Input pitches: {71, 71, 74, 72, 71}
- Order pitches in ascending order: {71, 71, 71, 72, 74}
- Scramble the order of pitches randomly: {74, 72, 71, 71, 71}
- Output pitches: {74, 72, 71, 71, 71}

In this case, the output would be rendered as {D5, C5, B4, B4, B4}. Rerunning the transformation a further three times would give further variants, for example {72, 74, 71, 71, 71}, {71, 74, 72, 71, 71} and {71, 74, 71, 72, 71}, rendered as {C5, D5, B4, B4, B4}, {B4, D5, C5, B4, B4} and {B4, D5, B4, C5, B4}, respectively, as illustrated in Fig. 12.17.

As with the pitch inversion algorithm, the value of fMRI_index can be used to create a control signal with which the amount of transformation can be varied. In

Fig. 12.17 The result from applying the pitch scrambling algorithm four times on the same input

order to illustrate this, let us assume an fMRI_index equal to 3. This gives a CS value of 0.4.

Considering the same input measure as before (Fig. 12.14) and the transformed values from the first pitch scramble shown in Fig. 12.17, the value of *d*, between the first scramble and the input signal, is calculated as follows:

$$d = \{(74{-}71), (72{-}71), (71{-}74), (71{-}72), (71 - 71)\}$$
$$d = \{3, 1, -3, -1, 0\}$$

The scaled modifier values are then calculated by multiplying the difference values by CS = 0.4:

$$\text{SMV} = \{(3 * 0.4), (1 * 0.4), (-3 * 0.4), (-1 * 0.4), (0 * 0.4)\}$$
$$\text{SMV} = \{1.2, 0.4, -1.2, -0.4, 0\}$$

Finally, the SMV values are summed with the values of the original input to give a transformed set of output values:

$$\text{New\_pitches} = \{(71 + 1.2), (71 + 0.4), (74 - 1.2), (72 - 0.4), (71 - 0)\}$$
$$\text{New\_pitches} = \{72.2, 71.4, 72.8, 71.6, 71\}$$

As before, pitch values are rounded up to the nearest whole number as per the MIDI standard, giving a transformed set of pitch values equal to {72, 71, 73, 72, 71}, which is rendered as {C5, B4, C#5, C5, B4}, as shown in Fig. 12.18. Note that the output is significantly closer in overall structure to the unscrambled input than the first scrambled transformation shown in Fig. 12.17, with only the first and third notes having changed here.

Fig. 12.18 Transformed output created by pitch scrambling algorithm assuming fMRI_index = 3

12.7.3.3 Pitch Delta Algorithm

Pitch delta represents the rate of change in the pitch values in each measure. The algorithm works by calculating the difference between the initial pitch and the successive pitch in each pair of notes. The pitch delta value is then used to transform the input pitch by summing the two values together (Eq. 12.4).

$$\text{New\_pitch} = \{(\text{successive\_pitch} - \text{initial\_pitch}) + \text{initial\_pitch}\} \quad (12.4)$$

Assuming the same input as for the previous examples (Fig. 12.14), with a string of note pitch values equal to {71, 71, 74, 72, 71}, the delta values for this transformation are calculated as follows:

1. $\text{Delta\_1} = \{(71-71) + 71\}$
   $\text{Delta\_1} = 0$
   $\text{New\_pitch\_1} = 71(\text{no change})$
2. $\text{Delta\_2} = \{(71-71) + 71\}$
   $\text{Delta\_2} = 0$
   $\text{New\_pitch\_2} = 71(\text{no change})$
3. $\text{Delta\_3} = \{(74-71) + 74\}$
   $\text{Delta\_3} = 3$
   $\text{New\_pitch\_3} = 77$
4. $\text{Delta\_4} = \{(74-72) + 74\}$
   $\text{Delta\_4} = 2$
   $\text{New\_pitch\_4} = 76$
5. $\text{Delta\_5} = \{(72-71) + 72\}$
   $\text{Delta\_5} = 1$
   $\text{New\_pitch\_4} = 73$

The transformed output would therefore comprise a pitch string of {71, 71, 77, 76, 73}, which is rendered as {B4, B4, F5, E5, C#5}, as shown in Fig. 12.19. Thus, the application of the pitch delta algorithm gives the effect of exaggerating the melodic intervals from a given measure; large intervals become even larger, while melodies with little or no interval between successive notes remain unchanged.

Fig. 12.19 Transformed output created by the pitch delta algorithm

As with the previous cases of transformations, the above example assumed a maximal fMRI_index value, but the effect of the transformation can be mediated by reducing the value of the fMRI_index. This is illustrated in the following example.

Let us assume the case of fMRI_index = 2. This gives a CS value of 0.3. With such a low value for the control signal, we should expect only a small amount of pitch delta transformation.

As before, we will use the input signal shown in Fig. 12.14, this time with the transformed values from the full pitch delta transformation shown in Fig. 12.19, to the difference *d*, as follows:

$$d = \{(71-71), (71-71), (77-74), (76-72), (73-71)\}$$
$$d = \{0, 0, 3, 4, 2\}$$

The scaled modifier values are then calculated by multiplying the difference values by the CS value, which is equal to 0.3:

$$\text{SMV} = \{(0 * 0.3), (0 * 0.3), (3 * 0.3), (4 * 0.3), (2 * 0.3)\}$$
$$\text{SMV} = \{0, 0, 0.9, 1.2, 0.6\}$$

Finally, the SMV values are summed to the original input to give a new sequence of pitch values:

$$\text{New\_pitches} = \{(71 + 0), (71 + 0), (74 + 0.9), (72 + 1.2), (71 + 0.6)\}$$
$$\text{New\_pitches} = \{71, 71, 74.9, 73.2, 71.6\}$$

As with the previous transformation examples, pitch values are rounded up to the nearest whole number, giving a transformed sequence of pitch values of {71, 71, 75, 73, 72}, which is rendered as {B4, B4, D#5, C#5, C5}, as shown in Fig. 12.20. The exaggerating effect of the pitch delta has been radically mediated by the value of CS, with a much smaller amount of change seen in the transformed output than in the full pitch delta transformation shown in Fig. 12.19.

The generative potential of a composition system that incorporates transformative processes, such as that offered by MusEng, is high.

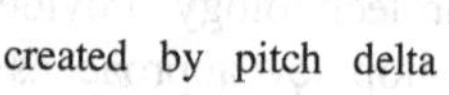

Fig. 12.20 Transformed output created by pitch delta algorithm with a relatively low fMRI_index value of 3

12.8 Concluding Remarks

Research into BCMI often is devoted to technical aspects of building BCMI systems, which is not surprising giving the plethora of technical difficulties that need to be addressed to implement a decent system. In this chapter, however, we ventured to explore the creative potential of the science and technology behind BCMI research: music neurotechnology.

We introduced an approach to music composition informed by the notion that the neural patterns and the corresponding mental images of objects and events around us are creations of the brain prompted by the information we receive through our senses. In the case of music, even though humans have identical mechanisms for processing the basics of sound, music as such is a construction of the brain. Indeed, there is increasing hard evidence that this construction differs from person to person. When we listen to music, sounds are deconstructed as soon as they enter the ear. Different streams of neuronally coded data travel through distinct auditory pathways towards cortical structures, such as the auditory cortex and beyond, where the data are reconstructed and mingled with data from other senses and memories, into what is perceived as music.

Metaphorically speaking, the compositional approach that we developed to compose *Symphony of Minds Listening* did to the Beethoven score what our hearing system does when we listen to music: sounds are deconstructed as they enter the ear and are relayed through various pathways towards cortical structures, where the data are then reconstructed into what is perceived as music.

We would like to highlight that the composition of the piece evolved in tandem with the development of the MusEng software. Some of MusEng's functionalities were firstly applied manually to compose a section of the piece, before they were implemented in software to aid the composition of other sections. The compositional practice therefore informed the design of the software, and the design of the software influenced the compositional practice. We believe that this is an important shift of paradigm from most scenarios of using computers in music. A piece of software is often developed from abstract specifications and tested only after it has been almost fully implemented. Moreover, composers are often confronted with software that does not always do what it needs to do. Our paradigm to systems development may not be as cost-effective as more standard methods, as it requires much more time to develop. However, it opens a significant opportunity for composers to actively participate in the design process. As we continue developing this work, more and more procedures emerging from the left-hand side of the block diagram in Fig. 12.5 will certainly make its way to the right-hand side.

We believe that Music Neurotechnology provides musicians with an unprecedented opportunity today to develop new approaches to music that would have been unthinkable a few years ago. This chapter unveiled only the tip of the iceberg.

12.9 Questions

1. How would you define the emerging field of Music Neurotechnology?
2. What do you understand by the 'mind as music' hypothesis?
3. Explain the metaphor that *Symphony of Minds Listening* is intended to express artistically.
4. What is the point of composing each movement of the symphony based on the fMRI scan of a different person?
5. Explain what ICA is and its role in the project presented in this chapter.
6. What are the approaches to using computer-generated materials in musical composition discussed in this chapter? Discuss the differences between them, and the advantages and disadvantages of each approach.
7. Can you envisage an approach to use computers in music beyond the ones discussed in this chapter?
8. What is the rational for dividing Beethoven's piece into 13 sections before processing it with MusEng?
9. Why does MusEng apply transformations to the music?
10. Create a new kind of transformation for MusEng and explain it in detail.

References

Ashburner J, The FIL Methods Group at UCL (2013) SMP8 Manual. Insitute of Neurology, University College London, London. Available online at http://www.fil.ion.ucl.ac.uk/spm/. Assessed on 04 Nov 2013

Bremment JD (2005) Brain imaging handbook. W. W. Norton & Co, London

Churchland P (2007) Neurophilosophy at work. Cambridge University Press, Cambridge

Eysenck MW, Keane MT (2005) Cognitive psychology: a student's handbook. Psychology Press (Taylor and Francis), Hove and New York

Lloyd D (2011) Mind as music. Frontiers in psychology 2:63. doi:10.3389/fpsyg.2011.00063

MathWorks (2014) Matlab: The Language of Technical Computing. http://www.mathworks.co.uk/ . Assessed on 04 Nov 2013

Miranda ER (ed) (2000) Readings in music and artificial intelligence. Routledge, Abingdon

Miranda ER (2013) On computer-aided composition, musical creativity and brain asymmetry. In: Collins D (ed) The act of musical composition. Ashgate, Farnham

Miranda ER (2014) Thinking Music. University of Plymouth Press, Plymouth

Miranda ER, Gimenes M (2011) An ontomemetic approach to musical intelligence. In: Miranda ER (ed) A-life for music: music and computer models of living systems. A-R Editions, Middleton, pp 261–286

Stone JV (2004) Independent component analysis: a tutorial introduction. The MIT Press, Cambridge

Passive Brain–Computer Interfaces

13

Laurent George and Anatole Lécuyer

Abstract
Passive brain–computer interfaces (passive BCI), also named implicit BCI, provide information from user mental activity to a computerized application without the need for the user to control his brain activity. Passive BCI seem particularly relevant in the context of music creation where they can provide novel information to adapt the music creation process (e.g., user mental concentration state to adapt the music tempo). In this chapter, we present an overview of the use of passive BCI in different contexts. We describe how passive BCI are used and the commonly employed signal processing schemes.

13.1 Introduction

In traditional brain–computer interfaces (BCI), called *active BCI*, the user deliberately tries to control his/her brain activity in order to send desired commands to the applications. One limitation of this approach for the design of BCI systems that are not targeted for special or medical scenarios is its relatively low performance compared to other control devices such as a keyboard. Indeed, BCI systems usually provide an information transfer rate below 25 bits/min (Wolpaw et al. 2002).

The recent *passive BCI* approach is less affected by this transfer rate limitation as it does not require a high bit rate (Coffey et al. 2010). Indeed, passive BCI do not try to replace traditional motor inputs but act as a complementary input providing

L. George (✉) · A. Lécuyer
INRIA, Rennes, France
e-mail: laurent.f.george@gmail.com

A. Lécuyer
e-mail: anatole.lecuyer@inria.fr

E.R. Miranda and J. Castet (eds.), *Guide to Brain-Computer Music Interfacing*,
DOI 10.1007/978-1-4471-6584-2_13

valuable information that reflects the user mental states (Cutrell and Tan 2007; Girouard 2009; Zander et al. 2009).

In passive BCI, the user does not try to control his/her brain activity, and he/she can remain mainly concerned by his/her primary task. The brain activity is analyzed to read out the user mental state which is used to adapt the interaction rules or the content of the application.

In this chapter, we present an overview of the use of passive BCI in different contexts: adaptive automation, multimedia content tagging, video game adaptation, error correction, etc. We describe how implicit BCI are used and the commonly employed signal processing schemes.

13.2 Passive BCI Definition

Cutrell and Tan were the first to introduce the expression "passive BCI." In Cutrell and Tan (2007), they wrote "We think there is a potential to use brain sensing in a more passive context, looking beyond direct system control to make BCI useful to the general population in a wide range of scenarios." Girouard (2009) referred to the work of Cutrell and Tan and defined the term "passive BCI" as "Passive BCI are interfaces that use brain measurements as an additional input, in addition to standard devices such as keyboards and mice." By developing passive BCI, her aim was to use the brain activity information to create "applications that pay attention to the user" by adapting them to user's mental state. Another point of view is presented by Zander et al. (2009) who defined passive BCI as "BCI based not on intended actions of the user, but instead on reactive states of the user's cognition automatically induced while interacting in the surrounding system." Recently, Makeig et al. (2012) described passive BCI as "BCI that produce passive readout of cognitive state variables for use in human–computer applications without requiring the user to perform voluntary control that may restrict performance of and attention to concurrent tasks." To summarize brain–computer interface (or interaction) could be categorized as:

- **active BCI** (explicit brain–computer interaction): The user deliberately tries to control his/her brain activity;
- **passive BCI** (implicit brain–computer interaction): The user does not try to control his/her brain activity, he/she is mainly concerned by his/her primary task.

It should be noted that there is no consensus about the integration of passive BCI in the BCI definition. Indeed, some researchers use a somewhat restrictive definition of BCI. For example, Pfurtscheller and Scherer defined a BCI as a communication system where a "brain signal that the user can intentionally modulate" is used for sending control commands (Pfurtscheller and Scherer 2010). This definition does not include passive BCI. However, the BCI community seems to accept passive BCI more and more as a new extension of BCI (Makeig et al. 2012).

13.3 Mental States Used in Passive BCI

Brain activity measurements, specifically EEG techniques, provide signals that can be used to read out user mental state. These signals and related mental states could be categorized into signals related to user emotional state, signals related to user's task, signals related to error potentials, and signal related to user's mental workload. In the following section, the EEG markers that could be used to read these different mental states are described.

13.3.1 Emotional State

Some EEG patterns have been shown to be correlated with user's emotional state (Onton and Makeig 2009; Heraz and Frasson 2007; Molina et al. 2009). These can be patterns over time (e.g., EEG rhythms) but emotional state can also influence punctual patterns (e.g., P300). In the following, we mainly refer to the works reported in Molina et al. (2009).

Lang et al. show that some rhythmic EEG patterns can be correlated to emotional state; it seems possible to detect dominance, arousal, and pleasure[1] using EEG signals (Heraz and Frasson 2007). Lang et al. (1997) use the International Affective Picture System which is a set of pictures that are known to cause specific emotions and measure EEG activity when presenting the pictures to the user. A correlation between different EEG rhythms and the three emotional states has been observed:

- pleasure: negative correlation with delta, theta, and beta rhythms[2] (highly for beta);
- arousal: positive correlation with theta and beta rhythms (highly for theta) and positive correlation with alpha rhythms;
- dominance: negative correlation with delta, theta, and beta (highly negative for beta) and a positive correlation with alpha rhythms.

Numerous works also seem to reveal a relation between frontal activity and emotion. A difference of symmetry has, for example, been shown between the left and the right frontal hemisphere in the alpha frequency band during emotion (Coan and Allen 2004).

Gamma band and emotion have also been found to be related to emotional process (Gemignani et al. 2000, Onton and Makeig 2009). Studies revealed an increase in gamma band over left frontal hemisphere during negative emotional stimuli (Gemignani et al. 2000). Onton and Makeig (2009) also describe a negative correlation between pleasure of imagined emotion and gamma power in occipital region.

---

[1] Emotional state can be represented into a three-dimensional space where the axes are pleasure (from unpleasant to pleasant), arousal (from calm to excited), and dominance (i.e., control) (Lang et al. 1997).

[2] Beta is considered to be frequency above 12 Hz in this study.

The event-related potentials (ERP) also seem to be modulated by the emotional state of the user. Olofsson et al. provide a survey of this modulation in Olofsson et al. (2008). For example, pleasant stimulus helps to induce larger P300 amplitude, compared to an unpleasant one.

13.3.2 Task-Related State

The brain activity measurement could also provide information concerning the user mental state related to an interaction task. In the following text, works that show markers related to user's perceived difficulty of the task and differentiation of tasked based on EEG activity are presented.

Lee and Tan (2006) found some differences in EEG patterns between different kinds of video game interactions[3] (relaxation without playing, playing without enemies, and playing with enemies). This work does not induce any modification of the game based on the acquired data. It reveals the feasibility of this kind of task differentiation by providing an average classification accuracy of 92.4 %.

Girouard et al. (2009) have also explored the differentiation of task during interaction with video game. They measure the blood oxygenation in cortical region of a user playing a Pacman-like video game. Their BCI system was able to distinguish between difficulty level and also active state versus passive state (user not playing). We can note that the difficulty level measured, as explained by Girouard, reflects the difficulty sensed by the user, not the difficulty of the game (the user can work hard and performs poorly).

Reuderink and colleagues observed brain activity related to frustration during video game interaction. They record EEG during a modified version of Pacman called *affective Pacman* (Reuderink et al. 2009). In *affective Pacman,* the game voluntary induced frustration by adding errors into movements (the key pressed did not always induce the same character movement). This experiment and the preliminary analysis reveal differences between normal and frustrated states. The authors propose to continue to explore the effect of frustration on EEG activity. They also propose to use the level of frustration as an interaction input to switch, for example, between an easy mode (where the character is controlled by a keyboard) and a difficult mode (where the user uses an explicit BCI to control the character).

13.3.3 Error-Related Potentials

Error-related potentials (ErrP) are a certain type of event-related potentials that can be measured in the EEG activity when the user is aware of erroneous behavior. Four kinds of ErrP have been identified (Ferrez and Millán 2007):

[3] Halo, first person shooter game produced by Microsoft.

- "response ErrP" follow incorrect motor action (e.g., wrong key press). It is a negative potential (error negativity ERN) in the EEG 80–100 ms after the incorrect movement followed by a positive peak between 200 and 500 ms.
- "feedback ErrP" follow the presentation of a stimulus that reflects incorrect performance (e.g., system presents a feedback that tells the user he has done an error). It is a negative deflection in the EEG 250 ms after the stimulus.
- "observation ErrP" follow the observation of errors made by an operator during choice reaction task where the operator needs to respond to stimuli. It is a negative deflection 250 ms after an incorrect response of the operator was observed.
- "interaction ErrP" appear when the user interacts with an application that does not respond in the expected way. It is composed of a positive peak 200 ms after the feedback, a negative peak after 250 ms, a second larger peak after 320 ms, a larger negative peak after 450 ms, and a late peak after 600 ms.

Interaction ErrP can, for example, be used to detect error in a P300 speller (see Sect. 13.4.2).

13.3.4 Mental Workload

Real-time measurement of mental workload could present benefits in different contexts as suggested in Blankertz et al. (2010) and Coffey et al. (2010). For instance, it could be used for safety purpose (e.g., raising alert), for improving human effectiveness and reducing errors (e.g., modifying task demands or activating assistance in times of cognitive overload), or as an objective measure in usability evaluation of new products.

Two relevant markers are used to estimate mental workload: ERP and EEG oscillatory activity (Van Erp et al. 2010). ERP have been shown to be affected by the user's mental workload, whereas EEG rhythmic activity is correlated with mental workload levels.

There are no obvious best markers to estimate user's mental workload. As reported by Berka et al. (2007), the requirement of an electing stimulus into real-world tasks to elicit the potentials is a limitation of the ERP-based approach. Oscillatory rhythmic activity-based estimators show the advantage of being able to be used without disrupting the primary mode of interaction of the user. Indeed, unlike ERP-based estimators, they do not require any external stimulus. They could be used in a completely passive BCI context. The combination of the two techniques has shown improvement of the mental workload estimator accuracy (Brouwer et al. 2012).

13.4 Applications of Passive BCI

In this section, our aim is to present existing applications of passive BCI. These applications can be categorized into four categories: implicit multimedia content tagging, error correction, adaptive aiding and automation, and virtual environment applications as displayed in Table 13.1.

13.4.1 Implicit Multimedia Content Tagging

Passive BCI have been used for tagging multimedia contents (Shenoy and Tan 2008; Kapoor et al. 2008; Koelstra et al. 2009). Shenoy and Tan used EEG activity to classify images (Shenoy and Tan 2008). They used ERP that occur in EEG activity after image presentation. Their system was able to classify images matching to human faces versus inanimate objects with a 75.3 % accuracy. For a three-class classification (human faces vs. animals vs. inanimate objects), an average accuracy of 55.3 % was obtained. Kapoor et al. (2008) used these results and proposed to combine BCI with a more classical recognition system. The experiment yielded significant gains in accuracy for the task of object categorization. In the two aforementioned works, users were not aware of the classification task. They were assigned "distractors task" to force them to look at the display. No feedback about the classification task was provided. This reinforces here the *implicit* property of the *interaction*.

Video content tagging has also been explored (Koelstra et al. 2009). Koelstra et al. proposed to use EEG brain activity to implicitly validate video tags. They demonstrated that incongruent tags could be successfully distinguished by EEG analysis. Recently, Moon et al. (2012) proposed to automatically extract interesting parts of video clip by using EEG activity. They used the commercial Emotiv device and one of the proprietary EEG index related to user emotional state named long-term excitement. In another study related to multimedia content, Scholler et al. (2012) proposed to use EEG activity and specifically P300 components to determine whether a change in video quality of multimedia clip occurred (the process is done off-line).

13.4.2 Error Detection and Correction

The detection of ErrP provides a promising possibility to correct errors in different contexts. For instance, the detection of ErrP has been used to correct error during classical computer interaction task (Parra et al. 2003) and to increase performance of active BCI (Ferrez and Millán 2008; Dal seno et al. 2010).

Parra et al. (2003) use the detection of error potentials in brain activity to correct errors in a visual discrimination task. In this study, the users had to push buttons corresponding to visual stimuli. The user sometimes failed and perceived error

Table 13.1 Existing systems using passive BCI

Application	Task	Features	Goal	Usage of implicit interaction	Ref
Multimedia content tagging	Image looking task	ERP	Use brain capabilities for classification task	Analyze brain activity after image presentation	Shenoy and Tan (2008)
	Image looking task	ERP	Use brain capabilities to improve automatic classification	Analyze brain activity after image presentation	Kapoor et al. (2008)
	Video watching		Automatic video clip extraction based on user brain activity	extract interesting parts of a video	Moon et al. (2012)
Adaptive aiding and automation	Tracking task	Band power ratios	Enhance mental engagement	Adapt automation of the task	Pope et al. (1995)
	Driving and distractors	Band power	Maintain low reaction time	Disable a task when high workload	Kohlmorgen et al. (2007)
Error correction	BCI motor imagery task	Errp	Improve explicit BCI	Filter out erroneous system's responses	Ferrez and Millán (2008)
	BCI P300 speller	Errp	Improve explicit BCI	Correct errors made by P300 speller	Dal Seno et al. (2010)
	Visual discrimination task	Errp	Improve user performance	Correct user's perceived errors	Parra et al. (2003)
Virtual environments	Game (Bacteria Hunt)	Alpha band power	Challenge player to relax	Affect controllability of avatar	Mhl et al. (2010)
	Game (AlphaWow)	Alpha band power	Enhance immersion	Shift avatar's form	Bos et al. (2010)
	Game (Tetris)	Blood oxygenation (fNIRS)	Enhance immersion	Adapt music to the predicted user's task	Girouard et al. (2013)
	Game (RLR)	Errp	Detect user's feeling of losing control	Detect whether the user perceived a system's error	Zander et al. (2009)

shortly after the action. The system could then identify error potentials and correct user's actions.

The use of error potential was also proposed to correct errors in active BCI systems (Ferrez and Millán 2008; Dal seno et al. 2010). Ferrez and Millàn (2008) used error potential detection to filter out erroneous responses of a BCI based on motor imagery. Dal seno et al. (2010) addressed the automatic detection and correction of the errors made by a P300 speller. Schmidt et al. (2012) also used an online detection of ErrP to increase the spelling speed of a visual p300 speller (increase of 29 %).

13.4.3 Adaptive Aiding and Adaptive Automation

Passive BCI have been used in several studies for off-line monitoring of workload during different tasks such as reading, writing, surfing, programming, mathematical tasks, and memory tasks (Berka et al. 2007). A few studies aim at using passive BCI for monitoring mental workload in online context (Heger et al. 2010; Berka et al. 2007). A commercial application system (B alert) based on EEG activity has also been proposed (Berka et al. 2007). B alert is an online monitoring system of mental workload and alertness that can provide an index of mental workload.

Passive BCI systems based on mental workload (or similar information) were also used for online adaptation purpose. Pope et al. (1995) proposed a brain-based adaptive automation system based on EEG activity. In their system, the allocation between human and machine of a tracking task is done based on an engagement index calculated using user's EEG indices. Ratios between the beta, alpha, and theta band power were used. An experiment conducted with 6 subjects shows the operability of such a system.

More recently, Wilson et al. (2007) proposed to use EEG data (F7, Fz, Pz, T5, O2) coupled with electro-oculographic and electrocardograph data to adapt an aiding system based on an online index of mental workload (more precisely, task demand level) during a complex aerial vehicle simulation. Two different task difficulties (high and low) were used. The mental workload index model was computed during the first task using artificial neural network. The mental workload model was then used online on the same task to adapt the aiding system that consists in providing more time to the subject to evaluate target stimuli. The aiding system was enabled when the user presented a high workload. This system allowed to improve operator's performance by approximately 50 %. Randomly presented aiding does not show the same level of performance improvement (approximately 15 % of performance improvement in random aiding condition).

Passive BCI based on mental workload have also been used to reduce workload by interrupting secondary tasks. Kohlmorgen et al. (2007) presented a passive BCI that measures mental workload in the context of a real car-driving environment. The user is engaged in a task mimicking interaction with the vehicle's electronic warning and information system. This task is suspended when high mental workload is detected. This experiment showed better reaction times on average using the passive BCI.

13.4.4 Passive BCI and Virtual Environments

Passive BCI have also been used, but scarcely, in virtual environments and video games. Several video games that use implicit interaction have been already developed. Some of them use implicit information to adapt the way the system responds to commands. It is the case of the game "Bacteria Hunt" in which the controllability of the player's character is impaired by considering the level of alpha power which is correlated here to relaxed wakefulness (Mhl et al. 2010).

Some other games adapt the avatar's characteristics based on implicit information. In "AlphaWoW" (Bos et al. 2010, Nijholt et al. 2009), which is based on the famous game World of Warcraft, the user's avatar form is updated (from elf to wolf) according to the measured level of alpha activity. Another way to use implicit information for games consists in adapting the game environment (e.g., background music). Girouard et al. (2013) described an experiment in which the user is engaged in two successive tasks watching a video and playing a Tetris game. The application was able to predict in which task the user was engaged in, based on measurement of the brain activity. This allowed to adapt the background music accordingly (e.g., increasing tempo). This adaptation was found to lead to a positive impact on user's satisfaction (Girouard et al. 2013).

Last, some video games can use implicit information to check whether the user has perceived specific game information. In the game developed by Zander et al. (2009), the user has to rotate a letter correctly, as fast as possible. Errors are introduced by the system. A passive BCI is used to detect whether the user's mental state reveals that the user has perceived the errors. In this case, the speed of rotation is increased. A false positive (a perceived error when there is none) slows the rotation down.

We can notice that these games combine the use of classical devices (e.g., keyboard) with a passive BCI. One of them also uses an explicit BCI together with a passive BCI (Mhl et al. 2010).

13.5 Conclusion

In this chapter, we proposed an overview of related work on the use of passive BCI (also named implicit BCI) for interacting with computer applications. We discussed the definition of passive BCI; then, we presented the different brain patterns that seem to be relevant for this kind of interaction. Finally, we presented the different applications of passive BCI.

The passive BCI approach holds good potential for BCMI. For instance, the detection of auditory error response could be used in order to create music systems that would be aware of the user's perception of music. An other example could be the evaluation of the user's mental state in order to create or select a musical playlist that matches the user's emotional state.

13.6 Questions

1. What is the difference between active and passive BCI ?
2. List three contexts where passive BCI have been used.
3. List three mental states that could be used in passive BCI.
4. What kind of EEG patterns has been shown to be related to the user's emotional states?
5. What are ErrP?
6. How error-related potentials could be used to improve an active BCI system?
7. How a passive BCI could be used to enhance the immersion of the user in a video game?
8. The conclusion hinted on two possible musical applications of passive BCI. Can you envisage a passive BCMI for generating music? Please describe how your idea would work.

References

Berka C, Levendowski D, Lumicao M, Yau A, Davis G, Zivkovic V, Olmstead R, Tremoulet P, Craven P (2007) EEG correlates of task engagement and mental workload in vigilance, learning, and memory tasks. Aviat Space Environ Med 78(Supplement 1):B231–B244

Blankertz B, Tangermann M, Vidaurre C, Fazli S, Sannelli C, Haufe S, Maeder C, Ramsey L, Sturm I, Curio G, et al (2010) The Berlin brain-computer interface: non-medical uses of BCI technology. Front. Neurosci. 4

Bos DPO, Reuderink B, van de Laar B, Grkk H, Mhl C, Poel M, Nijholt A, Heylen D (2010) Brain-computer interfacing and games. In: Tan DS, Nijholt A (eds) Brain-computer interfaces, human-computer interaction series. Springer, Berlin, pp 149–178 (chap 10)

Brouwer A, Hogervorst M, Van Erp J, Heffelaar T, Zimmerman P, Oostenveld R (2012) Estimating workload using EEG spectral power and ERPs in the n-back task. J Neural Eng 9 (4):045008

Coan JA, Allen JJB (2004) Frontal EEG asymmetry as a moderator and mediator of emotion. Biol Psychol 67(1–2):7–50

Coffey E, Brouwer A, Wilschut E, Van Erp J (2010) Brain-machine interfaces in space: using spontaneous rather than intentionally generated brain signals. Acta Astronaut 67(1–2):1–11

Cutrell E, Tan D (2007) BCI for passive input in HCI. In: Proceedings of ACM CHI 2008 conference on human factors in computing systems workshop on brain-computer interfaces for HCI and games

Dal Seno B, Matteucci M, Mainardi L (2010) Online detection of P300 and error potentials in a BCI speller. Comput Intell Neurosci 11

Ferrez P, Millán JR (2007) Error-related EEG potentials in brain-computer interfaces. Towards Brain-Comput Interf 291–301

Ferrez PW, Millán JR (2008) Simultaneous real-time detection of motor imagery and error-related potentials for improved BCI accuracy. In: Proceedings of the international brain-computer interface workshop and training course

Gemignani A, Santarcangelo E, Sebastiani L, Marchese C, Mammoliti R, Simoni A, Ghelarducci B (2000) Changes in autonomic and EEG patterns induced by hypnotic imagination of aversive stimuli in man. Brain Res Bull 53(1):105–111

Girouard A (2009) Adaptive brain-computer interface. In: Proceedings of the 27th international conference extended abstracts on human factors in computing systems, ACM, pp 3097–3100

Girouard A, Solovey ET, Hirshfield LM, Chauncey K, Sassaroli A, Fantini S, Jacob RJ (2009) Distinguishing difficulty levels with non-invasive brain activity measurements. In: Proceedings of the 12th IFIP TC 13 international conference on human-computer interaction, INTERACT '09. Springer, Berlin, pp 440–452

Girouard A, Solovey ET, Jacob RJ (2013) Designing a passive brain computer interface using real time classification of functional near–infrared spectroscopy. Int J Auton Adaptive Commun Syst 6(1):26–44

Heger D, Putze F, Schultz T (2010) Online workload recognition from EEG data during cognitive tests and human-machine interaction. Adv Artif Intell 410–417

Heraz A, Frasson C (2007) Predicting the three major dimensions of the learners emotions from brainwaves. Int J Comput Sci

Kapoor A, Shenoy P, Tan D (2008) Combining brain computer interfaces with vision for object categorization. In: Proceedings of the IEEE conference on computer vision and pattern recognition

Koelstra S, Mhl C, Patras I (2009) EEG analysis for implicit tagging of video data. In: Proceedings of the international conference on affective computing and intelligent interaction and workshops

Kohlmorgen J, Dornhege G, Braun M, Blankertz B, Mller KR, Curio G, Hagemann K, Bruns A, Schrauf M, Kincses W (2007) Improving human performance in a real operating environment through real-time mental workload detection. In: Dornhege G, R del Milln J, Hinterberger T, McFarland D, Mller KR (eds) Toward brain-computer interfacing. MIT press, Cambridge, pp 409–422

Lang P, Bradley M, Cuthbert B (1997) International affective picture system (IAPS): technical manual and affective ratings. In: NIMH center for the study of emotion and attention

Lee J, Tan D (2006) Using a low-cost electroencephalograph for task classification in HCI research. In: Proceedings of the 19th annual ACM symposium on User interface software and technology, ACM, p 90

Makeig S, Kothe C, Mullen T, Shamlo NB, Zhang Z, Kreutz-Delgado K (2012) Evolving signal processing for brain-computer interfaces. In: Proceedings of the IEEE 100(Centennial-Issue):1567–1584

Molina GG, Tsoneva T, Nijholt A (2009) Emotional brain-computer interfaces. In: Cohn J, Nijholt A, Pantic M (eds) ACII 2009: affective computing and intelligent interaction. IEEE Computer Society Press, Los Alamitos, pp 138–146

Moon JY, Kim Y, Lee HJ, Bae CS (2012) My own clips: automatic clip generation from watched television programs by using EEG-based user response data. In: 2012 IEEE 16th international symposium on consumer electronics (ISCE), pp 1–2

Mhl C, Grkk H, Bos DPO, Thurlings ME, Scherffig L, Duvinage M, Elbakyan AA, Kang S, Poel M, Heylen DKJ (2010) Bacteria hunt: a multimodal, multiparadigm BCI game. In: Proceedings of the international summer workshop on multimodal interfaces, Genua

Nijholt A, Bos DPO, Reuderink B (2009) Turning shortcomings into challenges: brain computer interfaces for games. Entertain Comput 1(2):85–94

Olofsson JK, Nordin S, Sequeira H, Polich J (2008) Affective picture processing: an integrative review of ERP findings. Biol Psychol 77(3):247–265

Onton J, Makeig S (2009) High-frequency broadband modulations of electroencephalographic spectra. Front Hum Neurosci 3

Parra LC, Spence CD, Gerson AD, Sajda P (2003) Response error correction-a demonstration of improved human-machine performance using real-time EEG monitoring. IEEE Trans Neural Syst Rehabil Eng

Pfurtscheller G, Scherer R (2010) Brain-computer interfaces used for virtual reality control. Proc ICABB

Pope AT, Bogart EH, Bartolome DS (1995) Biocybernetic system evaluates indices of operator engagement in automated task. Biol Psychol 40(1):187–195

Reuderink B, Nijholt A, Poel M, Nijholt A, Reidsma D, Hondorp G (2009) Affective pacman: a frustrating game for brain-computer interface experiments. Intell Technol Interact Entertain 9:221–227

Schmidt N, Blankertz B, Treder M (2012) Online detection of error-related potentials boosts the performance of mental typewriters. BMC Neurosci 13(1):19

Scholler S, Bosse S, Treder M, Blankertz B, Curio G, Muller KR, Wiegand T (2012) Toward a direct measure of video quality perception using EEG. IEEE Trans Image Process 21 (5):2619–2629

Shenoy P, Tan D (2008) Human-aided computing: utilizing implicit human processing to classify images. In: Proceeding of SIGCHI conference on human factors in computing systems

Van Erp JBF, Veltman HJA, Grootjen M (2010) Brain-based indices for user system symbiosis. In: Tan DS, Nijholt A (eds) Brain-computer interfaces., Human computer interaction seriesSpringer, London, pp 201–219

Wilson GF, Caldwell JA, Russell CA (2007) Performance and psychophysiological measures of fatigue effects on aviation related tasks of varying difficulty. Int J Aviation Psychol 17 (2):219–247

Wolpaw JR, Birbaumer N, McFarland DJ, Pfurtscheller G, Vaughan TM (2002) Brain computer interfaces for communication and control. Clin Neurophysiol 113(6):767–791

Zander T, Kothe C, Welke S, Roetting M (2009) Utilizing secondary input from passive brain-computer interfaces for enhancing human-machine interaction. In: Schmorrow D, Estabrooke I, Grootjen M (eds) Foundations of augmented cognition. Neuroergonomics and operational neuroscience, lecture notes in computer science, vol 5638. Springer, Berlin, pp 759–771

Index

E.R. Miranda and J. Castet (eds.), *Guide to Brain-Computer Music Interfacing*,
DOI 10.1007/978-1-4471-6584-2

Printed in the United States
By Bookmasters